T0321165

Particles, Fields and Topology
Celebrating A P Balachandran

Particles, Fields and Topology
Celebrating A P Balachandran

Edited by

T R Govindarajan
The Institute of Mathematical Sciences, India

Giuseppe Marmo
Federico II University of Naples, Italy

V Parameswaran Nair
The City College of New York, USA

Denjoe O'Connor
Dublin Institute for Advanced Study, Ireland

Sarada G Rajeev
University of Rochester, USA

Sachindeo Vaidya
Indian Institute of Science, Bangalore, India

 **World Scientific**

NEW JERSEY · LONDON · SINGAPORE · BEIJING · SHANGHAI · HONG KONG · TAIPEI · CHENNAI · TOKYO

Published by

World Scientific Publishing Co. Pte. Ltd.

5 Toh Tuck Link, Singapore 596224

USA office: 27 Warren Street, Suite 401-402, Hackensack, NJ 07601

UK office: 57 Shelton Street, Covent Garden, London WC2H 9HE

Library of Congress Control Number: 2023932597

British Library Cataloguing-in-Publication Data
A catalogue record for this book is available from the British Library.

PARTICLES, FIELDS AND TOPOLOGY
Celebrating A P Balachandran

ISBN 978-981-127-042-0 (hardcover)
ISBN 978-981-127-043-7 (ebook for institutions)
ISBN 978-981-127-044-4 (ebook for individuals)

For any available supplementary material, please visit
https://www.worldscientific.com/worldscibooks/10.1142/13251#t=suppl

Typeset by Stallion Press
Email: enquiries@stallionpress.com

Printed in Singapore

A.P. Balachandran

Origins by physicist and sculptor Michele Bourdeau. The Lagrangian shown is related to the sigma model, one of the early topics of interest for Prof. Balachandran. The increasing lumpiness to the right symbolizes a gradual transition to quantum space-time, another topic of interest to Bal.

Preface

A.P. Balachandran started his research career in the early 1960s and has a long and impressive record of intellectual contributions, continuing to this day, spanning over 50 years. Working broadly in the area of particle physics and quantum field theory, Balachandran has brought concepts of geometry, topology and operator algebras to the analysis of physical problems, particularly in particle physics and condensed matter physics. Current algebra, pion–nucleon scattering, magnetic monopoles, coadjoint orbits and group theory, skyrmions, anomalies, spin–statistics theorems, non-commutative geometry, boundary and edge effects and quantum Hall systems are some of the areas to which he has made many contributions. Balachandran and his collaborators have published several books in addition to well over 200 research articles. He also has had an influential role within the physics community, not only in terms of a large number of students, research associates and collaborators but also serving on the editorial boards of important journals.

This book of chapters by students and associates is meant as a tribute and celebration on the occasion of Balachandran's 85th birthday. We expect that this will help crystallize a clear perspective of Balachandran's contributions and how they have influenced physics over the last 50 years or so. As with Festschrifts for scientists, most of the chapters in this book will be scientific in nature, describing new perspectives and new directions resulting from Balachandran's contributions to physics and highlighting the continuing potential for future research. A few of the chapters will also contain reminiscences of collaborating and working with Balachandran.

We thank everyone who contributed to this effort. Some of our invitees could not contribute due to the press of competing demands on their time, but we do acknowledge and pass on to Balachandran their good wishes as well.

Happy Birthday Bal!

T.R. Govindarajan
Giuseppe Marmo
V. Parameswaran Nair
Denjoe O'Connor
S.G. Rajeev
Sachindeo Vaidya

Contents

Chapter 4. Exceptional Fuzzy Spaces and Octonions   **37**

Denjoe O'Connor and Brian P. Dolan

Chapter 5. Understanding Quantum Physics:
Why Geometry and Algebra Matter   **51**

Elisa Ercolessi

Chapter 6. Near-Horizon Conformal Structure,
Entropy and Decay Rate of Black Holes   **59**

Kumar S. Gupta and Siddhartha Sen

Chapter 20. A New Approach to Classical Einstein–Yang–Mills Theory   213

Donald Salisbury

Chapter 21. Renormalization Group and String Theory   223

Balachandran Sathiapalan and Homi Bhabha

Chapter 22. An Inside View of the Tensor Product   235

Rafael D. Sorkin

Chapter 23. Noncommutative AdS_2 II: The Correspondence Principle 265

Allen Stern and Aleksandr Pinzul

Chapter 24. Quantum Probability as the Sum of Holonomies of the Canonical One-Form 277

C. G. Trahern

Chapter 25. Noncommutative Geometry and Super-Chandrasekhar White Dwarf 289

T. R. Govindarajan, Surajit Kalita and Banibrata Mukhopadhyay

Chapter 26.   Chiral Anomaly in $SU(N)$ Gauge Matrix Models and Light Hadron Masses

301

S. Vaidya, N. Acharyya and M. Pandey

Chapter 27.   The QM/NCG Correspondence

313

Badis Ydri

Chapter 1

My Life and Times

A. P. Balachandran

Physics Department, Syracuse University
Syracuse, NY 13244-1130, USA
apbal1938@gmail.com

Hegel wrote: "The owl of Minerva spreads its wings only with the falling of the dusk."

To this I may add Edward Hallett Carr, in *What Is History?*: "What is history?, our answer, consciously or unconsciously, reflects our own position in time, and forms part of our answer to the broader question, what view we take of the society in which we live."

I retired from Syracuse University in 2012 and my ex-students and colleagues are preparing a Festschrift for my 85th birthday. They have requested me to write a review of my career. Reflecting on this request, it has become clear that the implications of the past get better clarified with the passage of time. I have contributed a few significant results to quantum theory, but my enduring contributions have been to the training of young researchers. They have been well over 30, perhaps nearer 40. They are spread over several countries and have themselves become accomplished physicists and, in turn, trained students or created schools. I will discuss both these aspects of my career in the following. But first let me describe where I come from.

I was born in Salem, Tamil Nadu, on 25 January 1938. Those were colonial times, I do not know how the Salem Corporation functioned, or if there was one, and even my birth certificate cannot be traced. My immediate family comes from small farmers in Kerala. Financially, it was not doing well in the 1930s, and that saw the migration of male members to cities in search of jobs. My Father was one such migrant who ended up in the port city of Kochi. I grew up in Ernakulam, a suburb then of Kochi and now a huge city in Kerala. Those were the days of a resurgent Kerala with a strong communist movement. The school itself had remarkable teachers. I recall in particular a high school math teacher who taught us how to think and an impassioned talk by a poor student in one of our Thursday "meeting" sessions on communism. All this went to my own formation.

I was in this school during 1944–1953 when my Father got transferred to Kozhikode. My first two years of college were there. I studied 'Intermediate' (Group one: Mathematics, Physics and Chemistry with Hindi as the second language). This college, which has now grown, was then unremarkable intellectually. But there was a city library and I read a lot: popular science books, English novels, history of mathematics (especially by E.T. Bell), etc. These readings largely decided my future path. From Kozhikode, I went to Madras Christian College (MCC) for a three-year course (B.Sc. Hons. specializing in Physics, 1955–1958) and then to Madras University for M.Sc. (Nuclear Physics) and Ph.D. (Particle Physics) under Alladi Ramakrishnan (1958–1962). MCC has steadily grown from a school established by Scottish missionaries in 1837 to become an autonomous degree-granting college affiliated with Madras University in 1978. MCC educationally was largely a washout, with a science syllabus from colonial times and faculty without even a basic knowledge of quantum physics. We seemed to have been taught 19th-century optics and properties of matter. But I do recall learning the Maxwell distribution at that time. One important memory is that Tom Kibble's father taught us mathematics. Tom himself spent his childhood days at MCC. My senior there, G. Rajasekaran (Rajaji), survived in physics like me. After MCC, he went to the Tata Institute of Fundamental Research, was deputed to work under Dalitz in Oxford for Ph.D. and now is an Emeritus Professor at the Institute of Mathematical Sciences after a distinguished career.

But my training with Alladi was different and unusual. Alladi had a background in stochastic processes and was keen on creating a quantum field theory group in Chennai. But he did not know this subject nor did the students who joined him. So we, Alladi included, literally taught ourselves the field. We would meet in the hall in his house on Luz Church Road and one of us would present a topic, maybe a chapter of Dirac's book or maybe Feynman's lectures on his rules, and found our way to a thesis in this fashion. We emerged from this experience with partial knowledge of physics. Basics of quantum physics were not taught to us, as was statistical mechanics, and it took years for me to fill these gaps adequately. But it taught us intellectual independence. Still, these efforts of Alladi have had a strong impact on Indian science. His political influences and Nehru's help led to the creation of the Institute of Mathematical Sciences (IMSc) and indirectly also to the Chennai Mathematical Institute which are now major research and teaching centers. The owl of Minerva spreads its wings only with the falling of the dusk.

In 1962, I went to Vienna on leave from a position at the newly created IMSc to work under Walter Thirring. I resigned from this position in 1964. I was in Vienna from 1 May 1962 to the end of October. Walter was intellectually intimidating and his colleagues were not socially hospitable. I did make friends with Herbert Pietschmann and with L. Tenaglia from Bari (he was spending his sabbatical in Vienna) and wrote papers on current algebras with Wolfgang Kummer and Herbert. They were among the earliest on the topic. We also had a few other publications. Herbert and I went to Trieste for the 16 July–25 August 1962 meeting arranged by Abdus Salam as a prelude to ICTP. It was a fabulous experience. There were Schwinger, Wightman, Wigner and other big names. Notes for Schwinger's lectures which covered his soluble model and work on quantum gravity were taken by A.P. Balachandran, B. Jaksic, I. Saavedra and J. Kvasnica. Salam talked about Dynkin diagrams and Wigner, if I remember correctly, about C, P and T. The ambiance was friendly, and I was thrilled by the intellectual feast. I remember Claud Lovelace and his vegetarianism from this period. From Vienna, I went to the Fermi Institute in Chicago. The two years I spent there were pleasant socially. I lived in the International House where there was a diverse student body with whom I could make friends. There were also Indian friends at the Fermi Institute, Rajaji and Divakaran, both Dalitz's students and R. Ramachandran (R.R.), Gregor Wentzel's student. Divakaran is well known among historians of Indian Mathematics and has several scholarly publications on the subject. R.R. has had a distinguished career as an administrator. He was the Director of IMSc for several years after having taught at the Indian Institute of Science, Kanpur, for many years. I have kept contacts with all three.

Those were the days when dispersion relations held sway over high energy theory and I too wanted to shine on that topic. But my knowledge of that topic was minuscule. Reinhard Oehme was working on the mathematical aspects of this subject and I tried to work with him. Oehme was collaborating with Peter Freund. I did not fit in that collaboration at all. Frank von Hippel was also a post-doc at the Fermi Institute and he was very friendly with me. But his interests were those of Dalitz, who had shifted from Oxford to the Fermi Institute, with focus then on hypernuclei, while I had no interest in phenomenology. Chandrasekhar too was in the faculty and I recall visiting him and Lalitha with Divakaran, Rajaji and R.R. Chandra was a formidable physicist, but the couple were totally alienated from their roots in India. Lalitha could not speak Tamil. I did not learn from him or from Nambu.

Professionally, this was a tense period. I needed work to get a job. Finally, towards the end of my second year, I had the idea of applying moment problem techniques to dispersion relations and wrote decent papers in the *Annals of Physics*. Moment problems and orthogonal polynomials reappeared in large N matrix models and more recently in the work by X. Han, S.A. Hartnoll and J. Kruthoff and by D. Berenstein.

George Sudarshan moved from the University of Rochester to Syracuse University in 1964. He had made fundamental and imaginative contributions to weak interaction theory, quantum optics and quantum information theory (on completely positive maps) and was brought to Syracuse University (SU) by Jack Leitner. Jack wanted a strong particle theory group to support his experimental efforts and Department Chair Fredrickson and Chancellor Tolley were very supportive. It was George who brought me to Syracuse as a Visiting Assistant Professor and, later in 1966, got me promoted to an Assistant Professor.

Around my second year at SU, clever methods were found to combine LSZ formalism, current algebras and PCAC, and I had the good sense to enter that field. Mike Gundzig was my first student and his thesis was on this topic. Mike left physics after a post-doc stint in Austin, became an insurance agent and affluent via that profession. I met him about three years ago at SU when he visited us with the potential for creating endowed lectures. Nothing came of it. George knew Gianfausto dell'Antonio and through this contact, we had Neapolitan visitors, Alberto Simoni, Franco Zaccaria and Chicco Nicodemi, who was my second student. From then started my regular visits to Napoli, and conversely our getting students from there. These have been marvelous experiences for me. I learnt Italian, became friends with Bruno Vitale and Beppe Marmo and became a Marxist, or rather let us say "a Historical Materialist". And Fedele Lizzi, Gianni Sparano and Peppe Bimonte became my students. We had a lot of visitors too, with whom I worked, Gianni Landi for one. And Elisa Ercolessi from Bologna also became my student: she was a student of Giuseppe Morandi who was working with Beppe and others in Napoli.

It must have been during 1977–1978 that I spent a year in the theory group at the University of Madras. It was then that I became a close friend of Govindarajan and Jayaraman and came to know the student, B. Vijayalakshmi, a legend among Indian women scientists and written up in *Lilavati's Daughters: The Women Scientists of India*. She came from a conservative Brahmin family and had stomach cancer even when she joined the doctoral program. She got radicalized after interacting with her colleagues, married Jayaraman, became active in the CPM (Communist Party of India — Marxist) politics and died of cancer in 1985. She reminds me of another colleague, Mythili Sivaraman, an ex-political science student from the Maxwell School at Syracuse University, who joined CPM and was very active, especially in women's issues. She had Alzheimer's disease for the past several years of her life and died of COVID-19 on May 30, 2021.

Govindarajan and I have written many papers together and discuss physics and politics on a regular basis. He participates in our Friday 316 Zoominars.

In the early 1980s, there were also regular visits to Göteborg, and Jan Nilsson was my host. During those visits, I became friends with Karl-Erik Ericsson and his family who lived in Kungsbacka. Karl-Erik is a remarkable person, socially committed, and has been a regular visitor to Kumasi. I also was visiting ICTP and became friends with Abdus Salam. On an occasion, Karl-Erik and I spent several hours with Abdus discussing Pakistani politics.

George left SU in 1967 and was replaced by Kamesh Wali. The Vietnam war was agitating us and after the US invasion of Cambodia, seized our conscience at SU. With a resolution by the University Senate, SU had no classes in the Fall of 1970 or 1971. We organized study groups about the history of the war, imperialism in S.E. Asia, etc. with moderate attendance.

During 1966–1967, I went to IAS in Princeton on leave. It was a crucial year intellectually. I became friends with Jean Nuyts and started being immersed in group theory. Jean and I wrote nice papers on crossing symmetry for partial waves: they were recently brought up by Aninda Sinha while discussing his work on scattering amplitudes. IAS also had a bunch of young Indian mathematicians, Raghunathan, Raghavan Narasimhan, Raghavan and Rangachari, all from TIFR, with whom I became friends. I learnt a lot talking to them, especially about groups and their orbits, topics of importance in spontaneous symmetry breaking. I also wrote papers on current algebras, which I continued to work on after my return to SU. I also worked there with Jean-Jacque Loeffel and wrote a rather mathematical paper with him.

Events after my return to Syracuse were kaleidoscopic. There were lecture notes on the Poincaré group and analyticity of scattering amplitudes, with notes by Bill Meggs. I continued working with Maurice Blackmon until about 1974 when he left Syracuse. Joe Weinberg, a student of Oppenheimer and a victim of McCarthyist repression, joined SU in 1970, and it was enchanting to drown in his intellectual brilliance. Pierre Ramond was my student, along with Bill, and both have had brilliant careers after leaving Syracuse. Pierre invented SUSY, an all-pervasive topic in string theory, and wrote *Field Theory: A Modern Primer*. He will be a candidate for the Nobel Prize if SUSY is discovered. Cassio Sigaud also was an early student, he returned to Rio, settled in Petropolis and I have visited him many times, once Indra included.

The early 1970s were turbulent times in the US with agitating students protesting the draft and being used in Vietnam. We, many SU faculty members, were protesting and joining marches. New communal living was also emerging among students and we had hopes for the dawn of socialism. But the underlying cause for these agitations was the mandatory draft of young people into the army. Once it was withdrawn in 1973, all these movements did fade away.

Science for the People emerged in the late 1960s as a socialist organization, and in 1976, the book, *The Bee and the Architect* (*L'Ape e L'Architetto*, a quote from Marx, Capital 1), also appeared. It is interesting that Marcello Cini, Giovanni Jona-Lasinio and Giorgio Parisi contributed to the second edition of this book. The role of science in a capitalist society was a matter of concern for many of us.

Science for the People is still published from Washington. In the late 1960s, it took the lead in exposing the Jason project, consisting of elite scientists advising the US government on sensitive matters. It was the source of the McNamara Line electronic barrier in Vietnam. The founders of JASON included John Wheeler and Charles H. Townes. Other early members included Murray Gell-Mann, Murph Goldberger, Hans Bethe, Sam Treiman, Steven Weinberg and Freeman Dyson. The exposures started an avalanche of criticisms of these people. Jan Nilsson, Bruno Vitale and I collected all the articles we could find at that time and made a book, *The War Physicists*, and Bruno distributed it as a preprint. There one sees Gell-Mann discussing cutting off the ears of Thai village headmen to control insurgency, and Dyson claiming that he was in Jason to influence the US government ! *The War Physicists* can be accessed at https://science-for-the-people.org/wp-content/uploads/2014/02/The-War-Physicists.pdf. The Jason group taken together was a vicious imperialist bunch. Now?

After the Vietnam days, the research at Syracuse began to take a turn towards topological issues in quantum physics. The practice of meeting regularly in Room 316 in the afternoons for free-wheeling discussions also became a habit then, although there were such meetings involving Mike Parkinson and Nick Papastamatiou during 1966–1968. This was a wonderful practice and is now continuing via Zoom under the title Room 316 on Fridays. We also used to go regularly for Friday dinners together, but alas, that cannot be done by Zoom.

My papers with Heinz Rupertsberger who was visiting us from Vienna during 1973–1976 show the steady shift of interest towards monopole physics. Our first paper on coadjoint orbits was with Steve Borchardt and Al Stern in 1978 although we did not know about them or, nor did we know of the Borel–Weil–Bott theory or the work of the Marseille group (Souriau). From then onwards, attention turned to unusual aspects of quantum physics stimulated by exceptional students: Al Stern (mid- to late 1970s), Parameswaran Nair and S.G. Rajeev (early 1980s), Fedele Lizzi and Vincent Rodgers (around 1983–1984), Kumar Gupta and Ajit Mohan Srivastava (late 1980s), Aneziris ans Bourdeau (1989) Elisa Ercolessi (early 1990s), Arshad Momen, Peppe Bimonte (1994), Paulo Teotonio (around the same time), Arshad Momen (92–97), Sachin Vaidya (1994–98), Badis Ydri and Babar Qureshi (2000). Also Gianni Landi, who visited us often and whom we visited in Napoli, had an influential role on our research. And Andrea Barducci and Luca Lusanna from Firenze also visited us for extended periods. Those were thrilling times. In addition to Beppe, Denjoe O'Connor, Brian Dolan, Giorgio Immirzi and Manolo Asorey came to Syracuse. And there was Sang Jo, a post-doc and visitor: the group theory book was written with him, and Sasha Pinzul, a very gifted mathematical physicist, with whom I continue to work and who now is in Brasilia, and my last student Pramod, very clever and imaginative, now in IIT Bhubaneswar.

There is an incident involving Kumar and he still talks about it. I was in charge of admissions and the offer to Kumar was getting delayed. He decided to make, a collect call and got the telegram with the offer immediately thereafter. He has been the unique student in my experience to have had such a call accepted!

These are brilliant people, who taught me physics and mathematics and have been making creative contributions to knowledge. Beppe Marmo used to visit us regularly and stay with us in Syracuse. He was financially supported by Eugene Saletan at Northeastern University for teaching courses. During that period and during trips to Göteborg, we formulated the use of symplectic forms in classical actions, later rediscovered by Ed Witten. We found this formulation under the name of path space formulation. It is described in our 1983 book, *Gauge Theories and Fibre Bundles.*

There were other students too. Bill Case (early 70's), Balram Rai (late 80's), who had great potential but could not realize it because of personal problems, Gianni Sparano (late 80's), in the faculty at Salerno and gifted in mathematics, Chandar and Anosh Joseph (late 2000's) and Charilaos Aneziris and Michele Bourdeau (1989), who work in Wall Street.

Cassio Sigaud was my student in the early 70's. After the doctorate, he returned to Rio and commuted to UFRJ (Universidade Federal do Rio de Janeiro) from their house in UFRJ Petropolis. I have visited him and his family may times, also with Indra.

I would have forgotten to mention a few students and friends. My apologies to them, not intentional.

It was also then that we gave a proof of non-abelian anomalies using heat kernels, and we ignored Beppe's repeated exhortations of their connection to cohomology theory: later Zumino and collaborators gave such an interpretation using descent equations. We also revived Skyrme's ideas on nucleons as solitons with Parameswaran Nair, S.G. Rajeev and Al Stern during 1982–1983. When Ed Witten visited us for a seminar, I told him our results. He already had a conjecture from large N expansions that baryons were solitons, and he caught on immediately. There followed his brilliant papers on Skyrme solitons in four dimensions, which had his five-dimensional action for anomalies.

There were decisive developments of this work: dibaryons were predicted as chiral solitons in work with Fedele, Vince, Al and Andrea Barducci from Firenze, and we also applied these ideas to nematic crystals and excitations in a He-3 phase. In Göteborg too, we discovered that colored monopoles broke global color transformations. It is a domain problem, and as time progressed, we came to understand that this phenomenon happens in every non-abelian gauge theory, or 'symmetries', commuting with all observables continuous or discrete (like the permutation symmetry of identical particles or symmetry groups of molecules). This result seems important physically: it leads to a demonstration that the QCD color group is spontaneously broken. The basics of this result are simple: QCD gauge group is a superselected group so that all observables commute with it. Each such sector is labeled by the eigenvalues of a complete commuting set of its group algebra. But as this group is non-abelian, its action changes these eigenvalues and the superselection sector. Hence, by definition, the operators changing these eigenvalues are spontaneously broken. The ADM group in quantum gravity suffers a similar fate. These results are reported in a PDF of one of the 316 Zoominar videos.

With the progress of time, our attention turned to algebraic formulations. The trigger was a paper by Rafael Sorkin where he discretized a manifold by a finite open cover and showed that this cover had the topology of a poset with T^0 topology. It also reproduced the homotopy groups of the manifold and had a projective limit to recover the exact manifold. As in the Gelfan'd–Naimark theory, we found that there was a dual AF algebra to a poset, but alas, they are all infinite dimensional. Our attempts to write a Hamiltonian for these algebras failed. This work was with Elisa, Fedele, Gianni Landi, Gianni Sparano and Paulo. Gianni Landi wrote a nice book on noncommutative geometry which discusses this work. Their K-theory was discussed by Elisa, Gianni and Paulo. The shift to algebras was steadily overtaking our group.

The next focus was fuzzy spheres where we wrote many papers, also with Brian Dolan, Giorgio Immirzi, Denjoe O'Connor and Peter Presnajder. For discretization of two-dimensional field theories, it is a novel approach, with no fermion doubling, but gauge theories present problems, and so do generalizations to higher dimensions. A basic reason is that the two-sphere is symplectic, but the four-sphere is not. The book I wrote with Sachin and Seckin Kürkçüoglu contains many results of general interest, *-products, index theory, projective modules, etc. I use it often. Seckin by the way has been an impressive physicist and now is the chair of the physics department of METU, Ankara. He has already built a talented school of students there. I continue to interact with him on a regular basis.

The brilliant work of Doplicher, Fredenhagen and Roberts (DFR) on the quantum nature of space-time appeared in 1995. It took us until early 2005 to appreciate the importance of this work. There also appeared the work of the Helsinki group applying the Drinfel'd twist to the Moyal algebra, and this new idea attracted us a lot. We wrote several papers on this topic.

Recently, Sachin, Seckin and I have completed a review of this topic for EPJ.

The work of Doplicher *et al.* which involved also Dorothea Bahns and Gherardo Piacitelli has important results with lower bounds on the measurability of space-time areas, volumes, etc. One suspects that these bounds have profound implications for space-time at the Planck scale.

Around 2012, Amilcar Queiroz, Sachin and I started to look at non-abelian gauge theories again. I do not recall what triggered this interest.

In any case, Amilcar, Sachin and I got involved in topological aspects of gauge theories, especially the Gribov problem, and we studied the beautiful paper of Narasimhan and Ramadas on this topic. With Nirrmalendu Acharyya and Mahul Pandey, we had the happy idea of developing a matrix model from their work, and we did numerical work to derive the glueball spectrum therefrom. The agreement was good and became better with improved variational ansatze. Sachin and collaborators have detailed developments of this model. Despite being a finite-dimensional matrix model, it agrees remarkably with experiments.

Around 1994–1995, I had the good fortune to team up with Bala Sathiapalan and work with him and Chandar. Bala would visit us once a week and we wrote a number of papers on the quantum Hall effect and its edge states. We were happy with these papers. With Chandar and Elisa, we also investigated edge states and discovered nice results. Bala joined IMSc and has now retired after a distinguished career, while Chandar also returned to India, left physics and is having an affluent career in the IT industry.

Also in 2014, I was visiting IMSc for several months — I had retired from SU in 2012 — and Amilcar and Andrés Reyes-Lega (with Ximena) also visited IMSc. It was a very productive time. While returning from Villa de Leyva after a summer school, I had stopped at UniAndes for a few days and had a wonderful discussion with Alonso Botero on identical particles and their entropy. The literature had tremendous confusion, but it was easy to see that the GNS construction of the Hilbert space of the observables with the states transforming appropriately under S_N (or even their q-deformed variants) would work. So, from Chennai, we published several papers on this subject with examples. Rafael Sorkin pointed out to me that the entropy associated with these constructions is not unique in the presence of certain degeneracies. We developed this idea as well, and not so long ago, using the Tomita–Takesaki (TT) theory, Andrés, his two students and I proved that this was a generic feature. We are happy with this paper.

In 2009, from 18 January, I could visit Universidad de Carlos Tercero as a 'Catedra d'Excelencia' Professor for six months. The stay in Madrid was exceptionally pleasant and fruitful as well. Madrid is a beautiful city. It was so easy to travel and the museums and parks are fabulous. It was then that I enjoyed the intellectual stimulation of Alberto Ibort and also connected to Manolo Asorey and Juan Manuel Peres Pardo: the outcome was a stream of papers on quantum geons and edge states. These developments are far from over and are feeding into ongoing work in myriad ways.

The last few years have taken a new turn. It was triggered by a beautiful paper by Brunetti, Guido and Longo in 2002 constructing free quantum fields from Wigner representations using TT theory and localization ideas (except maybe for 'continuous spin representations'). I spent a lot of time on this paper and talked about it at a Bengaluru school in 2016. This was the beginning of an appreciation of the wonderful works of Doplicher, Haag, Roberts and others. In time, it dawned on us that all there is at the roots of physics is quantum physics with its algebra of observables and states thereon, giving a noncommutative probability theory. The emergence of classical physics, manifolds, etc. then is a central problem. Our recent papers with Andrés, Parameswaran, Sachin and Sasha Pinzul on measurement theory and also gauge theories were inspired from immersion in algebraic quantum field theory. There is much to understand for a physicist here.

That brings us to the current times. Algebraic Quantum Field Theory as formulated by Haag and Kastler requires a mass gap and hence does not accommodate gauge theories. It has been recognized for decades that infrared divergences seriously affect quantum field theory and spontaneously break even Lorentz invariance. We have contributed to this problem and with Parameswaran, Sachin, Seckin and Amilcar, generalized the considerations to QCD and pointed out that SU(3) of color is also spontaneously broken. Very recently, Jens Mund, Pepe Gracia-Bondia, Rehren and Schroer have clarified the situation a lot. My attention for the past weeks has been on these papers. My expectation is that their escort field provides the order parameter for these spontaneous breakdowns. With Arshad, Amilcar, Babar and Manolo, we have been discussing these ideas. We already have a model for Z^0 to two photons, allowed now by Lorentz breakdown. There are other effects as well, such as surface states in condensed matter, which may be readily observable. Let us see how it goes.

With colleagues, I have written a few books. The first was on *Gauge Theories and Fibre Bundles*. Then, there have been books on *Group Theory and Hopf Algebras, The Hubbard Model, Classical Topology and Quantum States* and *Fuzzy and Fuzzy SUSY Physics*. They were written as we were learning the subjects. The *Classical Topology and Quantum States* book merits special comments. While the awareness that classical topology influences quantum physics is already dominating the *Gauge Theories* book, the title itself and the preface of this book claim this connection explicitly. And the text has many examples in this direction, fiber bundles, multiply connected spaces, soliton physics, QCD theta states and quantum geons. It has been Rafael Sorkin who taught us much of this material. His research has led him towards causal sets while mine has been hovering on the margins of this approach and the algebraic approaches. It seems that space-time notions emerge from the latter, seen in the quantum setting as noncommutative probability theory. But there are many problems to be overcome to better understand this emergence. If a new book is started by our group, it may be on *An Introduction to Algebraic Quantum Theory: An Engagement with Reality*. Let me conclude by recognising and thanking my colleagues who put together the festschrift (Parameswaran, Denjoe) and for organising this meeting (Govindarajan, Sanatan, Sachin, Parameswaran, Ajit, Ayan, Alok and Sujay). These are not easy tasks. I am grateful to them, and also Ravindran, IMSc director who has readily supported these projects.

Let me conclude by recognising and thanking my colleagues in the overall organising committee (Govindarajan, Beppe Marmo, Denjoe O'Connor, Parameswaran Nair, S.G. Rajeev, Sachin Vaidya) who put together this Festschrift and organised the Bal-Fest meeting at IMSc', and also the committee which managed the local organisation (Govindarajan, Sanatan Digal, Sujay Ashok, Alok Laddha, Ajit Srivastava, Ayan Mukhopadhyaya). These are not easy tasks. I am grateful to them, and also to Ravindran, IMSc director, who had readily and generously supported this meeting and who, along with Sanatan Digal, facilitated my extended stay at IMSc. I thank IIT Madras and CMI as well for their support of the meeting.

I am also grateful to Dr. K.K. Phua and the World Scientific Publishing Company who have graciously undertaken the publication of this book.

Chapter 2

Energy Preserving Boundary Conditions for Scalar Fields

Manuel Asorey[*] and Fernando Ezquerro[†]

*Centro de Astropartículas y Física de Altas Energías,
Departamento de Física Teórica, Universidad de Zaragoza,
E-50009 Zaragoza, Spain*
[*] asorey@unizar.es
[†] fezquerro@unizar.es

The dynamics of field theories in domains with boundaries is governed not only by field equations but also by its boundary conditions. In this chapter, we analyze the most general energy preserving boundary conditions for scalar field theories and the global properties of the space defined by all of them. In general, it has two connected components. In the case of complex scalar fields, we show the compatibility of these boundary conditions with the ones that preserve electric charge.

1. Introduction

There is an increasing number of quantum effects generated by the presence of boundaries. In condensed matter, the new role of quantum boundary effects is boosting a new era of quantum technologies. Indeed, the presence of plasmons and other surface effects in metals and dielectrics,[1] the appearance of edge currents in the Hall effect[2-4] and the discovery of new edge effects in topological insulators[5-10] and Weyl semiconductors[11] have very rich potential implications.

Although boundary effects arise in quantum physics since the early days of the theory, boundary effects in quantum field theory have a later development. Starting with the discovery of the Casimir effect,[12] boundary effects appear today as an essential ingredient in fundamental physics. They appear in Hawking radiation, black hole horizon effects, topological defects, topology change[13–16] and holography of the AdS/CFT correspondence.

The increasing relevant role of boundary effects is demanding a comprehensive theory of boundary conditions. In spite of the fact that quite a lot of work has been devoted to establish the foundations of the quantum theory, a comprehensive theory of boundary conditions for quantum field theories is still lacking. A first attempt to fill the gap was initiated by Asorey, Ibort and Marmo[14] and was further developed by Asorey and Muñoz-Castañeda.[17] This first global analysis was based on the preservation of unitarity for time evolution. Another essential remark was made by Asorey, García-Alvarez and Muñoz-Castañeda[18] who pointed out that the generalization for relativistic field theories requires a change of the basic principles, from unitarity to the preservation of the $U(1)$ symmetry which guarantees electric charge conservation. However, this principle does not apply to neutral fields where this new approach does not provide any fundamental law to be preserved.

In this chapter, we address the analysis of the theory of boundary conditions in field theories based only on the requirement of conservation of energy. This method applies to any bosonic or fermionic relativistic field theory including neutral fields like gauge fields, with the only exception of topological field theories.

2. Real Scalar Fields in Half-space

The dynamics of a real scalar field is governed by the Lagrangian density

$$\mathcal{L} = \frac{1}{2}\partial_\mu\phi\partial^\mu\phi - \frac{1}{2}m^2\phi^2 - V(\phi), \tag{2.1}$$

where $V(\phi)$ is any arbitrary local potential function. The translation symmetry induces by Noether's theorem four conservation laws

$$\partial^\mu T_{\mu\nu} = 0, \tag{2.2}$$

given in terms of the energy–momentum tensor

$$T_{\mu\nu} = \frac{1}{2}\partial_\mu\phi\partial_\nu\phi - \frac{1}{2}\eta_{\mu\nu}\partial_\alpha\phi\partial^\alpha\phi + \frac{1}{2}m^2\phi^2\eta_{\mu\nu} + V(\phi)\eta_{\mu\nu}, \tag{2.3}$$

where $\eta_{\mu\nu}$ denotes the Minkowski metric. In particular, when $\nu = 0$, we get a conservation law for the energy

$$\partial^\mu T_{\mu 0} = \partial_t \mathcal{E} + \partial^i P_i = 0, \tag{2.4}$$

where

$$\mathcal{E} = \frac{1}{2}\partial_t\phi\partial_t\phi + \frac{1}{2}\partial_i\phi\partial_i\phi + \frac{1}{2}m^2\phi^2 + V(\phi) \tag{2.5}$$

is the energy density and

$$P_i = \frac{1}{2}\partial_0\phi\,\partial_i\phi \tag{2.6}$$

is the momentum density of the field (we assume that $c = 1$). Thus, for any bounded domain Ω with regular boundary $\partial\Omega$,

$$\frac{d}{dt}E_\Omega = \int_\Omega \partial_t\mathcal{E} = \int_\Omega \partial^i P_i = \int_{\partial\Omega} n^i\,P_i, \tag{2.7}$$

where $\mathbf{n} = (n^i)$ denotes the normal vector to the boundary surface $\partial\Omega$.

Let us consider a simple case where Ω is just a half-space $\Omega = \{\mathbf{x} = (x^1, x^2, x^3)|x^3 \geq 0\}$ whose boundary is $\partial\Omega = \{\mathbf{x} = (x^1, x^2, 0)\}$ the plane perpendicular to the vector $\mathbf{n} = (0, 0, -1) \in \mathbb{R}^3$. In that case, the conservation of energy implies that

$$\frac{d}{dt}E_\Omega = -\int_{\partial\Omega} T_{03} = -\int \dot{\varphi}\,\varphi'\,dx^1 dx^2 = 0, \tag{2.8}$$

where

$$\dot{\varphi} = \partial_t\phi\Big|_{\partial\Omega}$$

and

$$\varphi' = \partial_3\phi\Big|_{\partial\Omega}$$

are the boundary values of the time derivative and the derivative along the normal direction to the surface of the fields ϕ.[a] If we consider only

---

[a] In dimensions higher than 2, there are some technical difficulties concerning the regularity of boundary values.[17] In this chapter, we will restrict ourselves to cases of regular boundary conditions.

boundary conditions which are invariant under translations along the x^1, x^2 plane, any solution verifies the equation

$$\dot\varphi \varphi' = 0 \tag{2.9}$$

which has two solutions $\varphi' = 0$ and $\dot\varphi = 0$. The first solution is Neumann boundary condition which is the usual boundary condition for open strings in string theory, whereas the second solution are the Dirichlet boundary conditions which correspond to D-branes in that theory.[19]

3. Complex Scalar Fields in Half-space

Let us now consider the case of complex scalar fields with Lagrangian density

$$\mathcal{L} = \frac{1}{2}\partial_\mu \phi^* \partial^\mu \phi - \frac{1}{2}m^2|\phi|^2 - V(|\phi|), \tag{3.1}$$

where $V(\phi)$ is any arbitrary local density potential function. Repeating the same arguments as in the case of real scalar fields, we get that energy conservation requires the vanishing of

$$\dot\varphi^* \varphi' + \varphi^{*\prime}\dot\varphi = 0. \tag{3.2}$$

If we introduce the following change of variables

$$\psi_1 = \begin{pmatrix} \varphi' + \dot\varphi \\ \varphi^{*\prime} + \dot\varphi^* \end{pmatrix} \quad \psi_2 = \begin{pmatrix} \varphi' - \dot\varphi \\ \varphi^{*\prime} - \dot\varphi^* \end{pmatrix}. \tag{3.3}$$

The vanishing condition becomes

$$|\psi_1|^2 - |\psi_2|^2 = 4\left(\varphi^{*\prime}\dot\varphi + \dot\varphi^*\varphi'\right). \tag{3.4}$$

If we assume translation invariance along the boundary plane, the most general solution of this condition satisfies

$$\left|\begin{pmatrix} \varphi' + \dot\varphi \\ \varphi^{*\prime} + \dot\varphi^* \end{pmatrix}\right|^2 = \left|\begin{pmatrix} \varphi' - \dot\varphi \\ \varphi^{*\prime} - \dot\varphi^* \end{pmatrix}\right|^2 \tag{3.5}$$

and is given by

$$\begin{pmatrix} \varphi' + \dot\varphi \\ \varphi^{*\prime} + \dot\varphi^* \end{pmatrix} = U \begin{pmatrix} \varphi' - \dot\varphi \\ \varphi^{*\prime} - \dot\varphi^* \end{pmatrix}, \tag{3.6}$$

where U is an arbitrary 2x2 unitary matrix. Conjugating (3.6) leads to

$$\begin{pmatrix} \varphi^{*\prime} + \dot{\varphi}^* \\ \varphi' + \dot{\varphi} \end{pmatrix} = U^* \begin{pmatrix} \varphi^{*\prime} - \dot{\varphi}^* \\ \varphi' - \dot{\varphi} \end{pmatrix}; \quad \sigma_1 \psi_1 = U^* \sigma_1 \psi_2, \quad \psi_1 = \sigma_1 U^* \sigma_1 \psi_2.$$

This implies that the unitary matrix U has to satisfy an extra condition

$$U = \sigma_1 U^* \sigma_1. \tag{3.7}$$

The meaning of this restriction is that the matrix U also has to belong to the $O(1,1)$ rotation group because from (3.7), it follows that U leaves the $(1,1)$ metric

$$\sigma_1 = \begin{pmatrix} 0 & 1 \\ 1 & 0 \end{pmatrix} \tag{3.8}$$

invariant, i.e.,

$$U^\perp \sigma_1 U = \sigma_1. \tag{3.9}$$

Thus, the general solution of local boundary conditions satisfying that constraint is given by the one-dimensional subgroup of unitary matrices $G = O(1,1) \cap U(2) \subset U(2)^{\text{b}}$ which has two disjoint components $G = G_+ \cup G_-$ differing only by the sign of the determinant $\det U = \pm 1$. The component G_+ contains all the matrices of the form

$$U_+(a) = e^{ia\sigma_3} \quad a \in [0, 2\pi) \tag{3.10}$$

that are continuously connected with the identity, whereas the other component G_- is given by all the matrices of the form

$$U_-(b) = \sigma_1 e^{ib\sigma_3} \quad b \in [0, 2\pi). \tag{3.11}$$

The general solution of the first type (3.10) is

$$\varphi' = -i \cot \frac{a}{2} \dot{\varphi}, \tag{3.12}$$

whereas

$$\text{Re}(\dot{\varphi} + \varphi') + \text{Re}(\dot{\varphi} - \varphi') \cos b = \text{Im}(\dot{\varphi} - \varphi') \sin b \tag{3.13}$$

$$\text{Im}(\dot{\varphi} + \varphi') - \text{Im}(\dot{\varphi} - \varphi') \cos b = \text{Re}(\dot{\varphi} - \varphi') \sin b \tag{3.14}$$

is the general solution of the second type (3.11).

---

^bThis is one of the three maximal compact subgroups of $U(2)$.[20]

Let us consider some particular cases of physical interest.

(i) $U_D = \mathbb{I}$: **Dirichlet boundary conditions**. This is a boundary condition of the first type with $a = 0$

$$\dot{\varphi} = 0. \tag{3.15}$$

(ii) $U_N = -\mathbb{I}$: **Neumann boundary conditions**. This is a boundary condition of the first type with $a = \pi$

$$\varphi' = 0. \tag{3.16}$$

(iii) $U_c = \pm i\sigma_3$: **Chiral boundary conditions**. These are boundary conditions of the first type with $a = \pm\frac{\pi}{2}$

$$\varphi' = \mp i\dot{\varphi}. \tag{3.17}$$

(iv) $U_t = \pm\sigma_1$: **Twisted boundary conditions**. These are boundary conditions of the second type with $b = 0, \pi$

$$\mathrm{Im}\,\varphi' = 0 \quad \mathrm{Re}\,\dot{\varphi} = 0 \tag{3.18}$$

$$\mathrm{Re}\,\varphi' = 0 \quad \mathrm{Im}\,\dot{\varphi} = 0. \tag{3.19}$$

(v) $U_{tc} = \pm\sigma_2$: **Twisted chiral boundary conditions**. These are boundary conditions of the second type with $b = \frac{\pi}{2}, \frac{3\pi}{2}$

$$\mathrm{Re}\,\varphi' = \mp\,\mathrm{Im}\,\varphi' \quad \mathrm{Im}\,\dot{\varphi} = \pm\,\mathrm{Re}\,\dot{\varphi}. \tag{3.20}$$

4. Fields Confined between Two Parallel Plates

Let us consider a complex scalar field confined between two parallel plates $\Omega = \{(x^1, x^2, x^3)\big| - L \leq x^3 \leq L\}$. In this case, local boundary conditions which preserve energy have to satisfy that

$$\dot{\varphi}_1^{*\prime}\dot{\varphi}_1 + \dot{\varphi}_1^{*}\varphi_1' - \dot{\varphi}_2^{*\prime}\dot{\varphi}_2 - \dot{\varphi}_2^{*}\varphi_2' = 0, \tag{4.1}$$

where $\varphi_1(x^1, x^2) = \phi(x^1, x^2, -L)$ and $\varphi_2(x^1, x^2) = \phi(x^1, x^2, L)$. In terms of the following vectors

$$H_1 = \begin{pmatrix} \dot{\varphi}_1 + \varphi_1' \\ \dot{\varphi}_2 - \varphi_2' \\ \dot{\varphi}_1^* + \varphi_1^{*\prime} \\ \dot{\varphi}_2^* - \varphi_2^{*\prime} \end{pmatrix} \quad H_2 = \begin{pmatrix} \dot{\varphi}_1 - \varphi_1' \\ \dot{\varphi}_2 + \varphi_2' \\ \dot{\varphi}_1^* - \varphi_1^{*\prime} \\ \dot{\varphi}_2^* + \varphi_2^{*\prime} \end{pmatrix}, \tag{4.2}$$

the restriction (4.1) reads

$$|H_1|^2 - |H_2|^2 = 4 \left(\dot{\varphi}_1^* \varphi_1' + \varphi_1^{*\prime} \dot{\varphi}_1 - \dot{\varphi}_2^* \varphi_2' - \varphi_2^{*\prime} \dot{\varphi}_2 \right) = 0. \qquad (4.3)$$

This means that any solution has to be of the form

$$H_1 = U H_2 \qquad (4.4)$$

with U an unitary matrix of $U(4)$. There is another extra condition that this solution has to satisfy. If we conjugate (4.4), we get

$$H_1^* = U^* H_2^* \qquad (4.5)$$

that implies

$$H_1 = \begin{pmatrix} 0 & \mathbb{I}_2 \\ \mathbb{I}_2 & 0 \end{pmatrix} U^* \begin{pmatrix} 0 & \mathbb{I}_2 \\ \mathbb{I}_2 & 0 \end{pmatrix} H_2. \qquad (4.6)$$

This imposes a further restriction on the unitary matrix

$$U = \begin{pmatrix} 0 & \mathbb{I}_2 \\ \mathbb{I}_2 & 0 \end{pmatrix} U^* \begin{pmatrix} 0 & \mathbb{I}_2 \\ \mathbb{I}_2 & 0 \end{pmatrix}. \qquad (4.7)$$

The meaning of this restriction is that the matrix U also has to belong to the $O(2,2)$ rotation group because from (4.7), it follows that U leaves the $(2,2)$ metric

$$\begin{pmatrix} 0 & \mathbb{I}_2 \\ \mathbb{I}_2 & 0 \end{pmatrix} \qquad (4.8)$$

invariant, i.e.,

$$U^\perp \begin{pmatrix} 0 & \mathbb{I}_2 \\ \mathbb{I}_2 & 0 \end{pmatrix} U = \begin{pmatrix} 0 & \mathbb{I}_2 \\ \mathbb{I}_2 & 0 \end{pmatrix}. \qquad (4.9)$$

Thus, the general solution of the local boundary conditions is given by the six-dimensional subgroup of unitary matrices $G = O(2,2) \cap U(4) \subset U(4)$ which has two disjoint components $G = O(2,2) \cap U(4) = G_+ \cup G_-$ distinguished by the sign of the determinant $\det U = \pm 1$. The component

G_0 contains all the matrices of the form

$$U_+(\mathbf{a}, \mathbf{b}, \mathbf{c}) = \exp i \begin{pmatrix} a_1 & b_1 + ib_2 & 0 & c_1 + ic_2 \\ b_1 - ib_2 & a_2 & -c_1 - ic_2 & 0 \\ 0 & -c_1 + ic_2 & -a_1 & -b_1 + ib_2 \\ c_1 - ic_2 & 0 & -b_1 - ib_2 & -a_2 \end{pmatrix} \quad \mathbf{a}, \mathbf{b}, \mathbf{c} \in \mathbb{R}^6$$

that are continuously connected with the identity. The other component is given by the matrices of the form

$$U_-(\mathbf{a}, \mathbf{b}, \mathbf{c}) = \frac{1}{2} \begin{pmatrix} 1 & 1 & -1 & 1 \\ 1 & 1 & 1 & -1 \\ -1 & 1 & 1 & 1 \\ 1 & -1 & 1 & 1 \end{pmatrix} U_+(\mathbf{a}, \mathbf{b}, \mathbf{c})$$

that are disconnected of G_+, i.e., $\pi_0(G) = \mathbb{Z}_2$.

This group contains the solutions of the type considered in the previous case for each one of the plane boundaries. But the fact that there are two boundaries gives rise to other remarkable boundary conditions like periodic boundary conditions. Let us consider these and other interesting particular cases:

(i) $U_N = \mathbb{I}_4$: **Neumann boundary conditions.** i.e., Neumann boundary conditions for both walls $\varphi'_1 = \varphi'_2 = 0$.

(ii) $U_D = -\mathbb{I}_4$: **Dirichlet boundary conditions.** i.e., Dirichlet boundary conditions for both walls $\dot{\varphi}_1 = \dot{\varphi}_2 = 0$.

(iii) $U_p = \begin{pmatrix} \sigma_1 & 0 \\ 0 & \sigma_1 \end{pmatrix}$: **Periodic boundary conditions** connecting the two walls $\dot{\varphi}_1 = \dot{\varphi}_2$, $\varphi'_1 = \varphi'_2$.

(iv) $U_{ap} = \begin{pmatrix} -\sigma_1 & 0 \\ 0 & -\sigma_1 \end{pmatrix}$: **Antiperiodic boundary conditions** connecting the two walls $\dot{\varphi}_1 = -\dot{\varphi}_2$, $\varphi'_1 = -\varphi'_2$.

5. Compatibility with Charge Density Conservation

In the case of complex fields, there is another conserved quantity: the electric charge. This conservation law provides another condition to be preserved by boundary conditions.[18] However, in principle, the families of boundary conditions which preserve each quantity are different. In the case of scalar fields, both families of boundary conditions are compatible. This very relevant property is a consequence of the compatibility of gauge transformations and space-time translations. Indeed, the actions of the group $U(1)$ of gauge transformations

$$G(\alpha)\phi = e^{i\alpha}\phi \tag{5.1}$$

and the group of translations T_4

$$T_a\phi(x) = \phi(x - a) \tag{5.2}$$

do commute. Moreover, as a consequence, the Poisson bracket of the charge density[c]

$$\rho = \frac{i}{2}\left(\phi^*\dot{\phi} - \phi\dot{\phi}^*\right) = \frac{i}{2}\left(\phi^*\Pi^* - \Pi\phi\right) \tag{5.4}$$

and the energy density

$$\mathcal{E} = \frac{1}{2}\left(\dot{\phi}^*\dot{\phi} + \nabla\phi^*\nabla\phi\right) + \frac{1}{2}m^2\phi^2 + V(\phi) \tag{5.5}$$

$$= \frac{1}{2}\left(\Pi\Pi^* + \nabla\phi^*\nabla\phi\right) + \frac{1}{2}m^2\phi^2 + V(\phi) \tag{5.6}$$

vanishes, i.e.,

$$\{\rho, \mathcal{E}\} = 0. \tag{5.7}$$

The conservation of electric charge is given by the continuity equation

$$\partial_t\rho + \partial^i j_i = 0, \tag{5.8}$$

where

$$\mathbf{j} = \frac{i}{2}\left(\phi^*\boldsymbol{\nabla}\phi - (\boldsymbol{\nabla}\phi^*)\phi\right). \tag{5.9}$$

The conservation of charge requires the vanishing of the electric current flux through the boundary[18]

$$-\int_{\Omega}\dot{\rho}d^3x = \int_{\Omega}\partial^i j_i d^3x = \int_{\partial\Omega}\mathbf{j}\,d\boldsymbol{\sigma} = \frac{i}{2}\int_{\partial\Omega}\left(\varphi^*\boldsymbol{\nabla}\varphi - (\boldsymbol{\nabla}\varphi^*)\varphi\right)d\boldsymbol{\sigma} \tag{5.10}$$

---

[c]Π and Π^* are the canonical momenta

$$\Pi = \frac{\partial\mathcal{L}}{\partial\dot{\phi}} = \dot{\phi}^* \quad \Pi^* = \frac{\partial\mathcal{L}}{\partial\dot{\phi}^*} = \dot{\phi}. \tag{5.3}$$

which in the half-space case reduces to

$$- \int_\Omega \dot{\rho} d^3 x = \frac{i}{2} \int (\varphi^* \varphi' - \varphi^{*\prime} \varphi) \, dx^1 dx^2. \qquad (5.11)$$

Thus, the local boundary conditions must satisfy

$$\varphi^* \varphi' - \varphi^{*\prime} \varphi = 0, \qquad (5.12)$$

and the most general boundary condition that satisfies (5.12) is given by[14]

$$\varphi + i\varphi' = U_c(\varphi - i\varphi'), \qquad (5.13)$$

where U_c is an arbitrary unitary matrix of $L^2(\mathbb{R}^2)$.[d] Now, if ϕ is a steady state

$$\phi = e^{i\omega t} \chi(x^1, x^2), \qquad (5.14)$$

we have that

$$\dot{\phi} = i\omega\phi \quad \dot{\phi}^* = -i\omega\phi^*. \qquad (5.15)$$

Thus, for steady states, the vanishing condition associated with the conservation of energy (3.2) implies the conservation of charge (5.12)

$$\dot{\varphi}^* \varphi' + \varphi^{*\prime} \dot{\varphi} = -i\omega \left(\varphi^* \varphi' - \varphi \varphi^{*\prime} \right) = 0. \qquad (5.16)$$

Since (5.16) holds for any value of ω, it also holds for non-steady states, and viceversa, the conservation of charge implies the conservation of energy for all states.

However, in spite of the fact that U is independent of ω, we cannot infer from the boundary conditions

$$\begin{pmatrix} \varphi' + i\omega\varphi \\ \varphi^{*\prime} - i\omega\varphi^* \end{pmatrix} = U \begin{pmatrix} \varphi' - i\omega\varphi \\ \varphi^{*\prime} + i\omega\varphi^* \end{pmatrix} \qquad (5.17)$$

that the fields also satisfy a boundary condition U_c of the type (5.13). Only for boundary conditions where U has all eigenvalues, ± 1 define boundary conditions of the type U_c.[21] This includes Dirichlet, Neumann or periodic boundary conditions U.

The application to higher spin theories is straightforward specially in the case of gauge fields.[22]

---

[d]There is a technical subtility associated with the fact that in higher dimension, the boundary values φ of the fields ϕ are singular, but in can be solved with a slight modification of the boundary analysis.[17]

Acknowledgments

Manuel Asorey thanks A. P. Balachandran for so many discussions on the role of boundary conditions in field theory and topological insulators; and not only on technical aspects but also on fundamental aspects of life and society. We are partially supported by Spanish MINECO/FEDER Grant PGC2018-095328-B-I00 funded by MCIN/AEI/10.13039/501100011033 and by ERDF A way of making Europe and DGA-FSE grant 2020-E21-17R.

References

1. J. Zhang, L. Zhang and W. Xu, *J. Phys. D: Appl. Phys.* **45** (2012) 113001.
2. R. B. Laughlin, *Phys. Rev.* **B 23** (1981) 5632.
3. B. I. Halperin, *Phys. Rev. B* **25**(4) (1982) 2185.
4. Y. Hatsugai, *Phys. Rev. Lett.* **71** (1993) 3697.
5. C. L. Kane and E. J. Mele, *Phys. Rev. Lett.* **95** (2005) 146802.
6. L. Fu and C. L. Kane, *Phys. Rev.* **B 76** (2007) 045302.
7. M. Z. Hasan and C. L. Kane, *Rev. Mod. Phys.* **82** (2010) 3045–3067.
8. J. E. Moore, *Nature* **464** (2010) 194–198.
9. X.-L. Qi and S.-C. Zhang, *Rev. Mod. Phys.* **83** (2011) 1057–1110.
10. B. A. Bernevig and T. L. Hughes, *Topological Insulators and Topological Superconductors*, Princeton University Press (2013).
11. M. Zahid Hasan, S.-Y. Xu and M. Neupane, *Topological Insulators: Fundamentals and Perspectives*, Wiley and Sons, Weinheim, Germany (2015) 55–100.
12. H. B. G. Casimir, *Proc. K. Ned. Akad. Wet.* **51** (1948) 793.
13. A. P. Balachandran, G. Bimonte, G. Marmo and A. Simoni, *Nucl. Phys.* **B446** (1995) 299.
14. A. Asorey, A. Ibort and G. Marmo, *Int. J. Mod. Phys.* **A 20** (2005) 1001.
15. A. D. Shapere, F. Wilczek and Z. Xiong, *Models of Topology Change*, arXiv:1210.3545 (2012).
16. M. Asorey *et al.*, *Quantum Physics and Fluctuating Topologies: Survey*, arXiv:1211.6882 [hep-th] (2012).
17. M. Asorey and J. M. Muñoz-Castañeda, *Nucl. Phys. B* **874** (2013) 852.
18. M. Asorey, D. García-Alvarez and J. M. Muñoz-Castañeda, *Int. J. Geom. Meth. Mod. Phys.* **12**(06) (2015) 1560004.
19. J. Polchinski, *String Theory, Vol. 1*, Cambridge University Press (2005).
20. F. Antoneli, M. Forger and P. Gaviria, *J. Lie Theory*, **22**(4) (2012) 949–1024.
21. M. Asorey, D. García-Alvarez and J. M. Muñoz-Castañeda, *J. Phys. A* **40** (2007) 6767.
22. M. Asorey and F. Ezquerro, In preparation.

https://doi.org/10.1142/9789811270437_0003

Chapter 3

Our Trysts with 'Bal' and Noncommutative Geometry

Biswajit Chakraborty[*,‡], Partha Nandi[†], Sayan Kumar Pal[†]
and Anwesha Chakraborty[†]

[*]*Department of Physics, School of Mathematical Sciences,
Ramakrishna Mission Vivekananda Educational and Research Institute,
PO Belur Math, Howrah 711202, India*
[†]*S.N. Bose National Centre for Basic Sciences,
Block JD, Sec 3, Salt Lake, Kolkata - 700106, India*
[‡]*dhrubashillong@gmail.com*

This contributory chapter begins with our fond and sincere reminiscences about our beloved Prof. A.P. Balachandran. In the main part, we discuss our recent formulation of quantum mechanics on (1+1)D noncommutative space-time using Hilbert–Schmidt operators. As an application, we demonstrate how geometrical phase in a system of time-dependent forced harmonic oscillator living in the Moyal space-time can emerge.

1. Down the Memory Lane with Prof. A.P. Balachandran

It is indeed a great privilege for us to get the opportunity to dedicate this note to one of our role models and teacher, a doyen of theoretical physics: Professor A.P. Balachandran, on his 85th birthday. To start with, let us begin with some brief nostalgic reminiscences of the past. One of us, B.C., knows Professor A.P. Balachandran, or simply Bal as we fondly refer to him, for almost four decades now. Although I never had the privilege of co-authoring any paper with him, I share with him our common alma-mater; we both are alumni of IMSc (Institute of Mathematical Sciences, Chennai) where we have carried out our Ph.D. works, a quarter century apart. I had joined IMSc in 1985, soon after Prof. E.C.G. Sudarshan took charge of this institute as its director and Prof. C.S. Seshadri, Prof. R. Balasubhramanian joined the Mathematics department and IMSc started getting very distinguished visitors like Bal in a regular basis. It was this rejuvenated academic environment that stimulated our passion for pursuing high quality research work to the best of our individual capacity. Importantly, Bal used to visit IMSc at least for once/twice every year to collaborate and give seminars/talks on new and interesting developments happening in theoretical Physics. He also used to deliver some pedagogical lecture series for the graduate students in IMSc. And once we got ourselves familiarized with the basics of mathematical aspects like topology, differential geometry, group theory, etc., his lectures and/or papers started becoming accessible and I was fascinated by the beauty and depth of the whole approach, and eventually, he became a role model for me, both as a researcher and a teacher.

Among the many areas that he has contributed, I would, particularly, like to mention about his enormous and deep contribution for understanding the symmetry aspects of noncommutative (NC) spaces, as captured by quantum groups/Hopf algebras and he has also written several books on the subject.[1] In fact, in one of his colloquia, delivered in the Saha Institute of Nuclear Physics, Kolkata, in early 2000, he gave an introductory lecture on noncommutative geometry — a new emerging area at that time. Particularly, his elaborated explanation on Gelfand–Naimark theorem was absolutely fascinating to me and in retrospect, I feel that this single event played a pivotal role in triggering my persistent fascination towards this subject — an area of my current interest.

On the other hand, the other co-authors have got to know Bal personally only since 2018, when he participated as a keynote speaker in the international conference on noncommutative geometry held at S.N. Bose Centre, Kolkata. Although we were familiar with his works which served as our references and checkpoints for our previous papers, this was the first occasion that we got to meet him personally and carry out intense academic discussions. We were quite enthralled to see his passion for physics even at this ripe age and patience to answer our naive questions. Subsequently, we have maintained our regular academic contacts in the online mode where he makes himself always available for physics discussions. Besides, almost all of our recent papers were written after incorporating all of his deep suggestions. Finally, he was quite generous to give us (P.N., S.K.P. and A.C.) the opportunity to speak at his famous Room-316 meetings from Syracuse University, which are being re-organized these days in online mode after the pandemic got started.

2. Noncommutative Geometry: Quantum Space-time

The celebrated idea of noncommutativity in modern physics has attracted a lot of interest theoretically in particle physics and condensed matter physics. Additionally, noncommutativity has played an important role in the study of quantum gravity at the Planck scale over the previous two decades. To motivate the emergence of noncommutative space-time or more generally quantum space-time, one can think of a superposition of two mass distributions. As has been argued by Penrose in Ref. 2 that as a feedback through Einstein's GR, this will give rise to a superposed geometry/space-time. Now, such a quantum space-time is likely to loose its time-translational symmetry resulting in the uncertainty of energy δE and time δt, indicating a finite lifetime $\sim \frac{\hbar}{\delta E}$ of the system. This heuristic argument indicates that one needs to reformulate the quantum theory without classical time, rather time should be promoted as an operator-valued coordinate, along with other operator-valued spatial coordinates (see also Refs. 3, 4). And we mention, in this context, the status of time in quantum gravity is an age-old problem. In fact, its status in QM itself is a bit ambiguous. One can, in fact, recall Pauli's objection in this context[5] and this ambiguity can result in other allied problems (for example, see Ref. 6). And the physics of such quantum systems should have its manifestations, at least in principle, even in sub-Planckian energy scales which also can even be effectively non-relativistic, like non-classical features in primordial gravitational waves.[7] Again, here we can cite many of the seminal works of Bal in this direction.[8] In fact, we were motivated primarily through these papers to pursue such studies and take it forward, as we describe in the sequel.

In this chapter, we discuss about our recent investigations and formulation of QM in $(1+1)$D noncommutative space-time.[9, 10] Thereafter, as an application, we discuss about the possibility of obtaining emergent Berry phase in a system inhabiting NC space-time. We start by giving a brief outline of the Hilbert–Schmidt operator-based formulation of noncommutative quantum mechanics (NCQM) which will serve as the basic mathematical framework for the present treatment.

3. Space-time NCQM Using Hilbert–Schmidt Operator Formulation

Although initially the Hilbert–Schmidt (HS) operator-based formulation of noncommutative quantum mechanics was systematically devised, following Ref. 11, to formulate quantum mechanics on spatial 2D noncommutative

Moyal plane,[12, 13] here, in this chapter, we provide a brief review of our recent works on the formulation of QM on noncommutative (1+1)D Moyal space-time (based on the results in Refs. 9, 10) and how the HS operator formulation can be adapted to formulate or rather used to extract an effective, consistent and commutative quantum mechanical theory. Now, before considering the quantum theory, let us first discuss the appearance of non-commuting nature of space-time brackets just at the classical level itself. To this end, let us consider the following Lagrangian of a non-relativistic particle in (1+1)D written in the first-order form[14]:

$$L^{\tau,\theta} = p_\mu \dot{x}^\mu + \frac{\theta}{2} \epsilon^{\mu\nu} p_\mu \dot{p}_\nu - \sigma(\tau)(p_t + H), \quad \mu, \nu = 0, 1, \qquad (3.1)$$

where $x^\mu = (t, x)$ and $p_\mu = (p_t, p_x)$ are both counted as configuration space variables. The evolution parameter τ is chosen a bit arbitrarily, except that it should be taken as a monotonically increasing function of time 't'. The overhead dots indicate $\tau-$ derivatives. The middle term is analogous to the Chern–Simons term with θ being the corresponding parameter. On carrying out Dirac's analysis of constraints, one arrives at the following Dirac brackets between the phase space variables:

$$\{x^\mu, x^\nu\}_D = \theta \epsilon^{\mu\nu}; \quad \{p_\mu, p_\nu\}_D = 0; \quad \{x^\mu, p_\nu\}_D = \delta^\mu{}_\nu. \qquad (3.2)$$

Finally, the Lagrange multiplier $\sigma(\tau)$ enforces the following first-class constraint in the system

$$\Sigma = p_t + H \approx 0 \qquad (3.3)$$

and can be shown to generate the τ evolution of the system in the form of gauge transformation of the theory.

We now elevate the Dirac brackets in (3.2) to the level of noncommuting operators in order to initiate the quantum mechanical analysis for this (1+1)D non-relativistic quantum mechanical system in the presence of the space-time noncommutativity of Moyal type:

$$[\hat{t}, \hat{x}] = i\theta, \quad \text{where } \theta \text{ is the NC parameter.} \qquad (3.4)$$

3.1. HS operator formulation

A representation of NC coordinate algebra (3.4) can be readily found to be furnished by the Hilbert space,

$$\mathcal{H}_c = Span\left\{|n\rangle = \frac{(b^\dagger)^n}{\sqrt{n!}}|0\rangle; \ b|0\rangle = \frac{\hat{t} + i\hat{x}}{\sqrt{2\theta}}|0\rangle = 0\right\}. \qquad (3.5)$$

We now introduce the associative NC operator algebra $(\hat{\mathcal{A}}_\theta)$ generated by $(\hat{t}, \hat{x})$ or equivalently by $(\hat{b}, \hat{b}^\dagger)$ acting on this configuration space $\mathcal{H}_c$ (3.5) as

$$\hat{\mathcal{A}}_\theta = \{|\psi) = \psi(\hat{t}, \hat{x}) = \psi(\hat{b}, \hat{b}^\dagger) = \sum_{m,n} c_{n,m}|m\rangle\langle n|\} \tag{3.6}$$

which is the set of all polynomials in the quotient algebra $(\hat{\mathcal{A}}/\mathcal{N})$ in $(\hat{t}, \hat{x})$ or equivalently in $(\hat{b}, \hat{b}^\dagger)$, subject to the identification of $[\hat{b}, \hat{b}^\dagger] = 1$. Thus, $\hat{\mathcal{A}}_\theta = \hat{\mathcal{A}}/\mathcal{N}$ is essentially identified as the universal enveloping algebra corresponding to (3.4), where $\hat{\mathcal{A}}$ is the free algebra generated by $(\hat{t}, \hat{x})$ and $\mathcal{N}$ is the ideal generated by (3.4). This $\hat{\mathcal{A}}_\theta$ is not equipped with any inner product at this stage.

We can now introduce a subspace $\mathcal{H}_q \subset \mathcal{B}(\mathcal{H}_c) \subset \hat{\mathcal{A}}_\theta$ as the space of 'HS' operators, which are bounded and compact operators with finite HS norm $||.||_{HS}$, which acts on $\mathcal{H}_c$ (3.5), and is given by

$$\mathcal{H}_q = \left\{ \psi(\hat{t}, \hat{x}) \equiv |\psi(\hat{t}, \hat{x})) \in \mathcal{B}(\mathcal{H}_c); \ ||\psi||_{HS} := \sqrt{tr_c(\psi^\dagger \psi)} < \infty \right\} \subset \hat{\mathcal{A}}_\theta, \tag{3.7}$$

where tr_c denotes the trace over $\mathcal{H}_c$ and $\mathcal{B}(\mathcal{H}_c) \subset \hat{\mathcal{A}}_\theta$ is a set of bounded operators on $\mathcal{H}_c$. This space can be equipped with the inner product

$$\left(\psi(\hat{t}, \hat{x}), \phi(\hat{t}, \hat{x}) \right) := tr_c\left(\psi^\dagger(\hat{t}, \hat{x})\phi(\hat{t}, \hat{x}) \right) \tag{3.8}$$

and therefore has the structure of a Hilbert space on its own. Note that we denote the elements of $\mathcal{H}_c$ and $\hat{\mathcal{A}}_\theta$ by the angular ket $|.\rangle$ and round ket $|.)$, respectively. We now define the quantum space-time coordinates $(\hat{T}, \hat{X})$ (which can be regarded as the representation of $(\hat{t}, \hat{x})$ and must be distinguished as their domains of actions are different, i.e., while $(\hat{T}, \hat{X})$ act on $\mathcal{H}_q$, $(\hat{t}, \hat{x})$ act on $\mathcal{H}_c$) as well as the corresponding momenta $(\hat{P}_t, \hat{P}_x)$ by their actions on a state vector $|\psi(\hat{t}, \hat{x})) \in \mathcal{H}_q$ as

$$\hat{T}|\psi(\hat{t}, \hat{x})) = |\hat{t}\psi(\hat{t}, \hat{x})), \quad \hat{X}|\psi(\hat{t}, \hat{x})) = |\hat{x}\psi(\hat{t}, \hat{x})),$$

$$\hat{P}_x|\psi(\hat{t}, \hat{x})) = -\frac{1}{\theta}|[\hat{t}, \psi(\hat{t}, \hat{x})]), \quad \hat{P}_t|\psi(\hat{t}, \hat{x})) = \frac{1}{\theta}|[\hat{x}, \psi(\hat{t}, \hat{x})]). \tag{3.9}$$

Thus, the momenta $(\hat{P}_t, \hat{P}_x)$ act adjointly and their actions are only defined in $\mathcal{H}_q$ and not in $\mathcal{H}_c$. It may be easily verified now that

$$[\hat{T}, \hat{X}] = i\theta, \ [\hat{T}, \hat{P}_t] = i\theta = [\hat{X}, \hat{P}_x], \ [\hat{P}_t, \hat{P}_x] = 0 \tag{3.10}$$

which represents the total NC Heisenberg algebra, quantum version of the classical algebra (3.2). Note, here we are working in the natural unit $\hbar = 1$.

3.1.1. *Schrödinger equation and an induced inner product*

As $\theta \neq 0$, simultaneous space-time eigenstate $|x, t\rangle$ does not exist. Nevertheless, an effective commutative theory can be constructed by making use of maximally localized events, i.e., the Sudarshan–Glauber coherent states

$$|z\rangle = e^{-\tilde{z}\tilde{b} + z\tilde{b}^\dagger} |0\rangle \in \mathcal{H}_c; \quad \tilde{b}|z\rangle = z|z\rangle, \tag{3.11}$$

where z is a dimensionless complex number and is given by

$$z = \frac{t + ix}{\sqrt{2\theta}}; \quad t = \langle z|\hat{t}|z\rangle, x = \langle z|\hat{x}|z\rangle, \tag{3.12}$$

where t and x are effective commutative coordinate variables. We can now construct the counterpart of coherent state basis in $\mathcal{H}_q$ (3.7), made out of the bases $|z\rangle \equiv |x, t\rangle$ (3.11), by taking their outer product as

$$|z, \bar{z}) \equiv |z) = |z\rangle\langle z| = \sqrt{2\pi\theta} \, |x, t) \in \mathcal{H}_q \quad \text{fulfilling } B|z) = z|z), \tag{3.13}$$

where the annihilation operator $\hat{B} = \frac{\hat{T} + i\hat{X}}{\sqrt{2\theta}}$ is a representation of the operator $\tilde{b}$ in $\mathcal{H}_q$ (3.7). It can also be checked that the basis $|z, \bar{z}) \equiv |z)$ satisfies the overcompleteness property:

$$\int \frac{d^2 z}{\pi} |z, \bar{z}) \star_V (z, \bar{z}| = \int dt dx \, |x, t) \star_V (x, t| = \mathbf{1}_q, \tag{3.14}$$

where $\star_V$ represents the Voros star product and is given by

$$\star_V = e^{\overleftarrow{\partial_z} \overrightarrow{\partial_{\bar{z}}}} = e^{\frac{i\theta}{2}(-i\delta_{ij} + \epsilon_{ij})\overleftarrow{\partial_i} \overrightarrow{\partial_j}}; \quad i, j = 0, 1; \quad x^0 = t, x^1 = x. \tag{3.15}$$

Then, the coherent state representation or the symbol of an abstract state $\psi(\hat{t}, \hat{x})$ gives the usual coordinate representation of a state just like ordinary QM:

$$\psi(x, t) = \frac{1}{\sqrt{2\pi\theta}} \left(z, \bar{z} \middle| \psi(\hat{x}, \hat{t}) \right) = \frac{1}{\sqrt{2\pi\theta}} tr_c \left[|z\rangle\langle z| \psi(\hat{x}, \hat{t}) \right]$$

$$= \frac{1}{\sqrt{2\pi\theta}} \langle z| \psi(\hat{x}, \hat{t})|z\rangle. \tag{3.16}$$

The corresponding representation of a composite operator say $\psi(\hat{x}, \hat{t})\phi(\hat{x}, \hat{t})$ is given by

$$\left(z \middle| \psi(\hat{x}, \hat{t})\phi(\hat{x}, \hat{t}) \right) = \left(z \middle| \psi(\hat{x}, \hat{t}) \right) \star_V \left(z \middle| \phi(\hat{x}, \hat{t}) \right). \tag{3.17}$$

This establishes an isomorphism between the space of HS operators $\mathcal{H}_q$ and the space of their respective symbols. Using (3.14), the overlap of two arbitrary states $(|\psi), |\phi))$ in the quantum Hilbert space $\mathcal{H}_q$ can be written in the form

$$(\psi|\phi) = \int dt dx \; \psi^*(x,t) \star_V \phi(x,t). \qquad (3.18)$$

Therefore, to each element $|\psi(\hat{x},\hat{t})) \in \mathcal{H}_q$, the corresponding symbol $\psi(x,t) \in L^2_\star(\mathbb{R}^2)$, where the $*$ occurring in the subscript is a reminder of the fact that the corresponding norm has to be computed by employing the Voros star product. In order to obtain an effective commutative Schrödinger equation in coordinate space, we will introduce coordinate representation of the phase space operators. To begin with, note that the coherent state representation of the actions of space-time operators $\{\hat{X}, \hat{T}\}$, acting on $|\psi)$, can be written by using (3.17) as

$$\left(x,t\Big|\hat{X}\,\psi(\hat{x},\hat{t})\right) = \frac{1}{\sqrt{2\pi\theta}}\left(z,\bar{z}\Big|\hat{x}\psi\right) = \frac{1}{\sqrt{2\pi\theta}}\,\langle z|\hat{x}|z\rangle \star_V (z,\bar{z}|\psi(\hat{x},\hat{t})).$$
$$(3.19)$$

Finally, on using (3.16), we have

$$\left(x,t\Big|\hat{X}\psi(\hat{x},\hat{t})\right) = X_\theta\left(x,t\Big|\psi(\hat{x},\hat{t})\right) = X_\theta\,\psi(x,t); \quad X_\theta = x + \frac{\theta}{2}(\partial_x - i\partial_t).$$
$$(3.20)$$

Proceeding exactly in the same way, we obtain the representation of $\hat{T}$ as

$$T_\theta = t + \frac{\theta}{2}(\partial_t + i\partial_x) \qquad (3.21)$$

so that $[T_\theta, X_\theta] = i\theta$ is trivially satisfied. It is now trivial to prove the self-adjointness property of both X_θ and T_θ, w.r.t. the inner product (3.18) in $\mathcal{H}_q$ by considering an arbitrary pair of different states $|\psi_1), |\psi_2) \in \mathcal{H}_q$ and their associated symbols, just by exploiting associativity of Voros star product. Since momenta operators act adjointly, their coherent state representations are

$$\left(x,t\Big|\hat{P}_t\psi(\hat{x},\hat{t})\right) = -i\partial_t\psi(x,t); \quad \left(x,t\Big|\hat{P}_x\psi(\hat{x},\hat{t})\right) = -i\partial_x\psi(x,t). \quad (3.22)$$

The effective commutative Schrödinger equation in NC space-time is then obtained by imposing the condition that the physical states $|\psi_{phy}) = \psi_{phy}(\hat{x},\hat{t})$ are annihilated by the operatorial version of (3.3):

$$(\hat{P}_t + \hat{H})|\psi_{phy}) = 0; \quad \psi_{phy}(\hat{x},\hat{t}) \in \mathcal{H}_{phy} \subset \hat{\mathcal{A}}_\theta, \qquad (3.23)$$

where $\hat{H} = \frac{\hat{P}_x^2}{2m} + V(\hat{X}, \hat{T})$. We are now ready to write down the effective commutative time-dependent Schrödinger equation in quantum space-time by taking the representation of (3.23) in $|x, t\rangle$ basis. Using (3.20,3.21,3.22), we finally get

$$i\partial_t \psi_{phy}(x, t) = \left[-\frac{1}{2m}\partial_x^2 + V(x, t) \star_V \right] \psi_{phy}(x, t). \qquad (3.24)$$

One can now obtain the continuity equation as

$$\partial_t \rho_\theta + \partial_x J_\theta^x = 0, \qquad (3.25)$$

where

$$\rho_\theta(x, t) = \psi_{phy}^*(x, t) \star_V \psi_{phy}(x, t) > 0; \quad J_\theta^x = \frac{1}{m}\mathfrak{Im}\left(\psi_{phy}^* \star_V (\partial_x \psi_{phy}) \right).$$

The positive definite property of $\rho_\theta(x, t)$ indicates that it can be interpreted as the probability density at point x at time t. However, note that for a consistent QM formulation, we ought to have $\psi_{phy}(x, t)$ to be "well behaved" in the sense that it should be square integrable at a constant time slice:

$$\langle \psi_{phy} | \psi_{phy} \rangle_{*t} = \int_{-\infty}^{\infty} dx \ \psi_{phy}^*(x, t) \star_V \psi_{phy}(x, t) < \infty \qquad (3.26)$$

so that $\psi_{phy}(x, t) \in L^2_\star(\mathbb{R}^1)$ which is naturally distinct from $L^2_\star(\mathbb{R}^2)$. Equivalently, at the operator level, $\psi_{phy}(\hat{x}, \hat{t})$ should belong to a suitable subspace of $\hat{\mathcal{A}}_\theta$ (3.6) which is distinct from $\mathcal{H}_q$, as the associated symbol for the latter is obtained from inner product defined for $L^2_\star(\mathbb{R}^2)$ (3.18). This is the main point of departure from the standard HS operator formulation of NCQM in (1+2)D Moyal plane with only spatial noncommutativity where time is treated as a c-parameter and one works with $\mathcal{H}_q$ or equivalently with a Hilbert space $L^2_\star(\mathbb{R}^2)$ for the corresponding symbols.[12, 13]

4. Emergence of Berry Phase in a Time-Dependent NC Quantum System

A time-independent system like a harmonic oscillator has been shown to exhibit no modifications in the spectrum in a noncommutative space-time.[9] Therefore, the forced harmonic oscillator, having explicit time-dependence, is a good prototype system to study and look for any possible emergent geometric phases, as one of the possible signals of space-time noncommutativity. We therefore take up the Hamiltonian of the forced harmonic oscillator in the following hermitian form for carrying out our analysis:

$$\hat{H} = \frac{\hat{P}_x^2}{2m} + \frac{1}{2}m\omega^2\hat{X}^2 + \frac{1}{2}[f(\hat{T})\hat{X} + \hat{X}f(\hat{T})] + g(\hat{T})\hat{P}_x. \qquad (4.1)$$

The corresponding effective commutative Schrödinger equation can be obtained by taking overlap of (3.23) in coherent state basis (3.13, 3.16),

$$i\partial_t \psi_{phy}(x,t) = \left[\frac{P_x^2}{2m} + \frac{1}{2}m\omega^2 X_\theta^2 + \frac{1}{2}\{f(T_\theta)X_\theta + X_\theta f(T_\theta)\} + g(T_\theta)P_x\right]$$
$$\times \psi_{phy}(x,t). \qquad (4.2)$$

At this stage, it will be interesting to note that X_θ and T_θ can be related to commutative x and t, defined in (3.20, 3.21), by making use of similarity transformations,

$$X_\theta = SxS^{-1}, \quad T_\theta = S^\dagger t(S^\dagger)^{-1}; \quad S = e^{\frac{\theta}{4}(\partial_t^2 + \partial_x^2)}e^{-\frac{i\theta}{2}\partial_t \partial_x}. \qquad (4.3)$$

This S, a non-unitary operator, can be used to define the following map:

$$S^{-1}: L_*^2(\mathbb{R}^1) \to L^2(\mathbb{R}^1); \quad i.e. \ S^{-1}\big(\psi_{phy}(x,t)\big) := \psi_c(x,t) \in L^2(\mathbb{R}^1). \qquad (4.4)$$

Now, one can easily verify at this stage,

$$\Big\langle \psi_{phy}, \phi_{phy} \Big\rangle_{*,t} = \langle \psi_c, \phi_c \rangle_t \ \forall \psi_{phy}, \phi_{phy} \in L_*^2(\mathbb{R}^1). \qquad (4.5)$$

Thus, one can replace non-local Voros star product with the local pointwise multiplication as in the usual commutative QM but *only* within the integral when working in $L^2(\mathbb{R}^1)$. This is because one can establish the identity only by dropping many of the boundary terms indicating that $\psi_{phy}^*(x,t) \star_V \psi_{phy}(x,t) \neq |\psi_c(x,t)|^2$. Thus, here, one can't interpret $|\psi_c(x,t)|^2$ as the probability density at point x at time t, unlike $(\psi_{phy}^* \star_V \psi_{phy})(x,t)$. In this context, one can recall the so-called T-map connecting Moyal and Voros star products.[15] Now, using (4.3, 4.4) in (4.2) and retaining terms up to linear in θ, finally, (4.2) can now be recast as

$$i\partial_t \psi_c(x,t) = H_c \psi_c(x,t), \tag{4.6}$$

where the corresponding effective commutative Hamiltonian H_c is given by

$$\begin{aligned}
H_c &= \alpha(t)p_x^2 + \beta x^2 + \gamma(t)(xp_x + p_x x) + f(t)x + g(t)p_x \\
&= H_{GHO} + f(t)x + g(t)p_x,
\end{aligned} \tag{4.7}$$

where H_{GHO} stands for the Hamiltonian of a generalized time-dependent harmonic oscillator representing the first three terms. The last two terms represent perturbations linear in position and momentum in coordinate basis. Finally, all the various coefficients in (4.7) are given by

$$\alpha(t) = \frac{1}{2m} - \theta \dot{g}(t); \quad \beta = \frac{1}{2}m\omega^2; \quad \gamma(t) = -\frac{1}{2}\theta \dot{f}(t). \tag{4.8}$$

This Hamiltonian can be put into the diagonal form

$$\tilde{H}_c = \Omega(t)\left(a^\dagger a + \frac{1}{2}\right) = H_{GHO} \tag{4.9}$$

after a series of time-dependent unitary transformations. For details, we refer to Ref. 10. Carrying out the analysis in the Heisenberg picture with adiabatic approximation, we find after evolution through a cycle of time period $t = \mathcal{T}$,

$$a^\dagger(\mathcal{T}) = a^\dagger(0)exp\left[i\int_0^{\mathcal{T}} \Omega \, d\tau + i\int_0^{\mathcal{T}} \left(\frac{1}{\Omega}\right)\frac{d\gamma}{d\tau}d\tau\right]. \tag{4.10}$$

The second term in the exponential of (4.10) represents the additional phase over and above the dynamical phase $e^{i\int \Omega(t)dt}$. This extra phase can now be written in a more familiar form of Berry phase,[16] given as a functional of the closed loop Γ,

$$\Phi_G[\Gamma] = \oint_{\Gamma=\partial S} \frac{1}{\Omega} \nabla_{\mathbf{R}}\gamma . d\mathbf{R} = -\frac{\theta}{2}\int\int_S \nabla_{\mathbf{R}}\left(\frac{1}{\Omega}\right) \times \nabla_{\mathbf{R}}\left(\dot{f}(t)\right) . d\mathbf{S}. \tag{4.11}$$

It is crucial to take note of the $su(1,1)$ Lie-algebraic structure of the GHO part of the effective Hamiltonian (4.7) in noncommutative space-time that is responsible for this emergent geometrical phase.

Acknowledgements

We sincerely thank Prof. V.P. Nair and Prof. T.R. Govindarajan for giving us the opportunity to contribute to this celebratory Festschrift to honor the legacy of Prof. A.P. Balachandran on his 85th birthday.

References

1. A. P. Balachandran, S. G. Jo and G. Marmo, *Group Theory and Hopf Algebras: Lectures for Physicists*, World Scientific Publishing Co. Pte. Ltd. (2010); A. P. Balachandran, S. Kurkcuoglu, S. Vaidya, *Lectures on Fuzzy and Fuzzy Susy Physics*, World Scientific Publishing Co. (2007).
2. R. Penrose, *Gen. Rel. Grav.* **8** (1996) 5.
3. F. G. Scholtz, *Double Slit Experiment in the Noncommutative Plane and the Quantum-to-Classical Transition*, arXiv:2101.06108 [quant-ph].
4. S. Doplicher, K. Fredenhagen and J. E. Roberts, *Comm. Math. Phys.* **172** (1995) 187.
5. W. Pauli, *Handbook der Physics*, edited by S. Flugge, Vol. 5/1, p. 60 Berlin (1926).
6. R. Gambini and J. Pullin, *New J. Phys.* (2022) 114803.
7. S. Kanno, J. Soda and J. Tokuda, *Phys. Rev. D* **103** (2021) 044017.
8. A. P. Balachandran, T. R. Govindarajan, A. G. Martins and P. Teotonio-Sobrinho, *JHEP* **0411** (2004) 068; A. P. Balachandran, T. R. Govindarajan, C. Molina and P. Teotonio-Sobrinho, *JHEP* **0410** (2004) 072; A. P. Balachandran and A. Pinzul, *Mod. Phys. Lett. A* **20** (2005) 2023; A. P. Balachandran, A. G. Martins and P. Teotonio-Sobrinho, *JHEP* **05** (2007) 066.
9. P. Nandi, S. K. Pal, A. N. Bose and B. Chakraborty, *Ann. Phys.* **386** (2017) 305–326.
10. A. Chakraborty, P. Nandi and B. Chakraborty, *Nucl. Phys. B* **975** (2022) 115691.
11. V. P. Nair and A. P. Polychronakos, *Phys. Lett. B* **505** (2001) 267–274.
12. F. G. Scholtz, B. Chakraborty, J. Govaerts and S. Vaidya, *J. Phys. A* **40** (2007) 14581.
13. F. G. Scholtz, L. Gouba, A. Hafver and C. M. Rohwer, *J. Phys. A* **42** (2009) 175303.
14. A. Deriglazov and B. F. Rizzuti, *Am. J. Phys.* **79** (2011) 882.
15. A. P. Balachandran, A. Ibort, G. Marmo and M. Martone, *Phys. Rev. D* **81** (2010) 085017; A. P. Balachandran, A. Ibort, G. Marmo and M. Martone, *SIGMA* **6** (2010) 052.
16. M. V. Berry, *Proc. Roy. Soc. A* **392** (1984) 45.

Chapter 4

Exceptional Fuzzy Spaces and Octonions

Denjoe O'Connor[*] and Brian P. Dolan[†]

School of Theoretical Physics, 10, Burlington Rd., Dublin 4, Ireland
[*] *denjoe@stp.dias.ie*
[†] *bdolan@stp.dias.ie*

We construct the fuzzy spaces based on the three non-trivial coadjoint orbits of the exceptional simple Lie group, G_2.

1. Introduction

Bal has had a long-standing interest in coadjoint orbits, noncommutative geometry and quantum space-times. In this contribution, we pull from all these ingredients, spiced with a smidgen of octonions, to make a contribution that hopefully Bal and others find interesting.

We consider the coadjoint orbits of the exceptional group G_2 and construct fuzzy versions of these spaces. In many respects, the orbits are very simple. One has holonomy $U(1) \times U(1)$ while two others have $U(2)$ as their holonomy group and, as Bal has emphasised to us, conformally compactified Minkowski space-time is also $U(2)$. These two are both 10-dimensional orbits, one of which is an S^2 bundle over the quaternionic projective plane $\mathbf{HP}^2$ and the other is a $\mathbf{CP}^2$ bundle over S^6 which is diffeomorphic but not isospectral to the $SO(7)$ quadric $\mathbf{Q}^5$.

2. G_2

The 14-dimensional exceptional group G_2 has rank 2. Its Dynkin diagram is not symmetric and G_2 therefore has no complex representations. The smallest non-trivial irrep. is 7-dimensional and we shall use an explicit representation for the 14 generators,[1] here labeled $T_1, \ldots, T_{14}$:

$$
T_1 = \begin{pmatrix}
0 & 0 & 0 & 0 & 0 & 0 & 0 \\
0 & 0 & 0 & 0 & 0 & 0 & 0 \\
0 & 0 & 0 & 0 & 0 & 0 & 0 \\
0 & 0 & 0 & 0 & 0 & 0 & -1 \\
0 & 0 & 0 & 0 & 0 & -1 & 0 \\
0 & 0 & 0 & 0 & 1 & 0 & 0 \\
0 & 0 & 0 & 1 & 0 & 0 & 0
\end{pmatrix}
\qquad
T_2 = \begin{pmatrix}
0 & 0 & 0 & 0 & 0 & 0 & 0 \\
0 & 0 & 0 & 0 & 0 & 0 & 0 \\
0 & 0 & 0 & 0 & 0 & 0 & 0 \\
0 & 0 & 0 & 0 & 0 & 1 & 0 \\
0 & 0 & 0 & 0 & 0 & 0 & -1 \\
0 & 0 & 0 & -1 & 0 & 0 & 0 \\
0 & 0 & 0 & 0 & 1 & 0 & 0
\end{pmatrix}
$$

$$
T_3 = \begin{pmatrix}
0 & 0 & 0 & 0 & 0 & 0 & 0 \\
0 & 0 & 0 & 0 & 0 & 0 & 0 \\
0 & 0 & 0 & 0 & 0 & 0 & 0 \\
0 & 0 & 0 & 0 & -1 & 0 & 0 \\
0 & 0 & 0 & 1 & 0 & 0 & 0 \\
0 & 0 & 0 & 0 & 0 & 0 & -1 \\
0 & 0 & 0 & 0 & 0 & 1 & 0
\end{pmatrix}
\qquad
T_4 = \begin{pmatrix}
0 & 0 & 0 & 0 & 0 & 0 & 0 \\
0 & 0 & 0 & 0 & 0 & 0 & 1 \\
0 & 0 & 0 & 0 & 0 & 1 & 0 \\
0 & 0 & 0 & 0 & 0 & 0 & 0 \\
0 & 0 & 0 & 0 & 0 & 0 & 0 \\
0 & 0 & -1 & 0 & 0 & 0 & 0 \\
0 & -1 & 0 & 0 & 0 & 0 & 0
\end{pmatrix}
$$

$$
T_5 = \begin{pmatrix}
0 & 0 & 0 & 0 & 0 & 0 & 0 \\
0 & 0 & 0 & 0 & 0 & -1 & 0 \\
0 & 0 & 0 & 0 & 0 & 0 & 1 \\
0 & 0 & 0 & 0 & 0 & 0 & 0 \\
0 & 0 & 0 & 0 & 0 & 0 & 0 \\
0 & 1 & 0 & 0 & 0 & 0 & 0 \\
0 & 0 & -1 & 0 & 0 & 0 & 0
\end{pmatrix}
\qquad
T_6 = \begin{pmatrix}
0 & 0 & 0 & 0 & 0 & 0 & 0 \\
0 & 0 & 0 & 0 & 1 & 0 & 0 \\
0 & 0 & 0 & -1 & 0 & 0 & 0 \\
0 & 0 & 1 & 0 & 0 & 0 & 0 \\
0 & -1 & 0 & 0 & 0 & 0 & 0 \\
0 & 0 & 0 & 0 & 0 & 0 & 0 \\
0 & 0 & 0 & 0 & 0 & 0 & 0
\end{pmatrix}
$$

$$
T_7 = \begin{pmatrix}
0 & 0 & 0 & 0 & 0 & 0 & 0 \\
0 & 0 & 0 & -1 & 0 & 0 & 0 \\
0 & 0 & 0 & 0 & -1 & 0 & 0 \\
0 & 1 & 0 & 0 & 0 & 0 & 0 \\
0 & 0 & 1 & 0 & 0 & 0 & 0 \\
0 & 0 & 0 & 0 & 0 & 0 & 0 \\
0 & 0 & 0 & 0 & 0 & 0 & 0
\end{pmatrix}
\qquad
T_8 = \frac{1}{\sqrt{3}} \begin{pmatrix}
0 & 0 & 0 & 0 & 0 & 0 & 0 \\
0 & 0 & -2 & 0 & 0 & 0 & 0 \\
0 & 2 & 0 & 0 & 0 & 0 & 0 \\
0 & 0 & 0 & 0 & 1 & 0 & 0 \\
0 & 0 & 0 & -1 & 0 & 0 & 0 \\
0 & 0 & 0 & 0 & 0 & 0 & -1 \\
0 & 0 & 0 & 0 & 0 & 1 & 0
\end{pmatrix}
$$

$$
T_9 = \frac{1}{\sqrt{3}} \begin{pmatrix}
0 & -2 & 0 & 0 & 0 & 0 & 0 \\
2 & 0 & 0 & 0 & 0 & 0 & 0 \\
0 & 0 & 0 & 0 & 0 & 0 & 0 \\
0 & 0 & 0 & 0 & 0 & 0 & 1 \\
0 & 0 & 0 & 0 & 0 & -1 & 0 \\
0 & 0 & 0 & 0 & 1 & 0 & 0 \\
0 & 0 & 0 & -1 & 0 & 0 & 0
\end{pmatrix}
\qquad
T_{10} = \frac{1}{\sqrt{3}} \begin{pmatrix}
0 & 0 & -2 & 0 & 0 & 0 & 0 \\
0 & 0 & 0 & 0 & 0 & 0 & 0 \\
2 & 0 & 0 & 0 & 0 & 0 & 0 \\
0 & 0 & 0 & 0 & 0 & -1 & 0 \\
0 & 0 & 0 & 0 & 0 & 0 & -1 \\
0 & 0 & 0 & 1 & 0 & 0 & 0 \\
0 & 0 & 0 & 0 & 1 & 0 & 0
\end{pmatrix}
$$

$$
T_{11} = \frac{1}{\sqrt{3}} \begin{pmatrix}
0 & 0 & 0 & -2 & 0 & 0 & 0 \\
0 & 0 & 0 & 0 & 0 & 0 & -1 \\
0 & 0 & 0 & 0 & 0 & 1 & 0 \\
2 & 0 & 0 & 0 & 0 & 0 & 0 \\
0 & 0 & 0 & 0 & 0 & 0 & 0 \\
0 & 0 & -1 & 0 & 0 & 0 & 0 \\
0 & 1 & 0 & 0 & 0 & 0 & 0
\end{pmatrix}
\qquad
T_{12} = \frac{1}{\sqrt{3}} \begin{pmatrix}
0 & 0 & 0 & 0 & -2 & 0 & 0 \\
0 & 0 & 0 & 0 & 0 & 1 & 0 \\
0 & 0 & 0 & 0 & 0 & 0 & 1 \\
0 & 0 & 0 & 0 & 0 & 0 & 0 \\
2 & 0 & 0 & 0 & 0 & 0 & 0 \\
0 & -1 & 0 & 0 & 0 & 0 & 0 \\
0 & 0 & -1 & 0 & 0 & 0 & 0
\end{pmatrix}
$$

$$T_{13} = \frac{1}{\sqrt{3}} \begin{pmatrix} 0 & 0 & 0 & 0 & 0 & -2 & 0 \\ 0 & 0 & 0 & 0 & -1 & 0 & 0 \\ 0 & 0 & 0 & -1 & 0 & 0 & 0 \\ 0 & 0 & 1 & 0 & 0 & 0 & 0 \\ 0 & 1 & 0 & 0 & 0 & 0 & 0 \\ 2 & 0 & 0 & 0 & 0 & 0 & 0 \\ 0 & 0 & 0 & 0 & 0 & 0 & 0 \end{pmatrix} \quad T_{14} = \frac{1}{\sqrt{3}} \begin{pmatrix} 0 & 0 & 0 & 0 & 0 & 0 & -2 \\ 0 & 0 & 0 & 1 & 0 & 0 & 0 \\ 0 & 0 & 0 & 0 & -1 & 0 & 0 \\ 0 & -1 & 0 & 0 & 0 & 0 & 0 \\ 0 & 0 & 1 & 0 & 0 & 0 & 0 \\ 0 & 0 & 0 & 0 & 0 & 0 & 0 \\ 2 & 0 & 0 & 0 & 0 & 0 & 0 \end{pmatrix}.$$

The quadratic Casimir is $-\frac{1}{4}\sum_{a=1}^{14} T_a^2 = 2\,\mathbf{I}$, where $\mathbf{I}$ is the 7×7 identity matrix. Reference 2 gives many useful tensor identities for G_2.

For a general 7×7 matrix, we have the G_2 decomposition into irreps.

$$\mathbf{7} \times \mathbf{7} = \mathbf{1} + \mathbf{7} + \mathbf{14} + \mathbf{27},$$

where the $\mathbf{7} + \mathbf{14}$ span all anti-symmetric 7×7 matrices and $\mathbf{27}$ spans all traceless symmetric 7×7 matrices.

Irreps of G_2 are labeled by two Dynkin labels $[n_1, n_2]$ and the $\mathbf{7}$ is $[1, 0]$, the $\mathbf{14}$ is $[0, 1]$ and the $\mathbf{27}$ is $[2, 0]$. In general, $[n_1, n_2]$ has dimension

$$dim[n_1, n_2] = \frac{(n_1 + 1)(n_2 + 1)(n_1 + n_2 + 2)(n_1 + 2n_2 + 3)(n_1 + 3n_2 + 4)(2n_1 + 3n_2 + 5)}{5!}$$

and the quadratic Casimir is

$$C_2([n_1, n_2]) = \frac{1}{3}n_1(n_1 + 5) + n_2(n_1 + n_2 + 3).$$

There are three different non-trivial coadjoint orbits in the Lie algebra of G_2,[3]

$$G_2/U(1) \times U(1), \quad G_2/U(2)_- \quad \text{and} \quad G_2/U(2)_+ \,.$$

$U(2)_-$ and $U(2)_+$ here reflect two different ways of embedding

$$U(2) = [U(1) \times SU(2)]/Z_2$$

in G_2 (made explicit in the following) and here denoted as[4,5]

$$G_2/U(2)_- = G_2/([SU(2) \times U(1)]/Z_2) \tag{2.1}$$

$$G_2/U(2)_+ = G_2/([U(1) \times SU(2)]/Z_2). \tag{2.2}$$

As coadjoint orbits, these three spaces all lend themselves to a fuzzy construction. $G_2/U(2)_-$ and $G_2/U(2)_+$ are both 10-dimensional manifolds and $G_2/U(2)_-$ is homeomorphic but not isometric to the complex quadric $\mathbf{Q}^5 = SO(7)/SO(5) \times SO(2)$ (page 503 of Boyer *et al.*[4]) A fuzzy version of $\mathbf{Q}^5$ was previously constructed in Dolan *et al.*[6] G_2, the automorphism group of the octonions, is a subgroup of $SO(7)$ and with the embedding

$G_2 \hookrightarrow SO(7)$, the homogeneous space $SO(7)/G_2$ has G_2 holonomy and admits a metric with $SO(7)$ isometry, it is a squashed 7-sphere $SO(7)/G_2 \approx S^7$. The tensor product of the 7-dimensional vector representation of $SO(7)$ with itself decomposes as

$$7 \times 7 = 1 + 21 + 27,$$

where the **21** decomposes as **7** + **14** under the embedding $G \hookrightarrow SO(7)$. Denoting the 21 generators of $SO(7)$ in the 7-dimensional representation by $X_1, \ldots, X_7, T_1, \ldots, T_{14}$, we can choose

$$X_1 = \begin{pmatrix} 0 & 0 & 0 & 0 & 0 & 0 & 0 \\ 0 & 0 & 1 & 0 & 0 & 0 & 0 \\ 0 & -1 & 0 & 0 & 0 & 0 & 0 \\ 0 & 0 & 0 & 0 & 1 & 0 & 0 \\ 0 & 0 & 0 & -1 & 0 & 0 & 0 \\ 0 & 0 & 0 & 0 & 0 & 0 & -1 \\ 0 & 0 & 0 & 0 & 0 & 1 & 0 \end{pmatrix} \quad X_2 = \begin{pmatrix} 0 & 0 & -1 & 0 & 0 & 0 & 0 \\ 0 & 0 & 0 & 0 & 0 & 0 & 0 \\ 1 & 0 & 0 & 0 & 0 & 0 & 0 \\ 0 & 0 & 0 & 0 & 0 & 1 & 0 \\ 0 & 0 & 0 & 0 & 0 & 0 & 1 \\ 0 & 0 & 0 & -1 & 0 & 0 & 0 \\ 0 & 0 & 0 & 0 & -1 & 0 & 0 \end{pmatrix}$$

$$X_3 = \begin{pmatrix} 0 & 1 & 0 & 0 & 0 & 0 & 0 \\ -1 & 0 & 0 & 0 & 0 & 0 & 0 \\ 0 & 0 & 0 & 0 & 0 & 0 & 0 \\ 0 & 0 & 0 & 0 & 0 & 0 & 1 \\ 0 & 0 & 0 & 0 & 0 & -1 & 0 \\ 0 & 0 & 0 & 0 & 1 & 0 & 0 \\ 0 & 0 & 0 & -1 & 0 & 0 & 0 \end{pmatrix} \quad X_4 = \begin{pmatrix} 0 & 0 & 0 & 0 & -1 & 0 & 0 \\ 0 & 0 & 0 & 0 & 0 & -1 & 0 \\ 0 & 0 & 0 & 0 & 0 & 0 & -1 \\ 0 & 0 & 0 & 0 & 0 & 0 & 0 \\ 1 & 0 & 0 & 0 & 0 & 0 & 0 \\ 0 & 1 & 0 & 0 & 0 & 0 & 0 \\ 0 & 0 & 1 & 0 & 0 & 0 & 0 \end{pmatrix}$$

$$X_5 = \begin{pmatrix} 0 & 0 & 0 & 1 & 0 & 0 & 0 \\ 0 & 0 & 0 & 0 & 0 & 0 & -1 \\ 0 & 0 & 0 & 0 & 0 & 1 & 0 \\ -1 & 0 & 0 & 0 & 0 & 0 & 0 \\ 0 & 0 & 0 & 0 & 0 & 0 & 0 \\ 0 & 0 & -1 & 0 & 0 & 0 & 0 \\ 0 & 1 & 0 & 0 & 0 & 0 & 0 \end{pmatrix} \quad X_6 = \begin{pmatrix} 0 & 0 & 0 & 0 & 0 & 0 & 1 \\ 0 & 0 & 0 & 1 & 0 & 0 & 0 \\ 0 & 0 & 0 & 0 & -1 & 0 & 0 \\ 0 & -1 & 0 & 0 & 0 & 0 & 0 \\ 0 & 0 & 1 & 0 & 0 & 0 & 0 \\ 0 & 0 & 0 & 0 & 0 & 0 & 0 \\ -1 & 0 & 0 & 0 & 0 & 0 & 0 \end{pmatrix}$$

$$X_7 = \begin{pmatrix} 0 & 0 & 0 & 0 & 0 & -1 & 0 \\ 0 & 0 & 0 & 0 & 1 & 0 & 0 \\ 0 & 0 & 0 & 1 & 0 & 0 & 0 \\ 0 & 0 & -1 & 0 & 0 & 0 & 0 \\ 0 & -1 & 0 & 0 & 0 & 0 & 0 \\ 1 & 0 & 0 & 0 & 0 & 0 & 0 \\ 0 & 0 & 0 & 0 & 0 & 0 & 0 \end{pmatrix}.$$

These X_i generators satisfy

$$[X_i, X_j] = c_{ij}{}^a T_a + c_{ij}{}^k X_k.$$

For $i \neq j$, the anti-commutators satisfy

$$\{X_i, X_j\} = X_i X_j + X_j X_i = e_{ij} + e_{ji},$$

where e_{ij} is the matrix with 1 in the $i - j$ position and zeros elsewhere and the anti-commutator is in the **27**. We can define a projector $\mathcal{P}$ that projects onto the **7** of $SO(7)$ (or equivalently projecting out the **14** and the **27** of G_2),

$$\mathcal{P}([X_i, X_j]) = c_{ij}{}^k X_k, \quad \mathcal{P}(\{X_i, X_j\}) = 0.$$

For $i = j$,

$$X_i^2 = -\mathbf{I} + e_{ii} = -\frac{6}{7}\mathbf{I} - \frac{1}{7}\begin{pmatrix} 1 & 0 & 0 & 0 & 0 & 0 & 0 \\ 0 & \ddots & 0 & 0 & \cdots & 0 & 0 \\ 0 & \cdots & 1 & 0 & 0 & \cdots & 0 \\ 0 & \cdots & 0 & -6 & 0 & \cdots & 0 \\ 0 & \cdots & 0 & 0 & 1 & \cdots & 0 \\ 0 & \cdots & 0 & 0 & 0 & \ddots & 0 \\ 0 & 0 & 0 & 0 & 0 & 0 & 1 \end{pmatrix} \quad \text{(no sum over } i\text{)},$$

where $\mathbf{I}$ is the 7×7 identity matrix and the -6 is in the $i - i$ position of the diagonal traceless matrix. Projecting out the **27** again,

$$\mathcal{P}(X_i^2) = -\frac{6}{7}\mathbf{I}$$

and

$$\mathcal{P}(X_i X_j) = -\frac{6}{7}\delta_{ij}\mathbf{I} + \frac{1}{2}c_{ij}{}^k X_k.$$

In fact, the $c_{ij}{}^k$, which furnish a torsion tensor on the non-symmetric homogeneous space $SO(7)/G_2$, are the structure constants of the pure imaginary octonions and the non-associative algebra of octonions is obtained by matrix multiplication of the X_i followed by projection onto the $\mathbf{1} + \mathbf{7}$. The non-associativity of the octonion algebra in this construction arises from this projection. In our convention, $c_{ij}{}^k = \epsilon_{ijk}$ is a completely anti-symmetric tensor with value 1 when $ijk = 123, 145, 176, 246, 257, 347, 365$ and the matrix X_j has components $(X_i)_j{}^k = c_{ij}{}^k$.

3. $G_2/U(2)_-$

The generator T_8 commutes with $\{T_1, T_2, T_3; T_8\}$ which generate $SU(2) \times U(1)$. We denote the adjoint orbit of T_8, $g^{-1}T_8 g$ with $g \in G_2$, by $G_2/U(2)_-$, where $U(2)_- = [SU(2) \times U(1)]/Z_2$. This adjoint orbit is related to the regular embedding of $SU(3)$ into G_2, where $SU(3)$ is generated by $\{T_1, \ldots, T_8\}$. We can gain some insight into this structure by focusing on the point which is the origin of the orbit, $g = 1$, which we shall call the N pole by analogy with $S^2 \approx SU(2)/U(1)$. The 10 generators of G_2 that do not commute with T_8 are $\{T_4, T_5, T_6, T_7, T_9, T_{10}, \ldots, T_{14}\}$ and these span the tangent space to $G_2/U(2)_-$ at the N pole. $G_2/SU(3) \approx S^6$ is a reduction of $G_2/U(2)_-$ under $U(2)_- \hookrightarrow SU(3)$ in the sense that $G_2/U(2)_-$ is a fiber bundle over $G_2/SU(3)$ (which is a squashed 6-sphere, $G_2/SU(3) \approx S^6$) with a 4-dimensional fiber. The adjoint action of $SU(3)$ on T_8 generates $\mathbf{CP}^2$, with $\{T_4, T_5, T_6, T_7\}$ tangent to $\mathbf{CP}^2$ at the N pole,[11] so the fiber is $\mathbf{CP}^2$. The remaining 6 generators $\{T_9, \ldots, T_{14}\}$ are tangent to the S^6 base at the N pole. In fact, $G_2/U(2)_-$ is diffeomorphic, but not isometric, to the complex quadric $\mathbf{Q}^5$ (see Miyaoka[5]). In summary, we have

$$\mathbf{CP}^2 \longrightarrow G_2/U(2)_- \approx \mathbf{Q}^5$$

$$\downarrow$$

$$G_2/SU(3) \approx S^6,$$

where again the equivalences on the total space and the base space are merely diffeomorphisms.

We can construct a rank 1 projector that commutes with the holonomy group $U(2)_-$ generated by $\{T_1, T_2, T_3; T_8\}$,

$$P_1 = \frac{1}{2} \begin{pmatrix} 0 & 0 & 0 & 0 & 0 & 0 & 0 \\ 0 & 1 & -i & 0 & 0 & 0 & 0 \\ 0 & i & 1 & 0 & 0 & 0 & 0 \\ 0 & 0 & 0 & 0 & 0 & 0 & 0 \\ 0 & 0 & 0 & 0 & 0 & 0 & 0 \\ 0 & 0 & 0 & 0 & 0 & 0 & 0 \\ 0 & 0 & 0 & 0 & 0 & 0 & 0 \end{pmatrix} = \mathbf{I} + \frac{1}{2}\{X_2, X_3\}^2 + \frac{i}{6}X_1 - \frac{i}{2\sqrt{3}}T_8,$$

then P_1 commutes with T_1, T_2, T_3, T_8 but not with any linear combination of the other T_i. The metric and Kähler structure are encoded into these projectors.[11–13] Also, the orbit of P_1 under $SO(7)$ gives $\mathbf{Q}^5$, and at a generic point of the orbit, $P_1 = \frac{1}{2}(\mathbf{m} + i\mathbf{n})$, where $\mathbf{m} = -\mathbf{n}^2$ and $\mathbf{n}^3 = -\mathbf{n}$. Multinomials in the components of $\mathbf{n}$ are a basis for functions on $\mathbf{Q}^5$, in analogy with how the unit vector in $\mathbf{R}^3$ can be used to construct functions on S^2. The spectra of Laplacians on the orbits are given by the quadratic Casimirs of the representations of G_2 (and $SO(7)$ for $\mathbf{Q}^5$), so the spectra are not the same: though $\mathbf{n}$ provides the functions, the manifolds are not isospectral.

As a coadjoint orbit, $G_2/U(2)_-$ is a symplectic space that admits a fuzzy description. Under the chain of embeddings,

$$
\begin{array}{lll}
G_2 & \hookleftarrow \quad SU(3) \quad \hookleftarrow & U(2)_- \\
\mathbf{7} & \longrightarrow \mathbf{1} + \mathbf{3} + \bar{\mathbf{3}} \longrightarrow & \mathbf{1}_0 + (\mathbf{2}_1 + \mathbf{1}_{-2}) + (\mathbf{2}_{-1} + \mathbf{1}_2) \\
\mathbf{14} & \longrightarrow \mathbf{3} + \bar{\mathbf{3}} + \mathbf{8} \longrightarrow & (\mathbf{2}_1 + \mathbf{1}_{-2}) + (\mathbf{2}_{-1} + \mathbf{1}_2) + (\mathbf{1}_0 + \mathbf{2}_3 + \mathbf{2}_{-3} + \mathbf{3}_0).
\end{array}
$$

The $\mathbf{7}$ contains a neutral singlet of $G_2/U(2)_-$, and so $\mathbf{7} \times \mathbf{7}$ matrices provide a candidate for a matrix representation of functions on fuzzy $G_2/U(2)_-$. In terms of Dynkin labels, the $\mathbf{7}$ is $[1,0]$ and the symmetric product of n of these is the $[n,0]$ with dimension

$$
d_{n,0} = \frac{(n+1)(n+2)(n+3)(n+4)(2n+5)}{5!}.
$$

In analogy with the constructions in Refs. 6, 11–13, we propose $d_{n,0} \times d_{n,0}$ matrices as a fuzzy representation of functions on $G_2/U(2)_-$, with matrix multiplication giving a star product for multiplication of functions.

The Laplacian on a $d_{n,0} \times d_{n,0}$ matrix Φ representing a function on fuzzy space $G_2/U(2)_-$ is obtained from $-\nabla^2 \Phi = \frac{1}{4}[T_a[T_a, \Phi]]$, with T_a in the $d_{n,0}$ dimensional irrep. of G_2. Eigenvalues λ of the Laplacian on $G_2/U(2)_-$ are therefore given by the second-order Casimirs of the irreducible G_2 representations appearing in the product $[n,0] \times [n,0]$ representation. The first two of these, ordered with increasing λ, are

$$
\begin{array}{llllllllll}
[1,0] & \times & [1,0] & = & [0,0] & + & [1,0] & + & [0,1] & + & [2,0] \\
\mathbf{7} & \times & \mathbf{7} & & 1 & + & 7 & + & 14 & + & 27 \\
& \lambda: & & & 0 & & 2 & & 4 & & \frac{14}{3}
\end{array}
$$

$$
\begin{array}{l}
[2,0] \times [2,0] = [0,0] + [1,0] + [0,1] + 2[2,0] + 2[1,1] + [3,0] + [0,2] + [2,1] + [4,0] \\
\mathbf{27} \times \mathbf{27} = \quad 1 \quad + \quad 7 \quad + \quad 14 \quad + 2(27) + 2(64) + \quad 77 \quad + \quad 77' \quad + \quad 189 \quad + \quad 182 \\
\qquad \lambda: \quad\quad\quad 0 \quad\quad\quad 2 \quad\quad\quad 4 \quad\quad\quad \frac{14}{3} \quad\quad\quad 7 \quad\quad\quad 8 \quad\quad\quad 10 \quad\quad\quad \frac{32}{3} \quad\quad 12
\end{array}
$$

The bold numbers above the eigenvalues (the dimension of the relevant irrep.) are the degeneracies associated with the corresponding eigenvalue. The Laplacian has a largest eigenvalue which grows as n^2, and the total number of eigenvalues is $d_{n,0}^2$, so for large n,

$$d_{n,0} \sim \frac{1}{60} n^5,$$

and reading the dimension of the manifold from Weyl's law,[10] the space is of dimension 10, compatible with the fact that $G_2/U(2)_-$ is a 5-dimensional complex manifold.

4. $G_2/U(2)_+$

The generator T_3 commutes with $\{T_3; T_8, T_9, T_{10}\}$, which generates a $U(2) \subset G_2$ which we shall denote $U(2)_+$. The adjoint orbit of T_3, $g^{-1}T_3g$ with $g \in G_2$, this is $G_2/U(2)_+$. The 10 generators of G_2 that do not commute with T_3 are $\{T_1, T_2, T_4, T_5, T_6, T_7, T_{11}, T_{12}, T_{13}, T_{14}\}$ and these span the tangent space to $G_2/U(2)_+$ at its N pole. There is a regular embedding of $SO(4) = [SU(2) \times SU(2)]_{Z_2} \hookrightarrow G_2$, and $G_2/SO(4)$ is an 8-dimensional manifold that is a reduction of $G_2/U(2)_+$ under $U(2)_+ \hookrightarrow SO(4)$ in the sense that $G_2/U(2)_+$ is a fiber bundle over $G/SO(4)$ with a 2-dimensional fiber. At the N pole, the generators of $SO(4)$ that are not in $U(2)_+$ are $\{T_1, T_2\}$ and these must be tangent to the fiber. The orbit of the adjoint action of the $SU(2)$ generated by $\{T_1, T_2, T_3\}$ acting on T_3 is a 2-sphere, so the fiber is a 2-sphere. It is argued in Ref. 7 that the 8-dimensional base $G_2/SO(4)$ is in fact diffeomorphic, but not isometric, to the quaternionic projective plane $\mathbf{HP}^2 \approx Sp(3)/Sp(2) \times Sp(1)$, it admits a quaternionic Kähler structure[8] and is an example of a Wolf space.[9]

So, we have

$$S^2 \approx \mathbf{CP}^1 \longrightarrow \quad G_2/U(2)_+$$

$$\downarrow$$

$$G_2/SO(4) \approx \mathbf{HP}^2,$$

with the equivalences on the total space and the base space being merely diffeomorphisms not isometries, the metrics are not equivalent.

As a coadjoint orbit, $G_2/U(2)_+$ is a symplectic space that admits a fuzzy description. Under the chain of embeddings,

$$
G_2 \;\hookleftarrow\; SO(4) \approx SU(2) \times SU(2) \;\hookleftarrow\; U(2)_+
$$

$$
\mathbf{7} \;\longrightarrow\; (\mathbf{1},\mathbf{3}) + (\mathbf{2},\mathbf{2}) \;\longrightarrow\; \mathbf{3}_0 + \mathbf{2}_1 + \mathbf{2}_{-1}
$$

$$
\mathbf{14} \;\longrightarrow\; (\mathbf{1},\mathbf{3}) + (\mathbf{3},\mathbf{1}) + (\mathbf{2},\mathbf{4}) \;\longrightarrow\; \mathbf{3}_0 + \mathbf{1}_{1,0} + \mathbf{1}_{0,0} + \mathbf{1}_{-1,0} + \mathbf{4}_1 + \mathbf{4}_{-1}
$$

For a harmonic expansion of functions on $G_2/U(2)_+$, we need irreps of G_2 that contain a neutral singlet of the holonomy group $U(2)_+$, so a harmonic expansion should not contain a $\mathbf{7}$ and the smallest non-trivial irrep that can appear in a harmonic expansion of a function on $G_2/U(2)_+$ is the $\mathbf{14}$. Tensor products of the three lowest dimensional non-trivial G_2 irreps. are

$$
\mathbf{7} \times \mathbf{7} = \mathbf{1} + \mathbf{7} + \mathbf{14} + \mathbf{27}
$$

$$
\mathbf{14} \times \mathbf{14} = \mathbf{1} + \mathbf{14} + \mathbf{27} + \mathbf{77} + \mathbf{77}'
$$

$$
\mathbf{27} \times \mathbf{27} = \mathbf{1} + \mathbf{7} + \mathbf{14} + 2(\mathbf{27}) + 2(\mathbf{64}) + \mathbf{77} + \mathbf{77}' + \mathbf{189} + \mathbf{182}
$$

and the lowest dimensional non-trivial matrix representations for a candidate to provide functions on fuzzy $G_2/U(2)_+$ is that of $\mathbf{14} \times \mathbf{14}$ matrices. Higher-dimensional function expansions come from products of $\mathbf{14}$s. In terms of Dynkin labels, the $\mathbf{14}$ is $[0,1]$ and, taking n of these, the $[0,n]$ has dimension

$$
d_{0,n} = \frac{(n+1)(n+2)(2n+3)(3n+4)(3n+5)}{5!}.
$$

$d_{0,n} \times d_{0,n}$ matrix multiplication gives a star product for multiplication of functions on $G_2/U(2)_+$. For large n, the dimension scales as

$$
d_{0,n} \sim \frac{3}{20} n^5,
$$

compatible with the fact that $G_2/U(2)_+$ is a 5-dimensional complex manifold.[10]

We can construct a rank 2 projector that commutes with the holonomy group $U(2)_+$ generated by $\{T_3, T_8, T_9; T_{10}\}$,

$$P_2 = \frac{1}{2} \begin{pmatrix} 0 & 0 & 0 & 0 & 0 & 0 & 0 \\ 0 & 0 & 0 & 0 & 0 & 0 & 0 \\ 0 & 0 & 0 & 0 & 0 & 0 & 0 \\ 0 & 0 & 0 & 1 & -i & 0 & 0 \\ 0 & 0 & 0 & i & 1 & 0 & 0 \\ 0 & 0 & 0 & 0 & 0 & 1 & -i \\ 0 & 0 & 0 & 0 & 0 & i & 1 \end{pmatrix} = \frac{1}{2} \left(\{X_4, X_5\}^2 + \{X_6, X_7\}^2 \right) + iT_3,$$

then P_2 commutes with T_1, T_2, T_3, T_8 but not with any linear combination of the other T_i. On the $[0, 2]$ representation, we would use $P_2 \otimes_a P_2$ restricted to the **14**, where a denotes anti-symmetrisation. Again, the metric and Kähler structure are encoded into these projectors.

Eigenvalues λ of the Laplacian on $G_2/U(2)_+$ are given by the second-order Casimirs of the irreducible G_2 representations appearing in the product $[0, n] \times [0, n]$ representation. The first two of these, ordered with increasing λ, are

$$
\begin{array}{ccccccccccc}
[0,1] & \times & [0,1] & = & [0,0] & + & [0,1] & + & [2,0] & + & [3,0] & + & [0,2] \\
\mathbf{14} & \times & \mathbf{14} & = & \mathbf{1} & + & \mathbf{14} & + & \mathbf{27} & + & \mathbf{77} & + & \mathbf{77'} \\
& & \lambda: & & 0 & & 4 & & \frac{14}{3} & & 8 & & 10
\end{array}
$$

$$
\begin{array}{ccccccccccccccccc}
[0,2] & \times & [0,2] & = & [0,0] & + & [0,1] & + & [2,0] & + & [3,0] & + & 2[0,2] & + & [2,1] & + & [4,0] & + & [5,0] \\
\mathbf{77'} & \times & \mathbf{77'} & = & \mathbf{1} & + & \mathbf{14} & + & \mathbf{27} & + & \mathbf{77} & + & 2(\mathbf{77'}) & + & \mathbf{189} & + & \mathbf{182} & + & \mathbf{378} \\
& & \lambda: & & 0 & & 4 & & \frac{14}{3} & & 8 & & 10 & & \frac{32}{3} & & 12 & & \frac{50}{3}
\end{array}
$$

$$
\begin{array}{ccccccccccc}
& + & 2[3,1] & + & [0,3] & + & [2,2] & + & [6,0] & + & [3,2] & + & [0,4] \\
& + & 2(\mathbf{448}) & + & \mathbf{273} & + & \mathbf{729} & + & \mathbf{714} & + & \mathbf{1547} & + & \mathbf{748} \\
& & 15 & & 18 & & \frac{56}{3} & & 22 & & 24 & & 28.
\end{array}
$$

On inspection of the decomposition of the $[n, 0] \times [n, 0]$ and that of $[0, n] \times [0, n]$, one sees that the latter is a subset of the former, implying that the two spaces are not globally equivalent. A theorem of Nakata (see Boyer et al.[4] and Nakata[14]) establishes that in fact $\pi_3(G_2/U(2)_-) = \mathbb{Z}_3$ while $\pi_3(G_3/U(2)_+) = 0$ further emphasising the global inequivalence, though it is stated in Ref. 5 that they are both diffeomorphic to $\mathbf{Q}^5$.

5. $G_2/U(1) \times U(1)$

The combination $T_3 + T_8$ only commutes with T_3 and T_8 and so the orbit $g^{-1}(T_3 + T_8)g$ is $G_2/U(1) \times U(1)$. At the N pole, the 12 generators $\{T_1, T_2, T_4, T_5, T_6, T_7, T_9, T_{10}, T_{11}, T_{12}, T_{13}, T_{14}\}$ are tangent to $G_2/U(1) \times U(1)$. $G_2/U(1) \times U(1)$ is an S^2 bundle over $\mathbf{Q}^5$; at the N pole, the generators orthogonal to the base are $\{T_1, T_2, T_3, T_8\}$ and $\{T_1, T_2\}$ span the fiber while $\{T_4, \ldots, T_7, T_9, \ldots, T_{14}\}$ span the base, which is $G_2/U(2)_-$. As an adjoint orbit, $G_2/U(1) \times U(1)$ can be embedded in $\mathbf{R}^{14}$, spanned by the 14 generators. Reducing this to S^{13}, we have the bundle structure[5]

$$S^1 \quad \longrightarrow \quad S^{13}$$

$$\downarrow$$

$$S^2 \approx \mathbf{CP}^1 \longrightarrow G_2/U(1) \times U(1)$$

$$\downarrow$$

$$\mathbf{CP}^2 \quad \longrightarrow G_2/U(2)_- \approx \mathbf{Q}^5$$

$$\downarrow$$

$$G_2/SU(3) \approx S^6 \,.$$

Furthermore, since T_3 and T_8 are in $SU(3)$, the holonomy can be enlarged and $G_2/U(1) \times U(1)$ can also be viewed as a bundle over $G_2/SU(3) \approx S^6$ with fiber $SU(3)/U(1) \times U(1)$.

Under the chain of embeddings,

G_2	$\hookleftarrow$	$SU(3)$	$\hookleftarrow$	$U(1) \times U(1)$		

$\mathbf{7}$	$\longrightarrow \mathbf{1} + \mathbf{3} + \bar{\mathbf{3}}$	$\longrightarrow$	$(0,0) + (1,1) + (1,-1) + (0,-2)$	
			$+(-1,-1) + (-1,1) + (0,2)$	

$\mathbf{14}$	$\longrightarrow \mathbf{3} + \bar{\mathbf{3}} + \mathbf{8}$	$\longrightarrow$	$(1,1) + (1,-1) + (0,-2)$	
			$+(-1,-1) + (-1,1) + (2,0)$	
	$+ 2(0,0) + (0,2) + (0,-2) + (3,1) + (-3,-1) + (3,-1) + (-3,1),$			

where (p,q) represents the charges of $U(1) \times U(1)$.

However, we have already seen that symmetric products of $\mathbf{7}$ give $d_{n,0} \times d_{n,0}$ matrices, describing a space with 5 complex dimensions. A more likely candidate is the $[1,1]$, which is $\mathbf{64}$-dimensional. Taking n of these, the $[n, n]$-dimensional irrep. of G_2 has dimension

$$d_{n,n} = (n + 1)^6.$$

This large n behavior is suggestive of 6 complex dimensions which is compatible with $G_2/U(1) \times U(1)$. We therefore propose $d_{n,n} \times d_{n,n}$ matrices as the fuzzy representation of functions on $G_2/U(1) \times U(1)$.

There is no projector on the $\mathbf{7} \times \mathbf{7}$ representation that commutes with (T_3, T_8) only; we need to go to a higher-dimensional irrep. to construct the projector for $G_2/U(1) \times U(1)$. The $\mathbf{64} = [1, 1]$ is in the product of three $[1, 0]$s: two of which are anti-symmetrized, $\mathbf{7} \times \mathbf{14} = \mathbf{7} + \mathbf{27} + \mathbf{64}$, so a candidate for the projector in this case is $P_1 \otimes (P_2 \otimes_a P_2)$ on the $\mathbf{64}$.

Eigenvalues λ of the Laplacian on $G_2/U(1) \times U(1)$ are given by the second-order Casimirs of the irreducible G_2 representations appearing in the product $[n, n] \times [n, n]$ representation. The number of irreps in the tensor product rapidly becomes very large and we give only the $n = 1$ case

$$
\begin{array}{llllllllll}
[1, 1] & \times & [1, 1] & = & [0, 0] & + & [1, 0] & + & 2[0, 1] & + & 2[2, 0] & + & 2[1, 1] & + & 3[3, 0] & + & 2[0, 2] & + & 3[2, 1] \\
\mathbf{64} & \times & \mathbf{64} & = & \mathbf{1} & + & \mathbf{7} & + & 2(\mathbf{14}) & + & \mathbf{27} & + & 2(\mathbf{64}) & + & 3(\mathbf{77}) & + & 2(\mathbf{77'}) & + & 3(\mathbf{189}) \\
\lambda: & & & & 0 & & 2 & & 4 & & \frac{14}{3} & & 7 & & 8 & & 10 & & \frac{32}{3}
\end{array}
$$

$$
\begin{array}{lllllllll}
& & + & 2[4, 0] & + & [1, 2] & + & 2[3, 1] & + & [5, 0] & + & [0, 3] & + & [2, 2] \\
& & + & 2(\mathbf{182}) & + & \mathbf{286} & + & 2(\mathbf{448}) & + & \mathbf{378} & + & \mathbf{273} & + & \mathbf{729} \\
& & & 12 & & 14 & & 15 & & \frac{50}{3} & & 18 & & \frac{56}{3}.
\end{array}
$$

6. Conclusions

Our three families of fuzzy spaces are specified by matrices of dimension $d_{n,0}$ for $G_2/U(2)_-$, $d_{0,n}$ for $G_2/U(2)_+$ and $d_{n,n}$ for $G_2/U(1) \times U(1)$, where the associated Laplacians are specified by $-\nabla^2\Phi = \frac{1}{4}[T_a[T_a, \Phi]]$ and T_a are the generators of G_2 in the associated representations. The equivalence of the function algebras of $G_2/U(2)_-$ and $\mathbb{Q}^5$ is manifest in our construction, see Ref. 6. To tease out any relation of these with $G_2/U(2)_+$ would take more work and we have not pursued that here. Furthermore, one should be able to construct fuzzy Dirac operators and equivariant vector bundles over these spaces along the lines of Dolan et al.[12, 15]

Acknowledgements

We would like to thank Charles Nash for several helpful discussions.

Comments of Denjoe O'Connor: I am delighted to have the opportunity to contribute to this celebration of Bal's 85th year. It has been a pleasure knowing him for the past 35 years since when we first met in Syracuse.

Though we have few joint publications, he has been a significant influence over the years. I especially enjoyed our productive meetings in Cinvestav, Mexico, where the conditions and scientific atmosphere for such meetings were superb. More recently, Bal has been a lively and insightful participant in the new era of Zoom and especially in the DIAS, School of Theoretical Physics seminar series. I hope we have many future years of productive collaboration.

Comments of Brian Dolan: I first met Bal more than 30 years ago in 1991 when he visited Lochlainn O'Raifeartaigh's group in the Dublin Institute for Advances Studies (DIAS). People often gathered in the kitchen in Burlington Road and the conversation would wander over various topics, but Bal was usually very quiet unless the conversation was about either physics or politics. If we strayed from physics for too long, he would go silent for a while and then announce "We should discuss", which meant go to a blackboard and discuss physics. And his call was always answered.

I subsequently met Bal on a number of occasions: in Dublin, in Cinvestav in Mexico City and in Syracuse, as well as at various conferences elsewhere. His focus on physics was always intense, but he is also a very warm and helpful person. On one occasion, I was visiting Syracuse for a few days and stayed with Bal and Indra. It came out over breakfast one morning that I had no research grant and that I had paid for my flight to the US from Ireland myself. Bal immediately arranged that I be reimbursed for the travel from his own grant.

References

1. S. L. Cacciatori *et al.*, Euler angles for G_2, *J. Math. Phys.* **46** (2005) 083512, [arXiv:hep-th/0503106].
2. A. J. Macfarlane, Lie algebra and invariant tensor technology for g_2, [archiv:math-ph/0103021].
3. T. Miyasaka, Adjoint orbit types of compact exceptional Lie group G_2 in its Lie algebra, *Math. J. Okayama Univ.* **43** (2001) 17–23, [arxiv:1011.0048[math.DG]].
4. C. Boyer and K. Galicki, Sasakian Geometry Oxford Mathematical Monographs, Oxford Science Publications (2008).
5. R. Miyaoka, Geometry of G_2 orbits and isoparametric hypersurfaces, *Nagoya Math. J.* **203** (2011) 175–189.
6. B. P. Dolan, D. O'Connor and P. Presnajder, Fuzzy complex quadrics and spheres, *JHEP* **02** (2004) 055, [hep-th/0312190].
7. L. J. Boya, Symmetric spaces of exceptional groups, [arXiv:0811.0554[math-ph]].

8. D. Conti, T. Bruun Madsen and S. Salamo, Quaternionic geometry in dimension eight, [arXiv:1610.04833[math.DG]].

9. J. A. Wolf, Complex homogeneous contact manifolds and quaternionic symmetric spaces, *J. Math. Mech.* **14** (1965) 1033–1047.

10. H. Weyl, Über die asymptotische Verteilung der Eigenwerte, (1911) *Nachrichten der Königlichen Gesellschaft der Wissenschaften zu Göttingen*, 110–117.

11. A. P. Balachandran, B. P. Dolan, J. Lee, X. Martin and D. O'Connor, Fuzzy complex projective spaces and their star products, *J. Geom. Phys.* **43** (2002) 184–204, [hep-th/0107099].

12. B. P. Dolan, I. Huet, S. Murray and D. O'Connor, Noncommutative vector bundles over fuzzy CP(N) and their covariant derivatives, *JHEP* **0707** (2007) 007, [arXiv:hep-th/0611209].

13. B. P. Dolan and O. Jahn, Fuzzy complex Grassmannian spaces and their star products, *Int. J. Mod. Phys.* **A18** (2003) 1935–1958, [hep-th/0111020].

14. F. Nakata, Homotopy groups of $G_2/Sp(1)$ and $G_2/U(2)$, in *Contemporary Perspectives in Differential Geometry and its Related Fields*, World Scientific (2018).

15. B. P. Dolan, I. Huet, S. Murray and Denjoe O'Connor, A universal Dirac operator and noncommutative spin bundles over fuzzy complex projective spaces, *JHEP* **0803** (2008) 029, arXiv:0711.1347 [hep-th].

Chapter 5

Understanding Quantum Physics: Why Geometry and Algebra Matter

Elisa Ercolessi

Department of Physics and Astronomy,
University of Bologna, Bologna, Italy
INFN – Sezione di Bologna, Via Irnerio 46, 40127 Bologna, Italy
elisa.ercolessi@unibo.it

This short contribution is a review of some of the scientific results that were achieved in Syracuse during my stay as a PhD student under the supervision of A.P. Balachandran from 1990 to 1994. It was an exciting period during which we worked closely with other PhD students and many visiting researchers on the mathematical structures of quantum mechanics. The research focused on some fundamental issues that the physical community managed to understand but put in a full coherent theory only much later.

1. Quantum Physics

The first time I met Professor A.P. Balachandran, who quickly became simply Bal, was at the legendary School on *Anomalies, Phases, Defects*, that was held in Ferrara (Italy) in 1989. At that time, I was a Master's student who had recently started working on the dissertation, under the supervision of one of the organizers of the School, Giuseppe Morandi. The experimental discovery of High-Tc superconductivity dated just a few years back and I was investigating how the two-dimensional Hubbard model — and variants of it — could provide insight into the problem. Equipped with a solid background in field theory and many-body techniques, I embarked on the project to classify the quantum phases of the Hubbard model on a square lattice, to soon discover that analytical tools to which I had been trained in my scholastic career were not sufficient to understand the physics of strongly correlated low-dimensional quantum systems and that geometry not only mattered but was indeed essential.

At the School, I was introduced to the fabulous world of geometry and physics by some outstanding speakers[1]: among others, Michael Berry who taught us about applications of the adiabatic phase, David Mermin who introduced homotopy to classify defects in matter and Wilczek who explained how to obtain fractional statistics. Bal was one of the speakers and gave a series of lectures on "Classical Topology and Quantum Phases".

I was very impressed by the approach he used: pragmatical and founded on very concrete physical situations but at the same time very rigorous. There was no space for implicit assumptions: any basic concepts as well as any sophisticated mathematical tools that one needed to introduce had to be justified, interpreted and checked on examples. He presented a very clear and unifying picture of topological excitations in all fields of physics, with instances from anyon superconductivity to gravity. However, in preparing those lectures, I believe that Bal's fundamental goal was to give an answer to the fundamental question of why and how the quantum description of a system forces us to take into account geometry.

In Section 2, I elaborate a bit on this, by recalling some early work I did with Bal, as soon as I joined Syracuse University as a PhD student more than 30 years ago, and some ideas that he also developed with other collaborators. I consider these contributions to be forerunners of some very modern topics, such as topological matter, that the physics community was ready to understand and put in a full coherent theory only much later.

In Section 3, I move on to describe our joint work on posets and noncommutative geometry. I do not remember how and why Bal was attracted by the latter subject: I just recall a long discussion about Connes' paper in Rm. 316, where Bal used to reunite all his PhD students and collaborators every afternoon. At that time, I was working on quantum mechanics on discrete spaces, struggling with the well-known problem of losing information about the global properties of the continuum. Our investigation reached a turning point when we realized that the algebraic approach by Connes on noncommutative C^*-algebras[2] provided an alternative perspective. Surely, some of these techniques had been used in axiomatic approach to quantum field theory, but their role in quantum mechanics had not been fully investigated. So, it soon became clear to us that geometry was not the only mathematical structure hidden behind quantum mechanics and that the algebraic approach offered by noncommutative geometry was allowing for an additional framework to consider.

2. Why Geometry Matters

The focus of Bal's lecture in Ferrara was on one fundamental aspect that is rarely taught in introductory quantum mechanics classes, namely the fact that, when the configuration space has a non-trivial fundamental group, many different and inequivalent (i.e., non-unitarily related) quantizations are possible. The simplest example of this is a particle on a circle $S^1 = \{\phi \in [0, 2\pi]\}$ with square-integrable wavefunctions $\psi(\phi)$ that satisfy twisted boundary conditions: $\psi(2\pi) = e^{i\theta}\psi(0)$. Each value of the angle θ defines a different quantization. The non-equivalence can be seen, for example, by looking at the spectrum $E_n = (n + \theta)^2$ ($n \in \mathbf{Z}$) of the Laplacian operator $\Delta = -d^2/d\phi^2$, which turns out to explicitly depend on it. It is well known[3] that this problem can be recast in that of a charged particle that goes around an infinite thin solenoid, outside which there is no magnetic field but a non-zero gauge potential, whose flux is indeed proportional to θ: in other words, the Aharonov–Bohm effect. It is not difficult to identify less trivial and interesting situations that can be cast in an analogous framework, and one can find instances in all fields, ranging from the asymmetric rotor (to describe, e.g., nuclei with distinct moments of inertia) to non-abelian gauge theories that admit θ-vacua.[4]

The other prototypical and fundamental situation in which that happens is when studying statistical properties of identical particles. If we denote with $q_j \in M$ the coordinate of the jth particle moving in the classical space M, the total configuration space of N identical particles is

$$\mathcal{M} = \{(q_1, q_2, \ldots, q_N) : q_i \neq q_j\}/\mathbb{P}_N,$$

i.e., by the space M^N, with the exclusion of the points of coincidence of two (or more) particles ($q_i = q_j$), modulo the permutation group $\mathbb{P}_N$ that acts on the N particles. Whenever the first homotopy group $\Pi_1(\mathcal{M})$ is non-trivial, we have non-equivalent statistics. While (scalar) particles in $\mathbb{R}^3$ can be either bosons or fermions, being described by totally symmetric or anti-symmetric wavefunctions, particles in $\mathbb{R}^2$ admit more general statistics, being described by wavefunctions that, under an exchange of two particles, can pick up *any* phase $\phi \in [0, 2\pi)$, not just ± 1. These particles are called *anyons*. Similar to bosons and fermions, anyons satisfy a spin–statistics theorem that gives a connection between such statistical phase and that acquired under a 2π-rotation of the particle. The situation can be more complex if the configuration space is provided by a generic manifold or if we are considering vector-valued wavefunctions or, also, if our elementary objects are not point-like such as strings or geons, a type of excitations that were introduced in quantum gravity in Ref. [5].

By 1990, several examples of quasi-particle excitations with exotic statistics started to pile up. After the first example of fractional Hall effect, several authors noticed similar behaviors for excitations on the so-called resonating valence bond ground state in spin systems and started to describe the unusual properties of spin liquid states as well as to discuss the possibility of anyon superconductivity in two-dimensional systems that could be described via the Hubbard model. This was indeed the subject of my first collaboration with Bal (and with G. Morandi and A.M. Srivastava) that resulted in a book.[6]

To conclude this section, I would like to recall that, once in Syracuse, I got also partly involved in the work that Bal was developing with other PhD students (namely G. Bimonte and P. Teotonio Sobrinho) to analyze the role of boundary conditions in quantum field theories on manifolds with spatial boundaries. Again inspired by the Hall effect, the idea was to identify models with gauge fields that admit states that live at the boundaries (edge states) and/or are carried by topological excitations. Thirty years later, it is evident that one was trying to shed some light on phenomena connected with what we now call topological phases of matter, a subject that is considered of great relevance, for example, in the field of novel materials that can be used for quantum computation and other quantum technological applications. At that time, however, we had to sustain many discussions to reply to objections insisting on the fact that physics should be dictated purely by bulk properties.

3. Why Algebra Matters

Despite the fact that all separable Hilbert spaces are equivalent and we can choose any equivalent representation of the canonical commutation relations, in conventional quantum theory, the primary role is played by the (commutative) C^*-algebra $C(M)$ of continuous functions on the configuration space M. Actually, the role of $C(M)$ is two-fold.

First, $C(M)$ represents a fundamental subclass of observables, corresponding to classical observables which depend on the coordinates only. Immediately, one question arises: Does $C(M)$ contain all information about the configuration manifold M, or — in other words — can we reconstruct M from $C(M)$? As it is well known, the answer is given by the Gelfand–Naimark theorem, that establishes a one-to-one correspondence between Hausdorff topological spaces and commutative C^*-algebras. Indeed, the topological space M can be constructively built from $C(M)$ by looking at its structure space $\widehat{C(M)}$, i.e., the space of all unitary equivalence classes of irreducible *-representations: thus, $M = \widehat{C(M)}$ as topologic spaces.

Second, $C(M)$ is dense in the Hilbert space of square-integrable functions, and (eventually restricting to the dense subset of smooth functions) it constitutes the common domain of all relevant physical operators, such as derivative operators that define momentum and free Hamiltonian. As we have seen before, these operators carry information about the mathematical properties of the configuration space M: both local ones, such as the differential structure that is needed to define the conjugate momenta or derivative operator, and global ones, such as connections on a suitable bundle over M, which is necessary to introduce because of non-trivial homological or homotopic properties of the classical configuration space. It is natural to ask oneself whether these analytical and geometrical structures have a counterpart in the dual picture, where we actually study the C^*-algebra $C(M)$ instead of M. Such a question, which finds a clear motivation and a definitive answer in modern mathematics, was somewhat known to the (few) physicists working on rigorous aspects of quantum field theory but entered explosively theoretical physics only at the beginning of the 1990s, when Alain Connes extended the problem to noncommutative C^*-algebras, introducing the field of *Noncommutative Geometry* and showing how this picture can be relevant to describe field theories of fundamental interactions.[7]

Connes' ideas captured our attention in a period in which we were discussing how to approximate a continuum theory with a discrete one, without losing information about the geometry. Following Sorkin's original idea,[8] we were working on the Partially Order Set (or poset) approximation of a continuous Hausdorff space: given a topological space Q and a covering of it by open subsets, $Q = \cup_\lambda Q_\lambda$, one can construct the coset space $P(Q)$ of equivalence classes w.r.t. the relation

$$x \sim y \quad (x, y \in Q) \quad means \quad x \in Q_\lambda \Leftrightarrow y \in Q_\lambda.$$

Such a quotient space, which is finite whenever we choose a finite covering, e.g., when Q is compact, has a natural partial order, by declaring that $[x] \leq [x']$ if every open set containing x' contains also x.

It turns out that, when endowed with the quotient topology, a (finite) poset is not Hausdorff. Still, it is a T_0 space of a very special type. The C^*-algebra of its continuous functions is noncommutative and approximately finite dimensional, meaning that it can be obtained as an (inductive) limit of a sequence of finite-dimensional (i.e., matrix) algebras. Actually, the converse is also true.[9] Thus, we get a one-to-one correspondence between finite T_0 topological spaces and approximately finite dimensional C^*-algebras, giving a remarkable connection between finite physics and noncommutative geometry. This is why we called posets noncommutative lattices.[10]

My PhD thesis was all dedicated to the development of quantum physics on such finite approximations of continuum spaces, capable of encoding the topological features of the continuum. More specifically, I worked out homology and homotopy theory on posets and their bundle theory. But the work on this subject went much beyond, in collaboration with other PhD students, more specifically G. Bimonte and P. Teotonio-Sobrinho, and researchers who visited Syracuse in those years: F.Lizzi, G. Landi and G. Sparano.

4. Final Remarks

I would like to finish with some personal remarks. For me, Syracuse was definitely too cold. Also, it was not so lively from a social and cultural point of view. But the Department of Physics of Syracuse University was an excellent place where to learn new physics and do research. From this point of view, the years I spent there were incredibly exciting.

Bal is a great teacher, both during lectures and in informal discussions. I am really grateful to him for being a careful supervisor and an attentive advisor. He taught me — as I believe to his many other PhD students — not to take hypotheses for granted, to boldly defend my ideas and to never give up.

I would define the discussions in Rm. 316 as a sort of collective stream of consciousness about quantum physics, which often turned into very polite but animated debates that ended only because it was dinner time and discussions about politics had to start, preferably in front of a dish of pasta.

References

1. M. Bregola, G. Marmo and G. Morandi. *Anomalies, Phases, Defects*, Bibliopolis (1990).
2. A. Connes, *Noncommutative Geometry*, Elsevier (1994).
3. G. Morandi, *The Role of Topology in Classical and Quantum Physics*, Springer-Verlag (1992)

4. A. P. Balachandran, G. Marmo, B. S. Skagerstam and A. Stern, *Classical Topology and Quantum States*, World Scientific (1991).

5. J. L. Friedman and R. D. Sorkin, *Phys. Rev. Lett.* **44** (1980) 11; **45** (1980) 148.

6. A. P. Balachandran, E. Ercolessi, G. Morandi and A. M. Srivastava, *Hubbard Model and Anyon Superconductivity*, World Scientific (1990).

7. A. Connes and J. Lott, *Nucl. Phys.* **B18** Suppl. (1990) 29.

8. R. D. Sorkin, *Int. J. Theor. Phys.* **30** (1991) 923.

9. O. Bratteli, *Trans. Amer. Math. Soc.* **171** (1972) 195.

10. A. P. Balachandran, G. Bimonte, E. Ercolessi, G. Landi, F. Lizzi, G. Sparano and P. Teotonio-Sobrinho, *J. Geom. Phys.* **18** (1996) 163.

Chapter 6

Near-Horizon Conformal Structure, Entropy and Decay Rate of Black Holes

Kumar S. Gupta[*,†,‡] and Siddhartha Sen[*,†,§]

*Theory Division, Saha Institute of Nuclear Physics,
Bidhannagar, Kolkata 700064, India*
†*CRANN, Trinity College Dublin, Ireland*
‡*kumars.gupta@saha.ac.in*
§*sen@tcd.ie*

It is shown that a mass scalar field probe can be used to reveal the near-horizon conformal structure of a large class of black holes. The logarithmic correction to black hole entropy follows from the algebraic properties of the near-horizon Klein–Gordon operator. The black hole decay can be analyzed within this formalism as a geodesic motion in an appropriate moduli space.

1. Introduction

Conformal filed theory (CFT) is known to play an important role in the quantum description of black holes, with the AdS/CFT duality as a celebrated example. The near-horizon CFT appears in a broader context, through the algebra of diffeomorphism generators, even without the requirement of supersymmetry.[1,2] In this short chapter, which is dedicated to Prof. A.P. Balachandran, we shall review an alternate algebraic approach to the near-horizon CFT,[3,4] which is applicable for a large class of black holes.[5] It will be shown that the self-adjoint extensions of the near-horizon Klein–Gordon (KG) operator leads to the logarithmic correction to the Bekenstein–Hawking entropy. In addition, the black hole decay can be understood as a geodesic motion in the appropriate moduli space.[6]

2. Scalar Field Probe of the Near-horizon Geometry of Black Holes

We begin by considering the BTZ black hole of mass M and spin J, and the discussion will be generalized to include a large class of black holes shortly. The BTZ black hole is given by the metric[7]

$$ds^2 = -f(r)dt^2 + f^{-1}(r)dr^2 + r^2 \left(d\phi - \frac{J}{2r^2}dt \right)^2, \qquad (2.1)$$

where $f(r) = -M + \frac{r^2}{l^2} + \frac{J^2}{4r^2}$, r and ϕ denote the radial and angular coordinates on the plane respectively and t denotes the time. This metric satisfies vacuum Einstein equations in $2 + 1$ dimensions, with a negative cosmological constant $\Lambda = -1/l^2$. The outer and inner horizons denoted by $r_\pm$ respectively are given by

$$r_\pm^2 = \frac{Ml^2}{2} \left(1 \pm \sqrt{1 - \frac{J^2}{M^2l^2}} \right). \qquad (2.2)$$

In the present approach, a massless scalar field is used as a probe of the near-horizon geometry. The Klein–Gordon (KG) equation for a massless scalar field ψ in a general space-time with a metric $g_{\mu\nu}$, where μ, ν run over the space-time indices, is given by

$$\frac{1}{\sqrt{-g}} \partial_\mu (\sqrt{-g} g^{\mu\nu} \partial_\nu \psi) = 0. \qquad (2.3)$$

For the BTZ metric (2.1), we use the ansatz $\psi(r, t, \phi) = R(r)e^{-i\omega t + im\phi}$. The periodicity of the coordinate ϕ in the BTZ construction leads to the quantization condition $m \in \mathbf{Z}$. Our main interest is to probe the region near the outer horizon of this black hole. To this end, we define a near-horizon coordinate $x \in [0, \infty)$ as

$$x = r - r_+. \qquad (2.4)$$

We now define a new radial wavefunction $\chi(r) \equiv \sqrt{x}R(r)$, in terms of which the near-horizon form of the KG equation is given by

$$H\chi(r) \equiv \left[-\frac{d^2}{dx^2} + \frac{a}{x^2} \right] \chi(r), \qquad (2.5)$$

where

$$a = - \left[\frac{1}{4} + \frac{\tilde\omega^2}{A^2} \right], \quad \tilde\omega^2 = \omega^2 - \frac{J\omega m}{r_+^2} + \frac{Mm^2}{r_+^2} - \frac{m^2}{l^2}. \qquad (2.6)$$

The parameter a contains the information specific to the geometry.

We shall now show that the near-horizon KG operator for a large class of black holes has the same form as in (2.5) with different values of the parameter a. Consider a metric in D space-time dimensions given by

$$ds^2 = -f(r)dt^2 + \frac{dr^2}{f(r)} + r^2 d\Omega_{D-2}^2, \tag{2.7}$$

where $f(r)$ is a function of the radial variable r and $d\Omega_{D-2}^2$ is the metric on unit S^{D-2}. The form of the functions $f(r)$ will depend on the choice of the specific black hole.

As before, we wish to probe the near-horizon geometry of the black hole using a massless scalar field. To that end, we again consider the dynamics of a scalar field ψ in the above general background, with the ansatz $\psi(t, r, \Omega) = e^{-i\omega t} R(r) Y_{lm}(\Omega)$. As in the case for BTZ, we now define a near-horizon coordinate $x \equiv r - r_h$, where r_h denotes the event horizon and $x \in [0, \infty)$. The near-horizon form of the function f for a large class of black holes is given by[5]

$$f(x) \sim Ax[1 + \mathcal{O}(x)], \tag{2.8}$$

where A is a constant which depends on parameters defining the black hole geometry. In terms of a new field $\chi \equiv \sqrt{x} R(r)$, the KG equation in the near-horizon region takes the form

$$H\chi = \left[-\frac{d^2}{dx^2} + \frac{a}{x^2} \right] \chi, \tag{2.9}$$

where

$$a = -\left[\frac{1}{4} + \frac{\omega^2}{A^2} \right]. \tag{2.10}$$

The structure of H here is the same as that for the BTZ black hole as given in (2.5). The constant a depends on the frequency of the probe and on the geometric details and distinguishes between the various black holes. For real and non-zero frequency of the scalar probe, the constant a is real and satisfies the condition $a < -\frac{1}{4}$. The above analysis applies to a large class of black holes having quite different geometric properties. Apart from asymptotically flat backgrounds, our analysis includes black holes with both signs of the cosmological constant. It also includes the Gauss–Bonnet case which is obtained from string-derived gravity going beyond the usual Einstein–Hilbert action.

3. Algebraic Properties of the Near-Horizon KG Operator

The operator H can be decomposed as[3,8]

$$H = C_+C_-, \quad \text{where } C_\pm = \pm\frac{d}{dx} + \frac{b}{x}, \tag{3.1}$$

and

$$b = \frac{1}{2} \pm \frac{\sqrt{1+4a}}{2}. \tag{3.2}$$

For the present cases of interest, $a < -\frac{1}{4}$, b is complex and C_+ and C_- are not formal adjoints of each other. Following, Ref. 10 we define the operators

$$L_n = -x^{n+1}\frac{d}{dx}, \quad n \in \mathbf{Z}, \tag{3.3}$$

$$P_m = \frac{1}{x^m}, \quad m \in \mathbf{Z}. \tag{3.4}$$

Using Eqs. (3.3) and (3.4), the operators $C_\pm$ and H can be written as

$$C_\pm = \mp L_{-1} + bP_1, \tag{3.5}$$

$$H = (-L_{-1} + bP_1)(L_{-1} + bP_1). \tag{3.6}$$

The operators L_m, P_m and H satisfy the commutation relations

$$[P_m, P_n] = 0, \tag{3.7}$$

$$[L_m, P_n] = nP_{n-m}, \tag{3.8}$$

$$[L_m, L_n] = (m-n)L_{m+n} + \frac{c}{12}(m^3 - m)\delta_{m+n,0}, \tag{3.9}$$

$$[P_m, H] = m(m+1)P_{m+2} + 2mL_{-m-2}, \tag{3.10}$$

$$[L_m, H] = 2b(b-1)P_{2-m} - (m+1)(L_{-1}L_{m-1} + L_{m-1}L_{-1}). \tag{3.11}$$

Note that Eq. (3.9) describes a Virasoro algebra with central charge c, while the algebra of the generators defined in Eq. (3.3) would lead to $[L_m, L_n] = (m-n)L_{m+n}$. However, this algebra admits a non-trivial central extension c and in any of its unitary irreducible highest weight representation $c \neq 0$.[9,10] That is why the central charge has been explicitly included in (3.9). We also note that the operator H is not an element of $\mathcal{M}$ but belongs to the corresponding enveloping algebra. This is due to the fact that the right-hand side of Eq. (3.11) contains the product of Virasoro generators. While such products are not elements of the algebra, they do belong to the corresponding enveloping algebra.

In order to study the quantum properties of H, we consider the equation

$$H\psi = \mathcal{E}\psi, \quad \psi(0) = 0, \tag{3.12}$$

where $\mathcal{E}$ is the eigenvalue and $\psi \in L^2[R^+, dr]$ is the associated wavefunction. H belongs to the class of unbounded linear operators on a Hilbert space and it admits a one-parameter family of self-adjoint extensions labeled by a $U(1)$ parameter e^{iz}, where z is real. The normalized bound state eigenfunction and eigenvalue of Eq. (5) are given by

$$\psi(x) = \sqrt{2Ex}K_0\left(\sqrt{E}x\right) \tag{3.13}$$

and

$$\mathcal{E} = -E = -\exp\left[\frac{\pi}{2}\cot\frac{z}{2}\right], \tag{3.14}$$

respectively, where K_0 is the modified Bessel function. In our formalism, this solution is interpreted as the bound state excitation of the black hole due to the capture of the scalar field.

In the analysis presented above, the information about the spectrum of the Hamiltonian in the black hole background is coded in the parameter z. The wavefunctions and the energies of the bound states depend smoothly on z. Thus, within the near-horizon region Δ, we propose to identify

$$\tilde{\rho}(z) = |\psi_0|^2 = 2A^2 \, e^{\frac{\pi}{4}\cot\frac{z}{2}} \tag{3.15}$$

as the density of states for this system written in terms of the variable z. $\tilde{\rho}(z)dz$ counts the number of states when the self-adjoint parameter lies between z and $z + dz$. As mentioned before, within the region, Δ z is positive and satisfies the consistency condition $z \sim 0$. Thus, within the region Δ, $\cot\frac{z}{2}$ is a large and positive number. We thus find that the density of states of a massive black hole is very large in the near-horizon region.

In order to proceed, we shall first provide a physical interpretation of the self-adjoint parameter z using the Bekenstein–Hawking entropy formula. To this end, recall that in our formalism, the capture of the scalar field probe gives rise to the excitations of the black hole which subsequently decay by emitting Hawking radiation. A method of deriving density of states and entropy for a black hole in a similar physical setting using quantum mechanical scattering theory has been suggested by 't Hooft.[11] This simple and robust derivation uses the black hole mass and the Hawking temperature as the only physical inputs and is independent of the microscopic details of the system. The interaction of infalling matter with the black hole is assumed to be described by Schrödinger's equation and the relevant emission and absorption cross-sections are calculated using Fermi's golden rule. Finally, time reversal invariance (which is equivalent to CPT invariance in this case) is used to relate the emission and absorption cross-sections. The density of states for a massive black hole of mass M obtained from this scattering calculation is given by[11]

$$\rho(M) = e^{4\pi M^2 + C'} = e^S, \tag{3.16}$$

where C' is a constant and S is the black hole entropy.

We are now ready to provide a physical interpretation of the parameter z. First note that the density of states calculated in Eqs. (3.15) and (3.16) corresponds to the same physical situation described in terms of different variables. In our picture, the near-horizon dynamics of the scalar field probe contains information regarding the black hole background through the self-adjoint parameter z. The same information in the formalism of 't Hooft is contained in the black hole mass M. It is thus meaningful to relate the density of states in our framework (cf. Eq. 3.15) to that given in Eq. (3.16). If these expressions describe the same physical situation, we are led to the identification

$$\frac{\pi}{4} \cot \frac{z}{2} = 4\pi M^2. \tag{3.17}$$

Note that the analysis presented here is valid only for massive black holes. We have also seen that in the near-horizon region, z must be positive and obey a consistency condition such that $\cot \frac{z}{2}$ is a large positive number. Thus, the relation between z and M given by Eq. (3.17) is consistent with the constraints of our formalism. We therefore conclude that the self-adjoint parameter z has a physical interpretation in terms of the mass of the black hole.

In view of the relation between z and M, we can write

$$\tilde{\rho}(z)dz \sim |J|\rho(M)dM, \tag{3.18}$$

where $J = \frac{dz}{dM}$ is the Jacobian of the transformation from the variable z to M. When $z \sim 0$, we get

$$\tilde{\rho}(z)dz \sim e^{4\pi M^2} \frac{1}{M^3} dM \sim e^{4\pi M^2 - \frac{3}{2}\ln M^2} dM. \tag{3.19}$$

The presence of the logarithmic correction term in the above equation is thus due to the effect of the Jacobian.

The entropy for the Schwarzschild black hole obtained from Eq. (3.19) can be written as[4]

$$S = S(0) - \frac{3}{2}\ln S(0) + C, \tag{3.20}$$

where $S(0) = 4\pi M^2$ is the Bekenstein–Hawking entropy and C is a constant. Thus, the leading correction to the Bekenstein–Hawking entropy is provided by the logarithmic term in Eq. (3.7) with a coefficient of $-\frac{3}{2}$.

4. Black Hole Decay

The requirement of near-horizon conformal symmetry places an important constraint on the self-adjoint extension parameter z. To see this, consider a band-like region $\Delta = [x_0 - \delta/\sqrt{E}, x_0 + \delta/\sqrt{E}]$, where $x_0 \sim \frac{1}{\sqrt{E}}$ and $\delta \sim 0$ is real and positive. When $z > 0$ and satisfies the condition $z \sim 0$, we see that $x_0 \approx 0$. Under this condition, Δ belongs to the near-horizon region of the black hole. At any point $x \in \Delta$, the leading behavior of the wavefunction is given by

$$\psi = A\sqrt{Ex}, \tag{4.1}$$

where $A = \sqrt{2}(\ln 2 - \gamma)$ and γ is Euler's constant. Taking a typical value of $x = x_0$, we can write the wavefunction as

$$\psi = AE^{\frac{1}{4}} \approx Ae^{\frac{\pi}{4z}} \tag{4.2}$$

for $z \sim 0$. Substituting the value of z from Eq. (1) in Eq. (9), the wavefunction can be written as a function of M as

$$\psi(M) = Ae^{cM^2} \equiv Ag(M^2). \tag{4.3}$$

The function $g(M^2) = e^{cM^2}$ captures the M dependence of the wavefunction and $c = \frac{\pi}{4a}$ is a positive constant.

The wavefunction of a system is a natural object to examine in order to understand any symmetry that might be present in the system. To this end, consider the set $G \equiv \{g(M^2)|M^2 \in R\}$. The elements of G are the functions g defined with different values of M corresponding to different elements of G. For any two elements of G given by $g(M_1^2) = e^{cM_1^2}$ and $g(M_2^2) = e^{cM_2^2}$, we can define a composition law as $g(M_1^2) \cdot g(M_2^2) \equiv g(M_1^2 + M_2^2) \in G$. Similarly, for any $g(M^2) \in G$, we can define the inverse element as $g^{-1}(M^2) \equiv g(-M^2) \in G$. With respect to the composition law defined above, the set G has the structure of a continuous abelian group.

The presence of the continuous abelian group G allows us to describe the way the black hole mass changes in a geometric fashion. To do this, we need to construct a group invariant metric on the space $\mathcal{M}$. There is a well-known procedure for doing this. We begin with the observation that the group G has a natural action on the space $\mathcal{M}$. Under the action of G, a point $M_0 \in \mathcal{M}$ transforms as $M_0 \to e^{cM^2} M_0 \in \mathcal{M}$. G therefore acts as a group of transformations on the space $\mathcal{M}$. On a continuous group G, the group invariant metric can be written as

$$ds^2 = \text{Trace } (g^{-1}dg)^2. \tag{4.4}$$

In our case, G is abelian and we obtain the expression of the metric on $\mathcal{M}$ as

$$ds^2 = [d(\log g)]^2 = 4c^2 M^2 (dM)^2 \equiv h_{MM}(dM)^2, \tag{4.5}$$

where $h_{MM} = 4c^2 M^2$.

We now have all the ingredients to calculate the geodesic equation of motion in $\mathcal{M}$. For this purpose, consider a parametrized curve $M(\lambda) \in \mathcal{M}$, where $\lambda \in R$ is taken as the affine parameter. Using the metric in Eq. (4.5), the geodesic equation of motion in $\mathcal{M}$ can be written as

$$\frac{d^2 M}{d\lambda^2} + \Gamma^M_{MM}\left(\frac{dM}{d\lambda}\right)^2 = 0 \quad (\text{no sum over } M), \tag{4.6}$$

where

$$\Gamma^M_{MM} = \frac{1}{2} h^{MM} \frac{dh_{MM}}{dM} = \frac{1}{2}\frac{d}{dM}(\log h_{MM}) = \frac{1}{M} \tag{4.7}$$

and $h^{MM} = h_{MM}^{-1} = (4c^2)^{-1} M^{-2}$. In terms of the variable $v = \frac{dM}{d\lambda}$, the geodesic equation can be expressed as

$$\frac{dv}{dM} + \frac{v}{M} = 0. \tag{4.8}$$

Integrating the above, we get

$$v = \frac{dM}{d\lambda} = \frac{\text{constant}}{M}. \tag{4.9}$$

The above analysis describes how the mass of the black hole changes with respect to the affine parameter λ up to an overall undetermined constant.

It is chosen to be given by $\lambda = M^{-1}t + b$, where t is the time and b is a constant, whereby Eq. (4.9) reduces to

$$\frac{dM}{dt} = -\frac{k}{M^2},\tag{4.10}$$

where $k > 0$ is a constant. The sign of the constant k has been chosen based on the expectation that the back-reaction effects will cause the mass of the black hole to decrease in order to compensate for this energy loss. With these identifications, Eq. (4.10) agrees with the standard result of black hole decay. The assumption of large black hole mass used in our formalism is a feature present in the usual approach to black hole decay as well. It may be noted that the decay rate obtained above is independent of the constant a which appears in the relation between z and M. We thus have a robust description of the decay of black holes which is independent of the microscopic details of our formalism.

We have seen that the near-horizon conformal structure leads to a logarithmic correction to the Bekenstein–Hawking entropy. We would now demonstrate that this logarithmic contribution to the entropy generates a corresponding correction term for the decay rate of black holes. In our formalism, density of states for the black hole was related to the modulus square of the wavefunction as $\rho \sim |\psi|^2$. The logarithmic correction to the Bekenstein–Hawking entropy is given by $-\frac{3}{2}\log M^2$. A change in the entropy due to this term would lead to a corresponding change in the density of states. Let $\chi(M)$ denote the effective wavefunction associated with this new density of states. In our formalism, we then have

$$|\chi(M)|^2 \sim e^{2cM^2 - \frac{3}{2}\log M^2},\tag{4.11}$$

which gives

$$\chi(M) \sim \frac{1}{M^{\frac{3}{2}}}e^{cM^2}.\tag{4.12}$$

Next, we observe that $\chi(M)$ can be written as

$$\chi(M) = p(M)g(M^2),\tag{4.13}$$

where $p(M) = \frac{1}{M^{\frac{3}{2}}}$. The set $P = \{p(M)|M \neq 0\}$ forms an abelian group with respect to the composition law $p(M_1) \cdot p(M_2) = p(M_1 M_2)$ with $p^{-1}(M) = p(M^{-1})$. The wavefunction $\chi(M)$ thus belongs to the direct product $P \otimes G$ which is again an abelian group. Following the analysis presented above, we can write the metric on $P \otimes G$ as

$$ds^2 = (gp)^{-1}d(gp), \tag{4.14}$$

where $g \in G$ and $p \in P$. In the limit of large black hole mass, this metric has the form

$$ds^2 \approx 4c^2M^2 \left[1 - \frac{3}{4c^2M^2}\right]. \tag{4.15}$$

The corresponding geodesic equation of motion gives

$$\frac{dv}{dM} + v\left[\frac{1}{M} + \frac{3}{4cM^3}\right] = 0, \tag{4.16}$$

where $v = \frac{dM}{d\lambda}$. This can be integrated to give

$$vMe^{-\frac{3}{8cM^2}} = \text{constant}. \tag{4.17}$$

Recall that the affine parameter λ has already been identified with $M^{-1}t+b$. With this identification, and for large black hole mass, the decay rate of black hole is obtained as

$$\frac{dM}{dt} = -k\left[\frac{1}{M^2} + \frac{3}{8cM^4}\right]. \tag{4.18}$$

Note that using the same logic as discussed before, the constant in the above equation has been written as $-k$ with $k > 0$.

5. Conclusion

To summarize, we note that the presence of a conformal structure in the near-horizon region of a black hole is a consequence of the holographic principle. In our formalism, the appropriate condition for realizing holography is encoded in the self-adjoint extension parameter z. The parameter z, or equivalently, the black hole mass M, then has the natural interpretation as a moduli, which leads to the logarithmic correction of lack hole entropy as well as produces a correction to the black hole decay rate. This analysis holds for a large class of black holes including the Gauss–Bonnet type.

References

1. J. D. Brown and M. Henneaux, Central charges in the canonical realization of asymptotic symmetries: An example from three-dimensional gravity, *Commun. Math. Phys.* **104** (1986) 207.

2. S. Carlip, *Quantum Gravity in 2+1 Dimensions*, Cambridge University Press (1998).
3. D. Birmingham, K. S. Gupta and S. Sen, Near horizon conformal structure of black holes, *Phys. Lett.* **B 505** (2001) 191.
4. K. S. Gupta and S. Sen, Further evidence for the conformal structure of a Schwarzschild black hole in an algebraic approach, *Phys. Lett.* **B526** (2002) 121.
5. S. K. Chakrabarti, K. S. Gupta and S. Sen, Universal near-horizon conformal structure and black hole entropy, *Int. J. Mod. Phys. A* **23** (2008) 2547.
6. K. S. Gupta and S. Sen, Black hole decay as geodesic motion, *Phys. Lett. B* **574** (2003) 93.
7. M. Banados, C. Teitelboim and J. Zanelli, *Phys. Rev. Lett.* **69** (1992) 1849.
8. K. S. Gupta and S. G. Rajeev, *Phys. Rev.* **D48** (1993) 5940.
9. P. Goddard and D. Olive, *Int. Jour. Mod. Phys.* **A1** (1986) 303.
10. Bombay Lectures on Highest Weight Representations of Infinite Dimensional Lie Algebras, V. G. Kac and A. K. Raina (World Scientific, Singapore, 1987).
11. G. 't Hooft, *Nucl. Phys.* **B256** (1985) 727.

© 2023 World Scientific Publishing Company
https://doi.org/10.1142/9789811270437_0007

Chapter 7

A Proposal for the Groupoidal Description of Classical and Quantum Fields

A. Ibort[*,†,‡] and G. Marmo[**,†,§]

*Departamento de Matemáticas, Universidad Carlos III de Madrid,
Leganés, Madrid, Spain*
†*ICMAT, Instituto de Ciencias Matemáticas (CSIC-UAM-UC3M-UCM),*
**Dipartimento di Fisica "E. Pancini", Università di Napoli Federico II,
Naples, Italy*
‡*albertoi@math.uc3m.es*
§*marmo@na.infn.it*

The recent groupoidal description of Schwinger's picture of quantum mechanics is used to provide a new approach to the notion of quantum field. Starting with Feynman's quantum mechanics, where classical paths play a key role, a groupoidal interpretation of them is provided by reckoning the basic notions involved on them, most singularly the notion of clocks. It is found that natural requisites about the use of clocks in describing quantum systems lead naturally to the notion of histories as a class of homomorphisms of groupoids, and such notion leads directly to Feynman's notion of classical paths. Similar arguments can be used when replacing clocks by reference frames. There the notion of universal histories of quantum systems and an associated notion of field emerge naturally. Moreover, the analysis of the properties of clocks in quantum systems leads to a characterization of semidirect products of groupoids.

1. Introduction

The development of the notion of quantum field is a forked road with many significant milestones like, among others, the second quantization scheme followed originally by Pauli and Dirac, Feynman's inspired path integral Lagrangian approach and the abstract operator approach devised by Schwinger to tackle the description of the interaction of matter with light.

The mathematical foundation of all this has been, to say the least, intricate as the various theories were plagued with ill-defined mathematical notions, all sorts of divergencies and cumbersome technical difficulties. Many of these circumstances were discussed in depth in the 1970s and 1980s of the past century and various renormalization/regularization schemes were invented to make sense of the subtle implications of the physical ideas involved in the cranky mathematical formalism that led to spectacular successes like the consolidation of the Standard Model (SM).

A somehow unexpected consequence of this activity was the construction of a rigorous mathematical background for quantum field theories by R. Haag (together with many others) based on abstract algebras of operators, where the notion of quantum field gets diluted on behalf of a deeper, perhaps more insightful notion of observables, known as Algebraic Quantum Field Theory (AQFT), whose ultimate aim will be to provide a more solid and rigorous mathematical background for such physical theories as the SM.

More recently, as an attempt to provide new answers to some of the mysteries surrounding the SM, A. Connes has proposed an appealing new Noncommutative (NC) algebra-based mathematical background for quantum field theories. Connes' approach, even if failing to supersede the standard treatments of quantum fields used in the standard presentation of the SM, points out towards some of the weak conceptual foundations of the standard descriptions of quantum fields, beginning with the notion of quantum field itself. For instance, has it a physical reality itself like in the old pictures of the various theories? Is there a natural dynamical description of quantum fields like the one the NC geometrical approach seems to suggest?

We will not pretend here to give an answer to such arduous questions but to advance another approach to the notion of quantum field which is inspired in J. Schwinger's original contributions and that is based on modern mathematical notions as categories and groupoids that we expect will help clarify and provide a sound mathematical background to some of the ideas discussed before.

1.1. *Quantum fields in the groupoid picture*

In the construction of a notion of quantum field derived from the groupoidal formulation of Schwinger's algebra of selective measurements, we will depart from the basic assumption that a quantum theory of fields is a theory describing a quantum system, hence it must be formulated by using the same abstract groupoidal description of quantum systems which is inspired in Schwinger's formulation of quantum mechanics.[16]

From this departing point, we will assume, following our previous work on the subject, that the notion of groupoid abstracts both Schwinger's notion of algebra of selective measurements and Heisenberg's notion of transitions (see, for instance, Refs. 1–6). More precisely, given an experimental setting used to study a quantum system, there is a groupoid Γ with space of outcomes Ω, whose morphisms $\alpha \colon a \to b$ correspond to the observed transitions of the system with outcomes $a, b, \in \Omega$. Thus, we may conceive such groupoid as an abstraction of the experimental setting used to "probe" the system, and it could consist, for instance, of families of test particles, their corresponding detectors, together with other auxiliary systems, like Stern–Gerlach devices, screens and cloud chambers, often used to interact with the system (an electron in a cavity, the electromagnetic field generated by an accelerated particle, a cosmic ray or a chunk of graphene, for instance). We may think, for instance, in the simple situation of a ferrita shard dropped in a magnetic field: its position and angle with respect to a given frame will provide the outcomes of the experiment. Now, an electron passing through the magnetic field would produce a change of the field that will modify the outcomes of the ferrita meter (this is a converse Stern–Gerlach experiment). The simultaneous use of many test particles (as the beam of electrons used in actual Stern–Gerlach experiments) would display characteristics of the field under study that will be reflected in the transitions of the used meters (or other auxiliary devices like interference devices).

Each transition $\alpha \colon a \to b$ has a "source" $s(\alpha)$ and a "target" $t(\alpha) = b$, and two transitions β, α can be composed if the source of β coincides with the target of α. The composition of them will be denoted by $\beta \circ \alpha$ and such composition law is associative, that is, $(\gamma \circ \beta) \circ \alpha = \gamma \circ (\beta \circ \alpha)$ whenever γ, β, α can be composed. For each outcome a of the system, there is a "trivial" transition denoted $1_a \colon a \to a$, called a unit, that leaves unaltered the system, that is, $\alpha \circ 1_a = 1_b \circ \alpha = \alpha$, for any $\alpha \colon a \to b$, and, finally, any transition of the system $\alpha \colon a \to b$ has its "inverse", that is, there is a transition $\alpha^{-1} \colon b \to a$, such that $\alpha^{-1} \circ \alpha = 1_a$ and $\alpha \circ \alpha^{-1} = 1_b$. We will often denote the groupoid Γ as $\Gamma \rightrightarrows \Omega$, to emphasise further its structural elements.

The simplest instance of this picture is provided by a system possessing just two outcomes, denoted say by "+" and "−", and just two transitions $\alpha \colon - \to +$ and $\alpha^{-1} \colon + \to -$ (together with the units $1_+, 1_-$). It is a

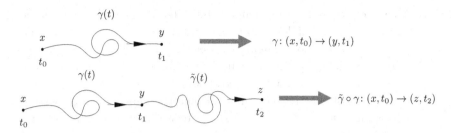

Fig. 1. Diagram representing the groupoidal interpretation of Feynman's classical paths and their composition.

straightforward observation that it defines a groupoid denoted A_2, whose algebra is isomorphic to the algebra of 2×2 matrices. Such groupoid provides the abstract description of a qubit (see Refs. 2, 3 for details).

Feynman's approach to quantum mechanics[13,14] can also be put in the previous framework. Consider a system whose classical space of configurations is Q. Now, given configurations $x, y \in Q$, a classical path is a continuous curve $\gamma: [t_0, t_1] \to Q$, such that $\gamma(t_0) = x$ and $\gamma(t_1) = y$. We can abstract such classical path as a "transition" from (x, t_0) to (y, t_1) and write it as $\gamma: (x, t_0) \to (y, t_1)$ (see Fig. 1). There is a natural composition of such transitions. If $\tilde{\gamma}: [t_1, t_2] \to Q$ is another classical path such that $\tilde{\gamma}(t_1) = y$, then we can concatenate both paths to get a new path $\tilde{\gamma} \circ \gamma: (x, t_0) \to (z, t_2)$, with $\tilde{\gamma}(t_2) = z$. Such composition is obviously associative and there are units given by the trivial paths $1_{(x,t)}: [t, t] = \{t\} \to Q$, with $1_{(x,t)}(t) = x$. We can formally introduce the inverse of a path γ as the path γ^{-1} that "runs backwards in time" and that erases the path γ in a Feynmanesque way of talking.

Certainly, classical paths constitute the "physical fields" in Feynman's description of quantum mechanics, however we would like to understand them in the more general (and abstract) background provided by the simplest description of quantum systems sketched above so that we can apply this notion to the description of the qubit, for instance. On the other hand, introducing fields as "classical paths" like in Feynman's formulation is not free from conceptual difficulties as Feynman's himself acknowledged when trying to determine the probability amplitude assigned to a path (see the discussion prior to state Postulate I in [Ref. 13, Section 3]).

There is also a striking difference between classical paths as fields in quantum mechanics and fields in a covariant description of a quantum system. In the later situation, classical fields become functions on space-time itself taking values on additional spaces that reflect additional properties of the fields (like its spinorial structure on the case of Dirac fields). Then, the question we will address in what follows is whether or not there is a natural groupoidal notion that would provide a common ground both for Feynman's description of quantum mechanics and the notion of quantum fields used in the standard description of Lagrangian quantum field theories.

2. Clocks and Histories

2.1. *Clocks*

Surprisingly, perhaps, it would be Feynman's approach to quantum mechanics what will provide the key to understand the notion that underlines both the notion of paths and the notion of quantum field. As it was discussed before, Feynman's primary notion to describe a quantum mechanical system is that of paths in space-time, that is, continuous parametrized classical paths $x(t)$ on the configuration space of the system, where the parameter t is interpreted as a physical time. We will agree with Einstein that the notion of time is an operational one, that is, time is what is measured by a clock, or in a language closer to the fundamental notions elucidated here, time corresponds to the outcomes of a physical reference device (satisfying some conditions that we will elucidate now).

What are the minimal requisites that a physical device C must satisfy to qualify as a clock? First, the outcomes of a clock are typically real numbers $s, t \in \mathbb{R}$ (or, a certain subset of the rationals $\mathbb{Q}$, even if it is also natural consider "periodic" clocks, as it has been historically the case). Thus, arguably, the simplest possible choice of the physical system acting as a clock would be the groupoid of pairs $C = P(\mathbb{R}) = \mathbb{R} \times \mathbb{R}$. It is of fundamental importance that the clock C is part of the experimental setting used to describe the given system, for instance, the clock could be "hanging in the wall of the laboratory", and it should be clearly identifiable as a subsystem of the total system, that is, it must be a subgroupoid of the given system Γ (see Ref. 5 for a discussion of subsystems and subgroupoids). Finally, the "clock" subsystem C, in order to work as a clock, must provide an accurate reading of the time for each transition $\alpha \colon a \to b$, that is, for each transition $\alpha \colon a \to b$, it is required a recording $s = s_C(\alpha) \in \mathbb{R}$ of the time at the beginning of the observation of the transition and a recording $t = t_C(\alpha)$ at the end of the observation. The difference $l(\alpha) = t_C(\alpha) - s_C(\alpha)$ will be called the 'lapse' of α (with respect to C). The map $T \colon \Gamma \to C$, defined in this way, i.e.,

$$T(\alpha) = (t_C(\alpha), s_C(\alpha)) = (t, s) \in C,$$

will be called the *clock map*. If two transitions β, α occur immediately, one after the other, then almost tautologically, the corresponding recordings ought to satisfy that $s_C(\beta) = t_C(\alpha)$, and $T(\beta) \circ T(\alpha) = (t_C(\beta), s_C(\beta)) \circ (t_C(\alpha), s_C(\alpha)) = (t_C(\beta), s_C(\alpha)) = T(\beta \circ \alpha)$. Moreover, we will assume that the clock is *perfect*, that is, the lapse of units is zero, i.e., if we consider the unit 1_a, then $t_C(1_a) = s_C(1_a)$. Then, the previous arguments show that the clock map T is a homomorphism from the groupoid Γ to the clock groupoid C (or, what is the same, a functor among them).

On the other hand, we have already argued that a clock must be a subgroupoid of the given system Γ, that is, there is an identification of the clock C with a subgroupoid of Γ. Such identification is provided by a injective homomorphism of groupoids, $\sigma \colon C \to \Gamma$. We can visualize this process by considering that before starting the experiment, we keep the clock C separated from the system. Then, we carry it into the laboratory, we start it up and, after checking that it works correctly, we use it to synchronize and calibrate the rest of the apparatuses used in the experiment. This process identifies the 'abstract' clock C with the 'clock' as a subsystem of the system under study.

There is a natural consistency condition associated with this interpretation and is that the readings of the clock itself provided by the clock map T must coincide with the outputs of the clock itself, i.e.,

$$T(\sigma(t, s)) = (t, s),$$

or, in other words, the groupoid homomorphism σ is a cross-section of the homomorphism T, that is, $T \circ \sigma = \mathrm{id}_C$. In addition, the consistency condition shows that T must be surjective, that is, T is a groupoid epimorphism.

The existence of the clock map, that is, of an unambiguous reading of time for the transitions of the system has strong implications. The first is that it determines another subsystem K_C, which is a normal subgroupoid of Γ. The subsystem K_C is defined as the kernel of T, that is, all transitions such that $T(\alpha) = (t, t)$ for some t. The space of outcomes of K_C is the same as the space of outcomes Ω of Γ, and the source and target maps of $K_C \rightrightarrows \Omega$ are denoted with the same symbols than those of Γ. It is most relevant to point out that the clock map T, being a functor from the groupoid Γ to the groupoid $C = P(\mathbb{R})$, induces a projection map $\pi \colon \Omega \to \mathbb{R}$, $\pi(a) = T(1_a) = t_C(1_a) = s_C(1_a)$, for any $a \in \Omega$. We will call the map π the canonical projection associated with the clock C and, whenever the space

Ω carries a differential structure, it will be assumed to be a locally trivial fibration.[a]

2.2. Direct and semidirect product of groupoids

The discussion in the previous section leads to consider the situation summarized by the following short exact sequence of groupoid homomorphisms:

$$1 \to K_C \to \Gamma \xrightarrow{\pi} C \to 1, \qquad (2.1)$$

and because there exists a cross-section homomorphism $\sigma\colon C \to \Gamma$, we will say that the sequence *splits*. In the particular instance that Γ would be the direct product of two groupoids $K \rightrightarrows Q$ and $C \rightrightarrows \mathbb{R}$, $\Gamma \cong K \times C \rightrightarrows Q \times \mathbb{R}$, then there is a canonical short exact sequence as above:

$$1 \to K_C \xrightarrow{i_1} K \times C \cong \Gamma \xrightarrow{\mathrm{pr}_2} C \to 1,$$

where $\mathrm{pr}_2\colon K \times C \to C$ is the canonical projection on the second factor, i.e., $\mathrm{pr}_2(\omega,(t,s)) = (t,s)$, $\omega\colon x \to y \in K$, $t,s \in \mathbb{R}$. Note that the normal subgroupoid K_C is not exactly K (or the image under a groupoid monomorphism $i\colon K \to \Gamma$) because $K_C = \{(\omega,(t,t)) \mid \omega \in K, t \in \mathbb{R}\}$. We call such groupoid the *pull-back* of the groupoid K along the projection $\pi_1\colon Q \times \mathbb{R} \to \Omega$ and denote it as $K_C = \pi_1^* K$.

We insist that there is no canonical inclusion homomorphism $i\colon K \to K \times C$ because $i(\omega)$ would have to have the form $(\omega,(t,s))$, and the times t,s must be specified consistently (see the discussion later on). Note again that the space of outputs of the direct product $K \times C$ is the Cartesian product of the space of outputs Q of K and $\mathbb{R}$, the space of outputs of C. Then, the space Ω of outputs of the system Γ will be isomorphic to $Q \times \mathbb{R}$ and the canonical projection associated with the clock map pr_2 will be the canonical projection onto the second factor $\pi_2\colon Q \times \mathbb{R} \to \mathbb{R}$.

A natural question we are facing here is under what conditions and in what sense the split exact sequence (2.1) determines a decomposition of the groupoid Γ as a direct product. This problem is the natural extension to the category of groupoids of the problem of describing group extensions that we succinctly review in what follows.

---

[a]In general, under much less restrictive conditions like, for instance, that Ω is a standard measurable space, then π would only we requested to define a measurable partition of Ω.

Consider two groups N, H; we will say that the group G is an extension of H by N if there exist group homomorphisms $i\colon N \to G$ and $\pi\colon G \to H$, such that the exact sequence of groups and homomorphisms $1 \to N \xrightarrow{i} G \xrightarrow{\pi} H \to 1$ is exact. Note that under these conditions, the homomorphism i must be a monomorphism and π must be an epimorphism. In addition, $i(N)$ must be a normal subgroup of G and there is a natural isomorphism between H and the quotient group $G/i(N)$. Two extensions G_a, $a = 1, 2$, of the groups N, H are isomorphic if there exist an isomorphism $g\colon G_1 \to G_2$ such that $i_2 = g \circ i_1$ and $\pi_2 \circ g = \pi_1$. More generally, we will say that two extensions of groups $1 \to N_a \to G \to H_a \to 1$, $a = 1, 2$, are isomorphic if there are group isomorphisms $h\colon H_1 \to H_2$, $g\colon G_1 \to G_2$ and $n\colon N_1 \to N_2$, such that the following diagram is commutative:

$$
\begin{array}{ccccccccc}
1 & \to & N_1 & \to & G_1 & \to & H_1 & \to & 1 \\
 & & n \downarrow & & g \downarrow & & h \downarrow & & \\
1 & \to & N_2 & \to & G_2 & \to & H_2 & \to & 1
\end{array}
$$

We will be interested in the classification of split extensions of groups, that is, the identification modulo, the equivalence relation defined above, of extensions $1 \to N \xrightarrow{i} G \xrightarrow{\pi} H \to 1$, such that there exists a homomorphism of groups $\sigma\colon H \to G$, satisfying $\pi \circ \sigma = \mathrm{id}_H$. It is well known that whenever a short exact sequence of groups $1 \to N \to G \xrightarrow{\pi} H \to 1$ splits, the middle group G is the semidirect product of the group H by the normal subgroup $N \triangleleft G$, with respect to the homomorphism of groups $w\colon H \to \mathrm{Aut}(N)$, defined as

$$
w_h(n) = \sigma(h) n \sigma(h)^{-1}, \quad n \in N, \quad h \in H.
$$

The group G is denoted as $N \rtimes_w H$. Finally, we observe that the group homomorphism w induces a group homomorphism, $[w]\colon H \to \mathrm{Out}(N)$, called the modular class of the extension, in the group of external automorphisms of N. It is simple to show that the semidirect product $G \cong N \rtimes_w H$ is trivial, that is, $G \cong N \times H$, iff the modular class $[w]$ vanishes.

We will turn now our attention to a similar situation in the category of groupoids. Let $C \rightrightarrows \tilde{\Omega}$, $N \rightrightarrows \Omega$ be two groupoids with spaces of outcomes Ω and $\tilde{\Omega}$, respectively. A groupoid $\Gamma \rightrightarrows \widehat{\Omega}$ will be said to be an extension of C by N if there exist a monomorphism of groupoids $i\colon N \to \Gamma$ and a epimorphism of groupoids $\pi\colon \Gamma \to C$, such that $\ker \pi = i(N)$, or, in other words, the following short sequence of groupoid homomorphisms:

$$
1 \to N \xrightarrow{i} \Gamma \xrightarrow{\pi} C \to 1, \tag{2.2}
$$

is exact. The monomorphism of groupoids $i\colon N \to \Gamma$ must satisfy that the induced map $i\colon \Omega \to \widehat{\Omega}$ is injective, which is just a consequence that the assignment $\omega \in N \mapsto i(\omega) \in \Gamma$ is injective. Hence, the groupoid N can be identified with the subgroupoid $i(N) \subset \Gamma$. Moreover, if $i(\Omega)$ were strictly contained in $\widehat{\Omega}$, we could always enlarge canonically the subgroupoid $i(N)$ in such a manner that its space of outcomes is $\widehat{\Omega}$ without altering the s.e.s. (2.2). In fact, all we have to do is to add the units 1_a to $i(N)$, $a \in \widehat{\Omega}$, with $a \notin i(\Omega)$. Thus, in what follows, we will assume that the groupoid N has the same space of outcomes as Γ, and we will denote it by Ω. With an abuse of notation, we will identify the groupoid N with its image $i(N) \subset \Gamma$, omitting the map i in the corresponding formulas.

The epimorphism of groupoids $\pi\colon \Gamma \to C$ induces a surjective map, denoted with the same symbol, $\pi\colon \Omega \to \tilde{\Omega}$. For each outcome $u \in \tilde{\Omega}$, we will denote by Ω_u the set $\pi^{-1}(u)$. Then, $\ker\pi$ consists of those transitions $\alpha\colon a \to b \in \Gamma$ such that $\pi(a) = \pi(b)$. Because the sequence (2.2) is exact, $\ker\pi = i(N)$, then the transitions $\omega\colon a \to b \in N$ are such that $\pi(a) = \pi(b) = u$, for some $u \in \tilde{\Omega}$. Hence, the groupoid N is disconnected and its connected components correspond to the fibers Ω_u of the map π. Then, if we denote by N_{Ω_u} the restriction of N to the fiber Ω_u, we get $N_{\Omega_u} = \{\alpha\colon a \to b \mid \pi(\alpha) = 1_{\pi(a)} = 1_{\pi(b)}\}$, and $N = \bigsqcup_{u \in \tilde{\Omega}} N_{\Omega_u}$ is the decomposition of N in connected components.

If we denote by N_0, Γ_0 and C_0, respectively, the canonical normal subgroupoids of the groupoids N, Γ and C, the short exact sequence (2.2) induces, in the obvious way, a short exact sequence:

$$1 \to N_0 \to \Gamma_0 \to C_0 \to 1. \tag{2.3}$$

The totally disconnected normal subgroupoids N_0, Γ_0 and C_0 consist of the union of isotropy subgroups of N, Γ and C, respectively. Thus, for every $a \in \Omega$ and $u = \pi(a)$, we get a short exact sequence of groups:

$$1 \to N(a) \to \Gamma(a) \xrightarrow{\pi_a} C(u) \to 1, \tag{2.4}$$

where $N(a) = N_a \cap N^a$ denotes the isotropy group of N and similarly for $\Gamma(a)$ and $C(u)$.

Moreover, as the isotropy groups $N(a)$, $N(b)$ with a, b in the same connected component of N are isomorphic (albeit not canonically), we can consider that for each $u \in \tilde{\Omega}$, there is an abstract extension of the corresponding isotropy groups. We conclude thus that an extension of the groupoid C by the groupoid N induces an extension of the isotropy group $C(u)$ by the group $H(a)$ with $\pi(a) = u$.

Finally, if the sequence (2.2) splits, that is, there is a homomorphism of groupoids $\sigma\colon C \to \Gamma$ such that $\pi \circ \sigma = \mathrm{id}_C$, then such functor induces a map $\sigma\colon \tilde{\Omega} \to \Omega$, $\sigma(u) \in \Omega_u$, i.e., a cross-section of the projection map $\pi\colon \Omega \to \tilde{\Omega}$. Moreover, it induces a family of group homomorphisms $\sigma_u\colon C(u) \to \Gamma(\sigma(u))$ which are cross-sections respectively of the epimorphisms of groups $\pi_u\colon \Gamma(\sigma(u)) \to C(u)$.

Hence, adapting the terminology from the theory of groups, if the groupoid Γ is the middle groupoid of a split short exact sequence (2.2), we will say that Γ is the semidirect product of the groupoids C and N with respect to the cross-section homomorphism $\sigma\colon C \to \Gamma$ that, in turn, defines a cross-section $\sigma\colon \tilde{\Omega} \to \Omega$ and a homomorphism of groupoids $w\colon C \to \mathrm{Aut}(N_0)$ defined as $w(\omega)(\gamma) = \sigma(\omega) \circ \gamma \circ \sigma(\omega)^{-1}$, for any $\omega \in C$, $\gamma \in N_0$.

A reconstruction theorem showing how the middle groupoid Γ can be recovered from these data will be discussed elsewhere.[10] It suffices for us now to note that the groupoid Γ will be the direct product of the groupoids C and K iff the space of objects Ω factorises trivially with respect to the projection π, i.e., $\Omega = Q \times \tilde{\Omega}$, and the induced modular classes $[w]_u$ vanish for all $u \in \tilde{\Omega}$. In such case, the normal subgroupoid N will be given by the pull-back of K along the projection $\pi_1\colon Q \times \tilde{\Omega} \to Q$.

2.3. *Histories*

We return now to analyze the situation of the split extension associated with a clock map $T\colon \Gamma \to C = P(\mathbb{R})$. Because the clock is a subgroupoid of Γ, determined by a synchronization homomorphism $\sigma\colon C \to \Gamma$, then the sequence $1 \to K_C \to \Gamma \xrightarrow{T} C \to 1$ splits and Γ is a semidirect product of K_C by C with respect to the homomorphism $w\colon C \to \mathrm{Aut}(K_0)$, defined as $w_{s,t}(\alpha) = \sigma(t,s) \circ \alpha \circ \sigma(s,t)$, where $\alpha\colon x(s) \to x(t)$, with $x(t) = \sigma(t)$, the curve in Ω defined by the cross-section of the projection $\pi\colon \Omega \to \mathbb{R}$ determined by the functor σ.

In such case, the induced sequence of extensions of isotropy groups, Eqs. (2.3) and (2.4), is trivial because the isotropy groups of $C = P(\mathbb{R})$ are trivial and, in consequence, the isotropy groups of the normal subgroupoid $\ker T = K_C$ coincide with the isotropy groups of Γ. Thus, because of the arguments exhibited in the previous section, the only obstruction for the semidirect product $\Gamma = K_C \rtimes_\sigma C$ being a direct product is determined by the projection map $\pi\colon \Omega \to \mathbb{R}$ induced by T. Hence, if the projection map

$\pi \colon \Omega \to \mathbb{R}$ is trivial, i.e., there is a space Q such that $\Omega = Q \times \mathbb{R}$ and $\pi = \mathrm{pr}_2$, then $\Gamma \cong K_C \times P(\mathbb{R})$. If this happens, K_C is isomorphic to the pull-back of a groupoid $K \rightrightarrows Q$ along the projection map pr_1, that is, K_C is isomorphic to the groupoid $\mathrm{pr}_1^* K = \{(\alpha, t) \mid t \in \mathbb{R}\}$.

In most applications, the space of outcomes Ω of the groupoid Γ carries some additional structure and the groupoid Γ is a groupoid belonging to the category determined by such structure. For instance, if the relevant structure were a Borel structure, the groupoids of interest would be measurable groupoids (see, for instance Ref. 9 and references therein); if the structure of interest were a topological structure, the groupoid Γ would be a topological groupoid.

It is often found that groupoids of interest are Lie groupoids, that is, the groupoid Γ is a smooth manifold such that all the relevant maps in the definition of the groupoid are smooth and the source and target maps are smooth submersions. In such case, if the clock map $T \colon \Gamma \to C = P(\mathbb{R})$ is smooth, then the induced map $\pi \colon \Omega \to \mathbb{R}$ is smooth too. It would be assumed that the projection π defines a locally trivial fibration over $\mathbb{R}$. The connected components of the groupoid $K_C = \ker T$ are the fibers of π. Thus, we will say that the bundle $\pi \colon \Omega \to \mathbb{R}$ has a K_C-structure. In general, there will not exist a smooth cross-section of such a bundle. For instance, if the groupoid K_C is the action groupoid of a group acting freely and transitively on the fibers of π, then such a section exists iff the bundle is trivial. Hence, in such case, the existence of a smooth clock map and a smooth synchronization map implies that the total groupoid Γ is the trivial direct product of the clock groupoid $C = P(\mathbb{R})$ and a groupoid $K \rightrightarrows Q$.

We will end this discussion by making the fundamental observation that the introduction of a clock in the description of a system allows us to define the "history" of the system, that is, the synchronization homomorphism $\sigma \colon C = P(\mathbb{R}) \to \Gamma$ can be regarded as defining a history on Γ as follows: Let us fix an initial instant of time t_0 and consider the family of transitions $\sigma(t, t_0) \colon a(t_0) \to a(t)$. Such family of transitions defines a path $a(t)$ on the space of outcomes and a path of transitions $\sigma_{t_0}(t) = \sigma(t, t_0)$. Moreover, such path satisfies that $\sigma_{t_0}(t_0) = 1_{a_0} = 1_{a(t_0)}$ and $\sigma_{t_0}(t+s) = \sigma(t+s, t_0) = \sigma(t+s, s) \circ \sigma(s, t_0) = \sigma_s(s+t) \circ \sigma_{t_0}(s)$. Such paths will be called Γ-paths (see Ref. 12 for the relation between Γ-paths and the corresponding paths on the Lie algebroid of the Lie groupoid Γ) and they are the main notion behind Feynman's interpretation of quantum mechanics. We will devote the coming section to discuss them in more detail.

3. Fields, Histories and Frames

Nothing constraints us to consider "clocks", in the sense discussed in the previous section, as reference systems. We may use more elaborated systems as reference systems. For instance, when describing a system on a certain spatial region, we often introduce a frame to locate it. We may consider that such frame consists of a system of detectors that assign specific spatial coordinates to the transitions of the system. For instance, a bubble chamber will be able to associate an actual trajectory in a three-dimensional frame to the sequence of transition describing the detection of a particle. In this sense, we may consider that a space-time frame is a system described by a groupoid F whose space of outcomes is some subset of the space-time itself, for instance, Minkowski space-time $\mathbb{M}$. Points in $\mathbb{M}$ will be denoted as $x = (x^\mu) = (t, x, y, z)$ and it will be assumed that $\mathbb{M}$ carries the standard Lorentzian metric $\eta = \eta_{\nu\nu}dx^\mu dx^\nu = -dt^2 + dx^2 + dy^2 + dz^2$. The simplest groupoid of this sort is the groupoid of pairs $F = P(\mathbb{M}) = \mathbb{M} \times \mathbb{M}$, and we will call it a global (abstract) space-time frame (system) and any of its subgroupoids of the form $F_O = P(O) = O \times O$, with $O \subset \mathbb{M}$ and open set, a local frame.

As in the previous discussion with clocks, there are some natural requirements to be put on the frame F to qualify as part of the description of our physical system. The first is that F itself is part of the experimental setting or, in groupoidal terms, that F is a subgroupoid of the overall groupoid $\Gamma \rightrightarrows \Omega$ describing the system. We will denote the monomorphism of groupoids describing it as $\sigma \colon F \to \Gamma$ and we will call it a synchronization homomorphism. In addition, the frame should be able to provide accurate readings of the positions and times of the transitions, thus there must be a homomorphism of groupoids $D \colon \Gamma \to F$ that describes such readings. Finally, the "detection" homomorphism D and the "synchronizing" homomorphism $\sigma \colon F \to \Gamma$ must be consistent in the sense that $D \circ \sigma = \mathrm{id}_F$, i.e., σ is a cross-section homomorphism of D and, in consequence, the short exact sequence defined by the homomorphism $D \colon \Gamma \to F$ splits. In other words, the groupoid Γ is a semidirect product of F and the normal subgroupoid $K_F = \ker D \lhd \Gamma$.

Note that, very much as in the previous discussion with clocks, the detection map D induces a projection $\pi \colon \Omega \to \mathbb{M}$ because it assigns a unit 1_x to any unit $1_a \in \Gamma$, i.e., $D(1_a) = 1_x = 1_{\pi(a)}$, and we denote such assignment as $\pi(a) = x$. If the projection map π is a locally trivial fibration, we conclude that the frame F fixes a bundle map $\pi \colon \Omega \to \mathbb{M}$. The fibers

$\Omega_x = \pi^{-1}(x)$ of such bundle are determined by additional outcomes of our system.

The synchronizing map $\sigma\colon F \to \Gamma$ determines a cross-section φ of the bundle $\pi\colon \Omega \to \mathbb{M}$, given by $\sigma(1_x) = 1_{\varphi(x)}$ and $\varphi(x) \in \Omega_x$. Hence, the different ways that our frame is synchronized with the experimental setting provides different cross-sections φ. We will interpret the functions $\varphi\colon \mathbb{M} \to \Omega$, $\varphi = \varphi(x)$, as the classical fields of the theory.

If we select two points $x, y \in \mathbb{M}$, they determine a causal domain $[y, x]$ in $\mathbb{M}$ that consists of all points z which lie in the causal future domain $J^+(x)$ of x and in the causal past domain $J^-(y)$ of y, i.e., $[y, x] = J^+(x) \cap J^-(y)$. The restriction of the groupoid $F = P(\mathbb{M})$ to the causal domain $[y, x]$ is just the groupoid of pairs $P(y, x) = [y, x] \times [y, x]$ which is a subgroupoid of F (see, for instance Ref. 8, for a discussion of causality and groupoids). As in the case of clocks, we can restrict the homomorphism σ to the subgroupoid $P(y, x)$ and we will get a concrete history of the system "starting" at x and "ending" at y. More concretely, fixing a point x_0, we define the maps

$$W(z) := \sigma(z, x_0)\colon \varphi(x_0) \to \varphi(z), \quad \varphi(z) = t(\sigma(z, x_0))$$

for all $z \in [y, x_0]$. The map $W = W(z)$ describes a "history" on the groupoid Γ that projects to a field φ on each causal domain $[y, x] \subset \mathbb{M}$. For this reason, in what follows, we will call the cross-sections $\sigma\colon F \to \Gamma$ *universal histories* (or *universes*), and the restrictions to causal domains $[y, x]$ *concrete histories* or *configurations*.

We can make this ideas concrete by considering a system described using Minkowski space as the reference system and whose additional outcomes have a linear structure, that is, they can be added and scaled. Then, the space of outcomes of the system will be a vector bundle $\pi\colon \Omega \to \mathbb{M}$. We can consider that the groupoid Γ is the groupoid of automorphisms of such bundle, that is, transitions are triples $(u_y; T_{y,x}; u_x)$ such that $u_x \in \Omega_x$, $u_y \in \Omega_y$, $T_{y,x}\colon \Omega_x \to \Omega_y$ is a linear map and $T_{y,x}(u_x) = u_y$. The source and target maps of Γ are the obvious ones and the composition is the composition of linear maps. A synchronization map $\sigma\colon P(\mathbb{M}) \to \Gamma$ defines a classical field $\varphi\colon \mathbb{M} \to \Omega$ and a family of transformations $\sigma(y, x)\colon \Omega_x \to \Omega_y$. Hence, the classical field φ can be written as $\varphi(x) = (\varphi^\alpha(x))$ (once a linear basis $e_\alpha(x)$ has been chosen on the fibers Ω_x) and the concrete histories $W(x) = (W(x)^\beta_\alpha)$ will be maps with values on the group of invertible matrices.

4. Conclusions and Discussion

The analysis of the role of clocks and space-time frames in the groupoidal description of quantum systems shows that the standard notions of paths and classical fields emerge from the concrete realization of the given reference system as a subsystem of groupoid describing the overall system. Such identifications, called universal histories, have concrete counterparts that are naturally identified with paths, in the case of clocks, and classical fields, in the case of space-time frames.

Any unitary representation of the groupoid describing the system will provide a quantum version of such fields, that is, a unitary representation U of the groupoid Γ (assumed discrete for simplicity) will associate an operator $U(\alpha)$ to any transition $\alpha\colon a \to b$ and a projector to any outcome $a \in \Omega$. Thus, if W is a history, such representation will define an assignment of an operator $\hat{\varphi}(x)$ to any $x \in \mathbb{M}$. Representations of groupoids are intimately related to states on their algebras, hence the standard notion of quantum fields as operator-valued distributions on $\mathbb{M}$ will be recovered once appropriate states are singled out in our groupoid. These and related topics will be discussed further in subsequent works.

Acknowledgements

The authors acknowledge financial support from the Spanish Ministry of Economy and Competitiveness through the Severo Ochoa Programme for Centres of Excellence in RD (SEV-2015/0554), the MINECO research project PID2020-117477GB-I00 and Comunidad de Madrid project QUITEMAD++, S2018/TCS-A4342. GM is also a member of the Gruppo Nazionale di Fisica Matematica (INDAM), Italy.

References

1. F. M. Ciaglia, A. Ibort and G. Marmo, A gentle introduction to Schwinger's picture of Quantum Mechanics, *Mod. Phys. Lett. A.* **33**(20) (2018) 1850122.
2. F. M. Ciaglia, A. Ibort and G. Marmo, G. Schwinger's picture of Quantum Mechanics I: Groupoids, *Int. J. Geom. Methods Modern Phys.* **16** (2019) 1950119.
3. F. M. Ciaglia, A. Ibort and G. Marmo, Schwinger's picture of Quantum Mechanics II: Algebras and observables. *Int. J. Geom. Methods Modern Phys.* **16**(9) (2019) 1950136.
4. F. M. Ciaglia, A. Ibort and G. Marmo, Schwinger's picture of Quantum Mechanics III: The statistical interpretation, *Int. J. Geom. Methods Modern Phys.* **16**(11) (2019) 1950165.

5. F. M. Ciaglia, F. Di Cosmo, A. Ibort and G. Marmo, Schwinger's picture of Quantum Mechanics IV: Composite systems, *Int. J. Geom. Methods Modern Phys.* **17**(4) (2020) 2050058.

6. F. M. Ciaglia, F. Di Cosmo, A. Ibort and G. Marmo, Schwinger's picture of Quantum Mechanics, *Int. J. Geom. Methods Modern Phys.* **17**(4) (2020) 2050054.

7. F. M. Ciaglia, F. Di Cosmo, A. Ibort and G. Marmo, Evolution of classical and Quantum states in the Groupoid picture of Quantum Mechanics, *Entropy* **22** (2020) 1292.

8. F. M. Ciaglia, F. Di Cosmo, A. Ibort, G. Marmo, L. Schiavone and A. Zampini, Causality in Schwinger's picture of Quantum Mechanics, *Entropy* **24** (2020) 75. https://doi.org/10.3390/e24010075.

9. F. M. Ciaglia, F. Di Cosmo, A. Ibort and G. Marmo, The type of a quantum system: Groupoids and their von Neumann algebras. In preparation.

10. F. M. Ciaglia, F. Di Cosmo, A. Ibort and G. Marmo, Extensions of groupoids and their algebras. In preparation (2022).

11. A. Connes, *Noncommutative Geometry*, Academic Press (1994).

12. M. Crainic and R. Loja Fernandes, Integrability of Lie brackets, *Ann. Maths.* **157** (2003) 575–620.

13. R. P. Feynman, Space-time approach to non-relativistic quantum mechanics, *Rev. Mod. Phys.* **20** (1948) 367–387.

14. R. P. Feynman, *Feynman's Thesis: A New Approach to Quantum Theory* (1942). Editor L. M. Brown, World Scientific (2005). Reprinted from R. P. Feynman, The principle of least action in Quantum Mechanics.

15. K. Mackenzie, General theory of Lie groupoids and Lie algebroids, London Mathematical Society Lecture Note Series **213** (2005) Cambridge University Press.

16. J. Schwinger, *Quantum Kinematics and Dynamics*, Advanced Book Classics, Frontiers in Physics Series, Perseus Books Group (1991).

Chapter 8

Galaxies without Dark Matter

Giorgio Immirzi

*Dipartimento di Fisica, Universit'a di Perugia
and INFN, Sezione di Perugia, Perugia, Italy*
giorgio.immirzi@pg.infn.it

It is an honour and a pleasure to be able to contribute to this Festschrift dedicated to Bal, but I want to write on a subject he does not like, and does not think much of, hoping to convince him that it is important and interesting, in spite of the fact that the work is not even finished: galaxies, which are sort of half way between elementary particles and cosmology but important for both. Galaxies are huge: for a typical one SGC1560 light takes ~25000 years to go from one end to the other;[1] it has a visible mass estimated to ~$8. \cdot 10^8 M_\odot$; it is more like a pancake than a sphere with an axis ratio ~5 : 1; it rotates, with an edge velocity ~36 km/s, the focus of our attention. Outside its visible part, there is a lot of ionized gas circling it. The famous astronomer Vera Rubin[2] discovered that the velocity of this gas does not fall down with distance as one might naively expect from Kepler law but stays roughly constant. Now, after 400 years arguing with Kepler's law, and with Newton's law that explains it, it is sort of sacrilegious, but what else can one do? Astrophysicists are resourceful: they invented 'dark matter'.[3] What you see, 'baryonic matter', is only a small percentage (5%) of the total mass; the rest is invisible and only interacts through gravitation, and it is distributed over a much larger volume that the visible mass. That explains the constant gas velocities. Humans are ignorant and pretentious, yet having to admit that we know nothing about what 95% of the Universe is <u>really</u> made of is a bit too much. Other physicists, mostly not astro-, are skeptical and try to find alternative explanations. I joined the "no dark matter" club, together with my colleague and long time friend Yogendra Srivastava.

I find the dark matter hypothesis extravagant; it reminds me of the 'cosmic aether' idea. I would like it to end in the same way. One weak point of the dark matter theory is that none has ever been found and not for lack of trying and ingenious attempts. Another weak point is the discovery of an empirical but very successful relation between the visible mass of the galaxy and the asymptotic velocity of the gas circling it: $v_{asym}^4 \sim M_{visible}$, the Tully–Fisher relation.[5] This creates a problem: if $M_{visible}$ is only 5% of the total, how comes it determines v_{asym}?

What we propose, in a paper unfinished but to be published any time soon, together with a few other authors,[6-9] is that general relativity, correctly applied, is perfectly capable of explaining the observed phenomena, if one takes into account the huge size of most galaxies and the basic fact that they rotate. Because of the rotation, the metric of the space surrounding it will have a non-diagonal term that produces gravitomagnetic effects, in its typical form, the Lense–Thirring effect. Here, I give the main points of our argument, with a minimum of equations. For the rest, you will have to wait.

The general form of the metric for a system with cylindrical symmetry has been given by Weyl[4] with coordinates (ct, ρ, φ, z):

$$ds^2 = -e^{2U}(cdt - ad\varphi)^2 + e^{-2U}\rho^2 d\varphi^2 + e^{2\nu - 2U}(d\rho^2 + dz^2),$$

where U, a, ν are only functions of $\rho = \sqrt{x^2 + y^2}$ and z. The function $U(\rho, z)$ is related to the Newtonian potential Φ through $e^{2U} = 1 + 2\frac{\Phi}{c^2}$; the function $a(\rho, z)$ is related to the angular momentum of the system and the gravitomagnetic field $A_\varphi = \frac{ca}{\rho}$, a vector potential $\sigma A = (A_r, A_\varphi, A_z) = (0, \frac{ca}{\rho}, 0)$. In the linearized approximation to the Einstein equations, it is convenient to use the analogy with electromagnetism to write the equations in terms of a 'gravitoelectric field' $\sigma E_g = -\nabla\Phi$ and a 'gravitomagnetic field' $\sigma B_g = \nabla \wedge \sigma A$, which we denote by GEM. This analogy was first noticed and used by Thirring in 1918; his initial purpose was to to compute the field inside a hollow rotating sphere; later, with Lense, he extended it to the analysis of the effect of proper rotation of a central body on the motion of other celestial bodies, which led to the discovery of the Lense–Thirring effect, finally verified in 2004. This work was recently extended by G. Ludwig in a set of beautiful papers[7-9] that in part we follow.

We assume that a galaxy can be modeled as a fluid with zero pressure, stationary and axially symmetric; its energy–momentum tensor can be written as $T^{\mu\nu} = \rho(\rho,z)_m u^\mu u^\nu$; additionally, we assume up-down symmetry with respect to its equatorial plane. The galaxy rotates around its $\rho = 0$ axis and the GEM field always produces a counter-rotating velocity field to material masses. The force is always attractive as can be seen in the equation for the GEM field

$$\nabla \times \mathbf{B} = -\left(\frac{G}{c^2}\right)\rho_m \mathbf{v} + \frac{1}{c^2}\frac{\partial \mathbf{E}}{\partial t}.$$

The minus sign in the first term on the right-hand side of this equation tells us that the induced magnetic field on the left side follows the *left-hand rule* always (Lenz's law). In terms of Φ and σA_g, the Einstein equations acquire the form of Gauss-like and Ampère-like laws; in cylindrical coordinates, assuming that $\sigma A_g = A_\varphi \hat\varphi, \sigma v = v\,\hat\varphi$ and that we have stationary conditions, the equations are

$$\frac{1}{\rho}\frac{\partial}{\partial\rho}\left(\rho\frac{\partial\Phi}{\partial\rho}\right) + \frac{\partial^2\Phi}{\partial z^2} = \nabla^2\Phi = 4\pi G\rho_m;$$

$$\frac{\partial}{\partial\rho}\left(\frac{1}{\rho}\frac{\partial(\rho A_\varphi)}{\partial\rho}\right) + \frac{\partial^2 A_\varphi}{\partial z^2} = \frac{4\pi G}{c^2}\rho_m\, v.$$

The assumption is that $v(\rho,z)$ describes continuously the motion of the rotating matter inside the galaxy and the motion of the ionized gas that circles round it. The geodesic equations for the (spatial) acceleration $\mathcal{A}^i$ of a particle are nonlinear and complicated, but we want to consider only equatorial circular motion around the z-axis with $\frac{d\varphi}{dt} = \frac{v}{\rho}$ and $\mathcal{A}^\varphi = \mathcal{A}^z = 0$, that to the lowest order give

$$\frac{\partial\Phi}{\partial\rho} - \frac{v^2}{\rho} = \frac{v}{\rho}\frac{\partial(ca)}{\partial\rho}, \qquad \frac{\partial\Phi}{\partial z} = \frac{v}{\rho}\frac{\partial(ca)}{\partial z}$$

$$\Longleftrightarrow E_z - vB_\rho = 0; \quad E_\rho + vB_z = -\frac{v^2}{\rho},$$

i.e., the Lorentz force equations. A simple qualitative argument for constant asymptotic velocity can be derived from these equations, written as $v^2 = \rho\frac{\partial\Phi}{\partial\rho} + v\rho(-B_z)$, a 'Newtonian' term plus a GEM term, $\beta^2 = g_N + \beta\beta_{mag}$. The Newtonian potential, roughly speaking, has two bumps and then goes down; you leave it intact and add the 'magnetic velocity' term, with the correct sign as dictated by Lenz's law, with a constant vector potential A_φ. For our galaxy, this produces quite a decent fit. The maximum of the Newtonian term more or less coincides with the asymptotic velocity, $\beta^2(\infty) = \frac{R_s}{2R_{max}}$. For a pillbox like galaxy $V = \pi R_{max}^2 h$, $M = \rho_m V$, so $\beta^2(\infty) \sim \frac{M}{M^{1/2}} \sim M^{1/2}$, which is the Tully–Fisher law.

Or we can use these equations to eliminate A_φ from the expression of the Ampère law that becomes

$$\frac{\partial}{\partial\rho}\left(\frac{1}{v}\frac{\partial\Phi}{\partial\rho} - \frac{v}{\rho}\right) + \frac{\partial}{\partial z}\left(\frac{1}{v}\frac{\partial\Phi}{\partial z}\right) = \frac{4\pi G}{c^2}\rho_m v.$$

This equation multiplied by v and subtracted from the expression of Gauss' law given above eliminates double derivatives and yields

$$4\pi G\rho_m\left(1 - \frac{v^2}{c^2}\right) = \left(\frac{1}{\rho} + \frac{1}{v}\frac{\partial v}{\partial\rho}\right)\frac{\partial\Phi}{\partial\rho} + \frac{1}{v}\frac{\partial v}{\partial z}\frac{\partial\Phi}{\partial z} + v\frac{\partial}{\partial\rho}\frac{v}{\rho},$$

a nonlinear first-order differential equation for $v(\rho, z)$ for given $\rho(\rho, z)_m, \Phi(\rho, z)$. In the equatorial plane $z = 0$ by the up-down symmetry, we can drop the $\frac{\partial\Phi}{\partial z}$, then

$$\left(\beta^2 + \rho\frac{\partial\tilde{\Phi}}{\partial\rho}\right)\rho\frac{\partial\beta}{\partial\rho} = \frac{\beta}{\rho}\left[\left(\beta^2 - \rho\frac{\partial\tilde{\Phi}}{\partial\rho}\right) + \frac{4\pi G\rho_m}{c^2}\rho^2(1 - \beta^2)\right];$$

$$\beta = \frac{v(\rho, 0)}{c}, \quad \tilde{\Phi} = \frac{\Phi}{c^2}.$$

Outside the galaxy, where $\rho(\rho, 0)_m = 0$, the equation becomes

$$\frac{\rho^2}{\beta}\frac{\partial\beta}{\partial\rho} = \frac{\beta^2 - \rho\frac{\partial\tilde{\Phi}}{\partial\rho}}{\beta^2 + \rho\frac{\partial\tilde{\Phi}}{\partial\rho}}$$

which shows the key role of the gravitomagnetic field that is now

$$\frac{B_z}{c} = \frac{\rho\frac{\partial\tilde{\Phi}}{\partial\rho} - \beta^2}{\beta\rho}.$$

One expects that at large ρ, it will agree with the Lense–Thirring metric, which takes into account the mass and the angular momentum of the galaxy, but at intermediate distances, the internal distribution of masses in the galaxy gives rise to a quite different behavior that produces the observed flattening of the rotation curves. We are late with the phenomenology; we have tried applying the Miyamoto–Nagai[11] field with adjusted parameters for NGC1350; it works fine but not for the luminosity curve, indicating that one needs a more careful analysis of the internal distribution of masses in the galaxy.

On the contrary, one could try a different approach, at the moment blocked by my incompetence with the use of Mathematica, in which the details of the internal distribution of masses are ignored: the idea is to look at the asymptotic behavior of the solution for the Weyl metric and to assimilate it to that of the Kerr metric; this metric describes a rotating black hole in terms of the mass M and a parameter a with the dimension of a length similar to our function $a(\rho, z)$; it is very complicated and there are lots of papers discussing it, but asymptotically, up to terms $O(1/r^2)$, it has in Cartesian coordinates the relatively simple form[10]:

$$g_{00} \simeq -1 + \frac{R_s}{r}, \quad g_{ij} \simeq \delta_{ij} + \frac{R_s}{r^3} x_i x_j, \quad g_{0i} \simeq \frac{R_s}{r^2} \left(x_i + \frac{1}{r} (\sigma a \wedge \sigma x)_i \right),$$

$$\sigma a = (0, 0, a), \quad i, j = 1, 2, 3,$$

where $R_s \equiv \frac{2GM}{c^2}$. From Weinberg's book,[10] it is easy to show that this is enough to determine mass, momentum and angular momentum of the source, using his equations integrated over a large sphere: $\sigma P = 0$, $P^0 = Mc$, $\sigma J = Mc\sigma a$. Vice versa, we can see that for the system to have a finite angular momentum, which a rotating galaxy certainly has, it is crucial that asymptotically the space-time part of $g_{\mu\nu}$ does not vanish.

The Kerr solution is not unique, it is singular at the origin, and an analysis of its relativistic multipole moments M_n shows that they are carefully arranged, with $a = 0$ giving the Schwarzschild solution. But this certainly does dot apply to the Weyl metric that we want with the same asymptotic behavior but non-singular at the origin. The asymptotic behavior chosen should also be sufficient to determine the velocity of a test body in circular, equatorial motion at large r. But we have not yet calculated it. Apologies.

References

1. A. H. Broeils, The mass distribution of the dwarf spiral NGC 1560, *Astron. Astrophys.* **256** (1992) 19.
2. V. C. Rubin, N. Thonnard and W. K. Fotd Jr, *ApJ* **225** (1978) L107. V. C. Rubin, One hundred years of rotating galaxies, *Publications of the Astronomical Society of the Pacific*, **112** (2000) 747–750.
3. P. J. E. Peebles, How the nonbaryonic dark matter theory grew. arXiv: 1701.05837.
4. H. Stephani, *An Introduction to Special and General Relativity*, Cambridge University Press; 3° edizione (21 agosto 2008).

5. R. B. Tully and J. R. Fisher, A new method for determining distances to galaxies, *Astron. Astrophys.* **54** (1977) 661; S. McGaugh, F. Lelli, J. M. Shombert...., The radial acceleration relation in rotationally supported galaxies, *Phys. Rev. Lett.* 117.201101, arXiv:1609.05917; S. McGaugh, Predictions and outcomes for the dynamics of rotating galaxies, *Galaxies* **8020035** (2020), arXiv:2004.14402.

6. H. Balasin and G. Grümiller, Non-Newtonian behaviour in weak field general relativity for external rotating sources, arXiv: astro-ph/0602519; M. T. Crosta, M. Gianmaria, M. G. Lattanzi and E. Poggio, Testing dark matter geometry sustained circular velocities in the Milky Way with Gaia DR2, arXiv:1810.04445; F. I. Cooperstock and S. Tieu, Perspectives on galactic dynamics via general relativity, astro-ph/0512048; D. Vogt and P. S. Letelier, Comments on perspectives on galactic dynamics via general relativity, astro-ph/0512553; D. J. Cross, Comments on the Cooperstock-Tieu galaxy model, astro-ph/0601191; A. G. Cornejo, The rotational velocity of spiral Sa galaxies in the general theory of relativity solution, *Int. J. Astron.* **10** (2021) 6.

7. G. O. Ludwig, Galactic rotation curve and dark matter according to gravitomagnetism, *Eur. Phys. J.* **C81** (2021) 186.

8. G. Ludwig, Extended gravitomagnetism. I. Variational formulation, *Eur. Phys. J. Plus* **136** (2021) 373.

9. G. O. Ludwig, Extended gravitomagnetism. II. metric perturbation, *Eur. Phys. J. Plus* **136** (2021) 465.

10. S. Weinberg, *Gravitation and Cosmology*, John Wiley & Sons (1972).

11. M. Miyamoto and R. Nagai, Three-dimensional models for the distribution of mass in galaxies, *Publ. Astron. Soc. Jpn.* **27** (1975) 533.

12. M. Visser, The Kerr spacetime: A brief introduction, arXiv: 0706.0622.

Chapter 9

Chaos in the Mass-Deformed ABJM Model

Seçkin Kürkçüoğlu

*Middle East Technical University, Department of Physics,
Dumlupınar Boulevard, Ankara 06800, Turkey*

kseckin@metu.edu.tr

Chaotic dynamics of the mass deformed ABJM model is explored. To do so, we consider spatially uniform fields and obtain a family of reduced effective Lagrangians by tracing over ansatz configurations involving fuzzy two-spheres with collective time dependence. We examine how the largest Lyapunov exponent, λ_L, changes as a function of E/N^2, where N is the matrix size. In particular, we inspect the temperature dependence of λ_L and present upper bounds on the temperature above which λ_L values comply with the MSS bound, $\lambda_L \leq 2\pi T$, and below which it will eventually be not obeyed.

1. Introduction

Research work exploring the structure of chaotic dynamics emerging from the matrix gauge theories has become quite abundant recently.[1-7] These studies are propelled by a result due Maldacena–Shenker–Stanford (MSS),[3] which briefly states that the largest Lyapunov exponent (which is a measure of chaos in both classical and quantum mechanical systems) for quantum chaos is controlled by a temperature-dependent bound given by $\lambda_L \leq 2\pi T$. Systems which are holographically dual to the black holes are conjectured to be maximally chaotic, i.e., saturate this bound. This is already proved for the Sachdev–Ye–Kitaev (SYK)[4] model and expected to be so for other matrix models which have a holographic dual such as the BFSS[8] model. In Ref. 2, classical chaotic dynamics of the Banks–Fischler–Shenker–Susskind (BFSS) model,[8] which provides a good approximation of the quantum theory in the high-temperature limit, is studied and it was found that the largest Lyapunov exponent scales as $\lambda_L = 0.2924(3)(\lambda_{'t\,Hooft}T)^{1/4}$ and therefore the MSS bound is violated only at temperatures below the critical temperature $T_c \approx 0.015$, while it remains parametrically smaller than $2\pi T$ for $T > T_c$. In Ref. 7, chaotic dynamics of massive deformations of the bosonic sector of the BFSS model was explored by exploiting matrix configurations involving fuzzy spheres, and upper bounds on the critical temperature, T_c, are estimated.

In a more recent paper,[9] we have examined chaos emerging from the massive deformation of the Aharony–Bergman–Jafferis–Maldacena (ABJM) model. Here, I will be reporting on a part of this work. In brief, our focus can be indicated as follows. ABJM model is a supersymmetric Chern–Simons (CS) gauge theory coupled to matter fields and describes the dynamics of N coincident $M2$-branes.[10,11] It possesses a massive deformation preserving all the supersymmetry but breaking the R-symmetry. The vacuum configurations in this model are fuzzy two-spheres described in terms of Gomis, Rodriguez-Gomez, Van Raamsdonk and Verlinde (GRVV) matrices.[12] For the purpose of studying the chaotic dynamics, we reduce this model from $2+1$ to $0+1$ dimensions by considering that the fields are spatially uniform and work in the 't Hooft limit. Tracing over an ansatz fuzzy 2-sphere matrix configuration with collective time dependence, we obtain a family of effective Hamiltonians. Solving the equations of motion numerically, we examine how the largest Lyapunov exponent, λ_L, changes as a function of E/N^2. Making use of the virial and equipartition theorems, we investigate the implications for the aforementioned MSS conjecture. The main outcomes of our work are the upper bounds we obtain on the temperatures above which largest Lyapunov exponents comply with the MSS bound and below which it will eventually not be obeyed.

2. Mass-Deformed ABJM Model with Spatially Uniform Fields

Bosonic part of the ABJM model[10] is an $SU(N)_k \times SU(N)_{-k}$ Chern–Simons gauge theory in $2+1$ dimensions. The subscripts $\pm k \in \mathbb{Z}$ label the level of the Chern–Simons terms associated with these gauge fields. The model involves the connections A_μ and $\hat{A}_\mu$ ($\mu : 0, 1, 2$) transforming in the standard manner under the $SU(N)_k$ and $SU(N)_{-k}$ gauge transformations, as well as the complex scalar fields (Q^α, R^α) which transform bifundamentally under the gauge symmetry, i.e., in the form $Q^\alpha \to U_L Q^\alpha U_R$, $R^\alpha \to U_L R^\alpha U_R$, where $(U_L, U_R) \in SU(N)_k \times SU(N)_{-k}$.

In order to dimensionally reduce S_{ABJM} to $0+1$ dimensions, we declare that both the gauge fields and the complex scalar are spatially uniform, i.e., independent of the spatial coordinates and depend on time only. Consequently, all partial derivatives with respect to the spatial coordinates vanish. We introduce the notation $A_\mu \equiv (A_0, X_i)$, $\hat{A}_\mu \equiv (\hat{A}_0, \hat{X}_i)$ with $(i = 1, 2)$. The action takes the form

$$S_{ABJM-R} = N \int dt \ -\frac{k}{4\pi}\text{Tr}(\epsilon^{ij}X_i\mathcal{D}_0X_j) + \frac{k}{4\pi}\text{Tr}(\epsilon^{ij}\hat{X}_i\hat{\mathcal{D}}_0\hat{X}_j)$$

$$+ \text{Tr}(|D_0Q^\alpha|^2) - \text{Tr}(|D_iQ^\alpha|^2) + \text{Tr}(|D_0R^\alpha|^2)$$

$$- \text{Tr}(|D_iR^\alpha|^2) - V, \tag{2.1}$$

where $D_iQ^\alpha = iX_iQ^\alpha - iQ^\alpha\hat{X}_i$, $D_iR^\alpha = iX_iR^\alpha - iR^\alpha\hat{X}_i$ and the covariant derivatives are $D_0Q^\alpha = \partial_0Q^\alpha + iA_0Q^\alpha - iQ^\alpha\hat{A}_0$, $D_0R^\alpha = \partial_0R^\alpha + iA_0R^\alpha - iR^\alpha\hat{A}_0$, $\mathcal{D}_0X_i = \partial_0X_i - i[A_0, X_i]$, $\hat{\mathcal{D}}_0\hat{X}_i = \partial_0\hat{X}_i - i[\hat{A}_0, \hat{X}_i]$. The potential term is given as $V = \text{Tr}(|M^\alpha|^2 + |N^\alpha|^2)$, where $M^\alpha = \mu Q^\alpha + \frac{2\pi}{k}(2Q^{[\alpha}Q_\beta^\dagger Q^{\beta]} + R^\beta R_\beta^\dagger Q^\alpha - Q^\alpha R_\beta^\dagger R^\beta + 2Q^\beta R_\beta^\dagger R^\alpha - 2R^\alpha R_\beta^\dagger Q^\beta)$ and $N^\alpha = -\mu R^\alpha + \frac{2\pi}{k}(2R^{[\alpha}R_\beta^\dagger R^{\beta]} + Q^\beta Q_\beta^\dagger R^\alpha - R^\alpha Q_\beta^\dagger Q^\beta + 2R^\beta Q_\beta^\dagger Q^\alpha - 2Q^\alpha Q_\beta^\dagger R^\beta)$. Here, μ stands for the masses of the fields (Q^α, R^α) and we have used the notation $Q^{[\alpha}Q_\beta^\dagger Q^{\beta]} = Q^\alpha Q_\beta^\dagger Q^\beta - Q^\beta Q_\beta^\dagger Q^\alpha$ and likewise for R^αs.

In (2.1), it is understood that all fields depend only on time. Let us also point out that this form of action is already written in the 't Hooft limit. The latter is defined as follows. Reducing from $2+1$ to $0+1$ dimensions, we have integrated over the two-dimensional space whose volume may be denoted, say, by V_2. Therefore, we may introduce $\lambda'_{t\,Hooft} := \frac{N}{V_2}$ and require that it remains finite in the limit $V_2 \to \infty$ and $N \to \infty$. In the action, S_{ABJM-R}, we have scaled $\lambda'_{t\,Hooft}$ to unity. If needed, $\lambda'_{t\,Hooft}$ may be restored back in S_{ABJM-R} by making the scalings $X_i \to \lambda^{-1/2}X_i$, $\hat{X}_i \to \lambda^{-1/2}\hat{X}_i$, $A_0 \to \lambda^{-1/2}A_0$, $\hat{A}_0 \to \lambda^{-1/2}\hat{A}_0$, $Q_\alpha \to \lambda^{-1/4}Q_\alpha$, $R_\alpha \to \lambda^{-1/4}R_\alpha$, $\mu \to \lambda^{-1/2}\mu$ and $t \to \lambda^{1/2}t$. It should be clear from (2.1) that S_{ABJM-R} is manifestly gauge invariant under the $SU(N)_k \times SU(N)_{-k}$ gauge symmetry and the reduced CS coupling $\frac{kV_2}{4\pi}$ is no longer level quantized. A more comprehensive discussion on the latter fact may be found in Ref. 9.

The ground states are given by configurations minimizing the potential V. Since the latter is positive definite, its minimum is zero and is given by the configuration $M^\alpha = 0 = N^\alpha$. There are two immediate solutions to this, which are given as $(R^\alpha, Q^\alpha) = (cG^\alpha, 0)$ and

$(R^\alpha, Q^\alpha) = (0, cG^\alpha)$, where G^α are GRVV matrices[11, 12] defining a fuzzy 2-sphere[13] at the matrix level N and $c = \sqrt{\frac{k\mu}{4\pi}}$. Explicitly, G^α are given as[12] $(G^1)_{mn} = \sqrt{m-1}\,\delta_{mn}$, $(G^2)_{mn} = \sqrt{N-m}\,\delta_{m+1\,n}$, $(G_1^\dagger)_{mn} = \sqrt{m-1}\,\delta_{mn}$, $(G_2^\dagger)_{mn} = \sqrt{N-n}\,\delta_{n+1\,m}$ with $m, n = 1, \ldots, N$, and they fulfill the relation $G^\alpha = G^\alpha G_\beta^\dagger G^\beta - G^\beta G_\beta^\dagger G^\alpha$.

In what follows, we make the gauge choice $A_0 = 0 = \hat{A}_0$ and therefore have the Gauss law constraints from variations of S_{ABJM-R} with respect to A_0 and $\hat{A}_0$. These are $\frac{k}{2\pi}[X_1, X_2] + \dot{Q}^\alpha Q_\alpha^\dagger - Q^\alpha \dot{Q}_\alpha^\dagger + \dot{R}^\alpha R_\alpha^\dagger - R^\alpha \dot{R}_\alpha^\dagger = 0$ and $-\frac{k}{2\pi}[\hat{X}_1, \hat{X}_2] - Q_\alpha^\dagger \dot{Q}^\alpha + \dot{Q}_\alpha^\dagger Q^\alpha - R_\alpha^\dagger \dot{R}^\alpha + \dot{R}_\alpha^\dagger R^\alpha = 0$.

Hamiltonian takes the form

$$H = \text{Tr}\left(\frac{1}{N}|P_Q^\alpha|^2 + \frac{1}{N}\text{Tr}|P_R^\alpha|^2 + N|D_iQ^\alpha|^2 + N|D_iR^\alpha|^2\right) + NV, \quad (2.2)$$

where $P_Q^\alpha = \frac{\partial L}{\partial \dot{Q}^\alpha} = N\dot{Q}^{\alpha\dagger}$ and $P_R^\alpha = \frac{\partial L}{\partial \dot{R}^\alpha} = N\dot{R}^{\alpha\dagger}$ are the conjugate momenta associated with Q_α and R_α, respectively.

Finally, we note that the scaling transformation $(Q_\alpha, R_\alpha) \rightarrow (\rho^{-1/2}Q_\alpha, \rho^{-1/2}R_\alpha)$, $(X_i, \hat{X}_i) \rightarrow (\rho^{-1}X_i, \rho^{-1}\hat{X}_i)$, $t \rightarrow \rho t$, where ρ is an arbitrary positive constant. Under this scaling, $(P_Q^\alpha, P_R^\alpha) \rightarrow (\rho^{-3/2}P_Q^\alpha, \rho^{-3/2}P_R^\alpha)$ and $V|_{\mu=0} \rightarrow \rho^{-3}V|_{\mu=0}$ indicating that the energy scales as $E \rightarrow \rho^{-3}E$. Since the Lyapunov exponent has the dimensions of inverse time, we see that it scales as $\lambda_L \propto E^{1/3}$ in the massless limit. In what follows, we will see that this scaling of the Lyapunov exponents with energy is essentially preserved after taking the mass deformations into account.

3. Ansatz Configuration and the Effective Action

We consider the matrices $X_i = \alpha(t)\text{diag}((A_i)_1, (A_i)_2, \ldots, (A_i)_N)$, $\hat{X}_i = \beta(t)\text{diag}((B_i)_1, (B_i)_2, \ldots, (B_i)_N)$, $Q_\alpha = \phi_\alpha(t)G_\alpha$, $R_\alpha = 0$, where $(A_i)_m$, $(B_i)_m$ are constants, $i = 1, 2$, $m = 1, 2, \ldots, N$ and $\alpha = 1, 2$. No sum over the repeated index α is implied. Here, $\phi_\alpha(t)$, $\alpha(t)$, $\beta(t)$ are real functions of time and the Gauss law constraint equations are easily seen to be satisfied by this choice of the matrices. Evaluating the equations of motion for $\alpha(t)$ and $\beta(t)$, we find that the emerging coupled equations have only one possible real solution and that is the trivial solution given simply as $\alpha(t) = \beta(t) = 0$.[9] Thus, setting X_i and $\hat{X}_i$ to zero from now on and performing the traces over the $N \times N$ GRVV matrices, we find the reduced

Hamiltonians

$$H_N(\phi_1, \phi_2\, p_{\phi_1}, p_{\phi_2}) = \frac{p_{\phi_1}^2}{2N^2(N-1)} + \frac{p_{\phi_2}^2}{2N^2(N-1)} + V_N(\phi_1, \phi_2), \quad (3.1)$$

where $V_N(\phi_1, \phi_2) = N^2(N-1)(\frac{1}{2}\mu^2(\phi_1^2 + \phi_2^2) + \frac{8\pi\mu}{k}\phi_1^2\phi_2^2 + \frac{8\pi^2}{k^2}\phi_1^4\phi_2^2 + \frac{8\pi^2}{k^2}\phi_2^4\phi_1^2)$. In the limit $\mu \to 0$, we have $H_N \to \rho^{-3}H_N$ under the scaling $(\phi_1, \phi_2) \to (\rho^{-1/2}\phi_1, \rho^{-1/2}\phi_2)$ and $t \to \rho t$, in view of the scaling properties of the matrix model given in the previous section. In what follows, we will explore the dynamics emerging from the Hamilton's equations at $\mu = 1$ at several different matrix levels N and the CS coupling k.

3.1. *Chaotic dynamics and the Lyapunov exponents*

We examine the chaotic dynamics of the models governed by the Hamiltonians H_N. For this purpose, we numerically evaluate the Lyapunov exponents of these models by solving the associated Hamilton's equations of motion. As it is well known (see Ref. 9 and references therein), the largest Lyapunov exponent is essentially a measure of the sensitivity of a system to given initial conditions. More precisely, it gives the exponential growth in perturbations and in this regard, it provides a quantitative means of detecting and examining chaos. The phase spaces for H_N are all 4-dimensional and their chaotic dynamics is governed by the largest (and only) positive Lyapunov exponent. Obtaining the solutions of the Hamilton's equations with 40 randomly selected initial conditions at a given energy value E and matrix level N, we calculate the mean of the time series for each and every Lyapunov exponent from all runs. In the simulation, we take a time step of 0.25 and run the code from time 0 to 1500. Results for the largest Lyapunov exponent, λ_L, as a function of E/N^2 for $N = 15, 25$ at several different values of the energy and $k = \pm 1, \pm 2$ are discussed in the following.

For $k = 1, 2$, the data and best fitting curves of the form $\lambda_L = \alpha_N(\frac{E}{N^2})^{1/3}$ are given in Figure 1 at $N = 15, 25$ and $k = 1, 2$. They clearly demonstrate the $\lambda_L \propto E^{1/3}$ dependence of the Lyapunov exponent anticipated by the scaling argument. Values of the coefficients α_N for the fitting curves in Figure 1 are $\alpha_{15} = 0.6092, 0.4788$ and $\alpha_{25} = 0.499, 0.3958$ for $k = 1, 2$, respectively.

For $k = -1, -2$, we seek best fitting curves of the form $\lambda_L \propto (\frac{E}{N^2} - \gamma_N)^{1/3}$ to the Lyapunov data, where γ_N is determined by N, k and μ and proportional to the minimum value of $\tilde{V}_N$ involving the quartic and

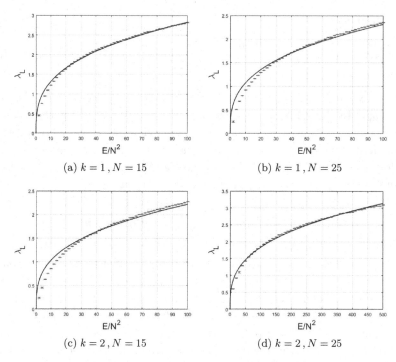

(a) $k = 1$, $N = 15$ (b) $k = 1$, $N = 25$

(c) $k = 2$, $N = 15$ (d) $k = 2$, $N = 25$

Fig. 1. Largest Lyapunov exponent and the best fitting curves in the form $\lambda_L = \alpha_N(\frac{E}{N^2})^{1/3}$.

sextic terms of V_N. This specific form of the γ_N is indeed motivated by the virial and equipartition theorems as will be shortly discussed in the following section. Data and the fitting curves are provided in Figure 2. The coefficients α_N for the fitting curves are $\alpha_{15} = 0.5539, 0.4522$ and $\alpha_{25} = 0.4971, 0.3660$ for $k = -1, -2$, respectively.

3.2. *Temperature dependence of the Lyapunov exponent*

In the massless limit, $\lambda_{'tHooft}$ and the temperature are the only dimensionful parameters and using dimensional analysis, we may easily see that $\lambda_L \propto (\lambda_{'tHooft}T)^{1/3}$. This is because, in $0 + 1$ dimensions $\lambda_{'tHooft} = \frac{N}{V_2}$, V_2 being the volume of the 2-dimensional space, we have integrated over in going from $2+1$ to $0+1$ dimensions and it has the dimension $[Length]^{-2}$ and hence λ_L has the dimension $[Length]^{-1}$. In view of the equipartition theorem, this is consistent with $\lambda_L \propto E^{1/3}$ which was independently already noted based on the scaling symmetry. Shortly, we will also discuss the

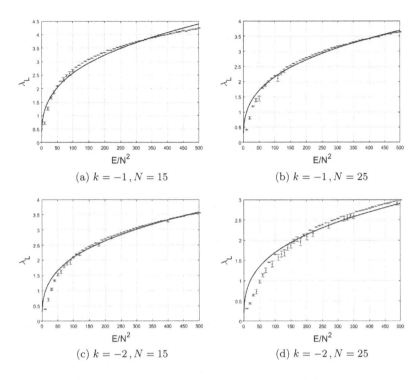

(a) $k = -1, N = 15$

(b) $k = -1, N = 25$

(c) $k = -2, N = 15$

(d) $k = -2, N = 25$

Fig. 2. Largest Lyapunov exponent and the best fitting curves in the form $\lambda_L = \alpha_N(\frac{E}{N^2} - \gamma_N)^{1/3}$ at $k = -1, -2$.

effects of mass parameter on the relation between the energy and temperature upon the application of the virial and the equipartition theorems. We may contrast these features with those of the BFSS model, in which $\lambda_L \propto (\lambda_{tHooft}T)^{1/4}$. Since the potential is purely quartic in this latter case, the system has a scaling symmetry implying that $\lambda_L \propto E^{1/4}$. Mass deformations in this model were examined via an ansatz involving fuzzy 2- and 4-spheres in Ref. 7.

In the model described by the ansatz configuration which is introduced at the beginning of this section (after setting X_i and $\hat{X}_i$ to zero), total number of real degrees of freedom is $4N^2$ before taking the global gauge symmetry and the Gauss law constraints into account. The latter imposes only N^2 real relations as the two equations comprising it reduce to the same equation upon integrating by parts and taking the Hermitian conjugate of one or the other, while the R-symmetry gives 8 real relations among the unconstrained degrees of freedom, therefore we have $n_{d.o.f.} \approx 3N^2$ at large N.

Applying the virial theorem to (3.1) gives $2\langle K\rangle = 2\langle V_N\rangle + \tilde{V}_N(\phi_1,\phi_2)$. Here, $\tilde{V}_N(\phi_1,\phi_2) = 2N^2(N-1)\left(\frac{8\pi\mu}{k}\phi_1^2\phi_2^2 + \frac{16\pi^2}{k^2}\phi_1^2\phi_2^4 + \frac{16\pi^2}{k^2}\phi_1^4\phi_2^2\right)$. It is positive definite if k and μ have the same sign, but for k and μ with opposite signs, we have $Min(\tilde{V}_N(\phi_1,\phi_2)) = N^2(N-1)\frac{k\mu^3}{27\pi}$.

Applying the equipartition theorem to the kinetic energy yields $\langle K\rangle \approx \frac{3}{2}N^2T$ at large N. In what follows, we consider $\mu = 1$.

For $k \geq 1$, we have the inequality $\langle K\rangle \geq \langle V_N\rangle$. Together with the results of the equipartition theorem, this implies that $\langle E\rangle = \langle K\rangle + \langle V_N\rangle \leq n_{d.o.f}T \approx 3N^2T$. We can express this inequality in the form $\frac{E}{N^2} \leq 3T$, where we have also dropped the brackets on energy for ease in notation. We may compare and relate this result to the MSS bound $\lambda_L \leq 2\pi T$ on the largest Lyapunov exponent for quantum chaos.[3] ABJM model has a gravity dual[11] via the AdS/CFT correspondence and we may expect the MSS conjecture to hold for quantum chaotic dynamics of the ABJM model too. Since our analysis here is confined to the classical regime, we should expect that the MSS bound be eventually not obeyed at sufficiently low temperatures. Indeed, from our results, we observe that there is a critical temperature, which we may denote as T_c and given by solving the equation $\alpha_N(3T)^{1/3} = 2\pi T_c$. This yields $T_c = \sqrt{3}\left(\frac{\alpha_N}{2\pi}\right)^{3/2}$ and it is an upper bound for the critical temperature at or below which MSS bound will eventually not be obeyed by our model. For $N = 15$, $k = 1, 2$, our estimates are $T_c = 0.0523, 0.0364$, respectively, while, for $N = 25$, we find $T_c = 0.0388, 0.0274$ for $k = 1, 2$. More comprehensive results, presented in our paper,[9] indicate that T_c values tend to decrease with increasing matrix size. This is in agreement with the fact that 't Hooft limit is better approximated with increasing values of N.

For $k \leq -1$, we proceed as follows. We may write $2\langle K\rangle = 2\langle V_N\rangle + \tilde{V}_N(\phi_1,\phi_2) + |Min(\tilde{V}_N)| - |Min(\tilde{V}_N)|$ and this implies that $\langle K\rangle \geq \langle V_N\rangle - \frac{1}{2}|Min(\tilde{V}_N)|$. Therefore, we have $E - \frac{1}{2}|Min(\tilde{V}_N)| = \langle K\rangle + \langle V_N\rangle - \frac{1}{2}|Min(\tilde{V}_N)| \leq n_{d.o.f}T$. Since $\langle K\rangle \approx \frac{3}{2}N^2T$ at large N, this leads to the inequality $\frac{E}{N^2} - \gamma_N \leq 3T$, where $\gamma_N := \frac{|Min(\tilde{V}_N)|}{2N^2}$. Hence, we now clearly observe the line of reasoning that led us in the previous section to consider the best fitting curves of the form $\lambda_N = \alpha_N\left(\frac{E}{N^2} - \gamma_N\right)^{1/3}$. These curves are already given in Figure 2 and they clearly represent the variation of the largest Lyapunov exponent with respect to E/N^2 quite well. Finally, by the same line of reasoning used earlier, we find that the critical temperatures are found to be $T_c = \sqrt{3}\left(\frac{\alpha_N}{2\pi}\right)^{3/2}$. Our numerical estimates are $T_c = 0.0609, 0.0497$ for $N = 15$, $k = -1, -2$, respectively, and $T_c = 0.0547, 0.0403$ for $N = 25$ and $k = -1, -2$.

4. Conclusions and Outlook

Here, we have reported on a part of our recent work.[9] Our main objective was to examine the structure of chaotic dynamics emerging from the massive deformation of the ABJM model. We have approached this problem by considering an ansatz configuration involving fuzzy 2-spheres with collective time dependence and obtained a family of effective actions parametrized by the matrix level N. We computed the largest Lyapunov exponent and presented its variation with respect to E/N^2 and demonstrated that $\lambda_L \propto (E/N^2)^{1/3}$ or $\lambda_L \propto (E/N^2 - \gamma_N)^{1/3}$ depending on the sign of the CS coupling. This allowed us to inspect the extent the largest Lyapunov exponent complies with the MSS bound upon the use of the virial and equipartition theorems.

5. Reminiscences and a Tribute

I met Bal in the late summer of 1999 at Syracuse. My recollection is that he was the first person in the States to ask me about relief efforts and the situation in the aftermath of the magnitude 7.6 earthquake that took place near Propontis, the sea of Marmara, in Turkey, at around the time. I joined his research group in the following year. At first, it was kind of discombobulating to work with Bal, trying to learn so many different concepts and techniques so fast, to adjust to the flow and exposure to the ideas in many diverse directions that were being discussed and debated with other students and collaborators and the frequent visitors, who were, at the time, Denjoe (O'Connor), Peter (Prešnajder) and Giorgio (Immirzi). These were the days, when the new directions in noncommutative geometry and fuzzy field theories were the main themes of interest in Bal's group.

After a while, with the persistent but always positive attitude of Bal in approaching his students, I found myself able to take up the challenge, to learn a particular problem pretty rapidly and contribute to the development and solution of research problems leading to novel publishable results. My scientific training with Bal, especially via discussions in room 316 in the physics department, which were already a classic before I arrived at Syracuse, allowed me to gain the confidence and assertiveness which helped me to build my own career path. My collaboration with Bal continued after my graduation and resulted in several papers and a book on fuzzy physics together with Sachin Vaidya. We still continue to collaborate today and we are entertaining ideas on how to make use of a coproduct that we have found long time ago for fuzzy spheres to model entanglement.

Meetings in room 316, at dinner tables or pot luck parties at Bal's house often expanded into conversations on literature, art house movies and music but most assuredly into heated debates in world politics. Working with Bal also led me to travel to diverse locations all over the world, to visit the pyramids of Sun and Moon in the Mesoamerican city of Teotihuacan, to the halls of the Prado museum in Madrid and perhaps most exhilaratingly, as I still vividly recall after so many years, to being exposed to mock charges of elephants in the monsoon forests on our way to the city of Mysore.

Wishing Bal a very happy 85th birthday with the final verses of the poem "Plea" by Nazım Hikmet:

> Do away with the enslaving of man by man,
> This plea is ours,
> To live! Like a tree alone and free
> Like a forest in brotherhood
> This yearning is ours.

Acknowledgements

I would like to thank the editors for the kind invitation to contribute to this volume. This work is supported by TÜBİTAK under the project number 118F100.

References

1. Y. Sekino and L. Susskind, *JHEP* **0810** (2008) 065; C. Asplund, D. Berenstein and D. Trancanelli, *Phys. Rev. Lett.* **107** (2011) 171602; S. H. Shenker and D. Stanford, *JHEP* **1403** (2014) 067; S. Aoki, M. Hanada and N. Iizuka, *JHEP* **1507** (2015) 029; Y. Asano, D. Kawai and K. Yoshida, *JHEP* **1506** (2015) 191.
2. G. Gur-Ari, M. Hanada and S. H. Shenker, *JHEP* **1602** (2016) 091.
3. J. Maldacena, S. H. Shenker and D. Stanford, *JHEP* **1608** (2016) 106.
4. J. Maldacena and D. Stanford, *Phys. Rev. D* **94**(10) (2016) 106002.
5. P. Buividovich, M. Hanada and A. Schäfer, *EPJ Web Conf.* **175** (2018) 08006; P. V. Buividovich, M. Hanada and A. Schäfer, *Phys. Rev. D* **99**(4) (2019) 046011.
6. Ü. H. Coşkun, S. Kürkçüoğlu, G. C. Toga and G. Ünal, *JHEP* **1812** (2018) 015.

7. K. Başkan, S. Kürkçüoğlu, O. Oktay and C. Taşcı, *JHEP* **10** (2020) 003.
8. T. Banks, W. Fischler, S. H. Shenker and L. Susskind, *Phys. Rev. D* **55** (1997) 5112.
9. K. Başkan, S. Kürkçüoğlu and C. Taşcı, To appear in Phys. Rev. D [arXiv:2203.08240 [hep-th]].
10. O. Aharony, O. Bergman, D. Jafferis and J. Maldacena, *J. High Energy Phys.* **2008** (2008, October) 091–091.
11. H. Nastase, *Introduction to the ADS/CFT Correspondence*, Cambridge University Press (2015).
12. J. Gomis, D. Rodríguez-Gómez, M. Raamsdonk and H. Verlinde, *J. High Energy Phys.* **2008**, (2008, September) 113–113.
13. A. P. Balachandran, S. Kürkçüoğlu and S. Vaidya, *Lectures on Fuzzy and Fuzzy SUSY Physics*, Singapore, World Scientific (2007), [hep-th/0511114].

Chapter 10

Quantum Groupoids and Gauge Transformations

Giovanni Landi

Università di Trieste, and INFN, Trieste, Italy

landi@units.it

We describe how to associate a Hopf algebroid to a noncommutative principal bundle. This is a quantization of the gauge groupoid of a classical bundle. The gauge group of the noncommutative bundle is isomorphic to the group of bisections of the algebroid. As examples, we give a monopole bundle over a quantum sphere and noncommutative bundles over a point associated with Taft algebras.

Per A.P. Balachandran con ammirazione e rispetto

1. Introduction

The study of groupoids and Lie algebroids on the one hand and gauge theories on the other hand is important in different areas of mathematics and physics. In particular, these subjects meet in the notion of the gauge groupoid of a principal bundle and associated Atiyah sequence. In view of the considerable amount of recent work on noncommutative principal bundles, it is desirable to come up with a noncommutative version of groupoids and study their relations to noncommutative gauge theories.

There is a natural bialgebroid (named Ehresmann–Schauenburg) associated with a noncommutative principal bundle that can be seen as a quantization of the classical gauge groupoid.[8] Classically, bisections of the gauge groupoid are closely related to gauge transformations. In parallel with this result, the gauge group of a noncommutative principal bundle is group isomorphic to the group of bisections of the corresponding bialgebroid. To illustrate the theory, we work out some details of the gauge group of the principal bundle and of the bialgebroid with corresponding group of bisections, for the noncommutative monopole bundle over the quantum Podleś 2-sphere. In fact, in this case, there is also an invertible antipode which satisfies the conditions for a Hopf algebroid.

An interesting class of examples comes from noncommutative principal bundles over a point (Galois objects), notably those associated with Taft algebras T_N. In contrast to the classical case, a noncommutative bundle over a point need not be trivial. The equivalence classes of T_N-Galois objects are in bijective correspondence with the abelian group $\mathbb{C}$. The corresponding bialgebroid has an antipode and is then a Hopf algebra.

This chapter is based on Ref. 8 where all the following statements are proven.

2. Noncommutative Principal Bundles

We start with a brief recall of Hopf–Galois extensions as noncommutative principal bundles. Then, we consider gauge transformations as equivariant automorphisms of the total space algebra which are vertical so that they leave invariant the base space algebra.

A Hopf–Galois extension is an algebra A with a coaction $\delta^A \to A \otimes H$ of a Hopf algebra H that we write as $\delta^A(a) = a_{(0)} \otimes a_{(1)}$ (with an implicit sum) and a canonical-defined map required to be invertible.[17]

Definition 2.1. Let H be a Hopf algebra and A be an algebra with a coaction δ^A. Take the subalgebra $B := A^{coH} = \{b \in A \mid \delta^A(b) = b \otimes 1_H\} \subseteq A$ of coinvariant elements with balanced tensor product $A \otimes_B A$. The extension $B \subseteq A$ is called an H-Hopf–Galois extension if the *map* (named canonical)

$$\chi : A \otimes_B A \longrightarrow A \otimes H, \quad a' \otimes_B a \mapsto a'a_{(0)} \otimes a_{(1)}$$

is an isomorphism.

Commutative Hopf–Galois extensions are obtained from principal bundles. The crudest situation is that of affine varieties. These are embedded in some $\mathbb{R}^n$ as the zero locus of polynomials. Let $\pi : P \to P/G$ be a principal bundle with G an affine algebraic group and P, P/G affine varieties. Let $H = \mathcal{O}(G)$, $A = \mathcal{O}(P)$ and $B = \mathcal{O}(P/G)$ be the corresponding algebras of coordinate functions. The group structure of G induces a Hopf algebra structure on H. Since $B \subset A$ is the subalgebra of functions constant on the fibers, one has $B = A^{coH}$. Also, $\mathcal{O}(P \times_{P/G} P) \simeq A \otimes_B A$ and bijectivity of the map $P \times G \to P \times_{P/G} P$, $(p, g) \mapsto (p, pg)$, that characterises principality of the bundle, is bijectivity of the canonical map $\chi : A \otimes_B A \to A \otimes H$, thus giving that $B = A^{coH} \subseteq A$ is a Hopf–Galois extension (see, e.g., Ref. [12, S8.5]).

Since the canonical map χ is left A-linear, its inverse is determined by the restriction $\tau := \chi^{-1}|_{1_A \otimes H}$, named *translation map*,

$$\tau = \chi^{-1}|_{1_A \otimes H} : H \to A \otimes_B A, \quad h \mapsto \tau(h) = h^{<1>} \otimes_B h^{<2>}. \qquad (2.1)$$

Thus, by definition, $h^{<1>} h^{<2>}{}_{(0)} \otimes h^{<2>}{}_{(1)} = 1_A \otimes h$.

In Ref. 4, gauge transformations for a noncommutative principal bundle were defined to be invertible H-equivariant maps, with no additional requirement such as being algebra maps. However, the resulting gauge group might be very big, even in the classical case. In contrast, in Ref. 1, gauge transformations were taken to be algebra homomorphisms. This property implies in particular that any such an algebra map is invertible.[8]

Prop 1. *Let $B = A^{coH} \subseteq A$ be a Hopf–Galois extension. Then, the collection $\mathrm{Aut}_H(A)$ of right H-equivariant unital algebra maps of A into itself which restrict to the identity on the subalgebra B is a group for map composition. For $F \in \mathrm{Aut}_H(A)$, its inverse $F^{-1} \in \mathrm{Aut}_H(A)$ is given by*

$$F^{-1}(a) = a_{(0)} F(a_{(1)}{}^{<1>}) a_{(1)}{}^{<2>}, \qquad (2.2)$$

for all $a \in A$, and notation (2.1) for the translation map.

Elements $F \in \mathrm{Aut}_H(A)$ preserve the (co)action of the structure quantum group since they obey $\delta^A \circ F = (F \otimes \mathrm{id})\delta^A$. And they also preserve the base space algebra B. This group reduces to the usual gauge group of a principal bundle in the classical case. However, for a general noncommutative principal bundle, this group may be quite small, due to the lack of enough algebra morphisms of a noncommutative algebra. Something in between, in the context of braided geometry, has been recently proposed in Ref. 2.

2.1. *A noncommutative monopole bundle*

Let us consider an explicit example of the above construction, that is, the U(1) principal bundle over the standard Podleś sphere S_q^2 of Ref. 14. With $q \in \mathbb{R}$ a deformation parameter, the coordinate algebra $\mathcal{O}(\mathrm{SL}_q(2))$ of the quantum group $SL_q(2)$ is generated by elements a, c and d, b with relations

$$ac = qca \quad \text{and} \quad bd = qdb, \quad ab = qba \quad \text{and} \quad cd = qdc,$$
$$cb = bc, \quad ad - da = (q - q^{-1})bc \quad \text{and} \quad da - q^{-1}bc = 1. \tag{2.3}$$

Then, the Hopf algebra $H = \mathcal{O}(U(1))$ coacts on the algebra $\mathcal{O}(SL_q(2))$ via

$$\delta(a) = a \otimes z, \quad \delta(d) = d \otimes z^{-1}, \quad \delta(c) = c \otimes z, \quad \delta(b) = b \otimes z^{-1}. \tag{2.4}$$

The subalgebra of coinvariant elements in $\mathcal{O}(SL_q(2))$ for this coaction is the coordinate algebra $B = \mathcal{O}(S_q^2)$ of the standard Podleś sphere $\mathcal{O}(S_q^2)$. As a set of generators for $\mathcal{O}(S_q^2)$, one may take

$$B_- := -q^{-1}ab, \quad B_+ := cd \quad \text{and} \quad B_0 := -q^{-1}cb, \tag{2.5}$$

for which one finds the relations

$$B_- B_0 = q^2 B_0 B_- \quad \text{and} \quad B_+ B_0 = q^{-2} B_0 B_+,$$
$$B_- B_+ = q^2 B_0 \left(1 - q^2 B_0\right) \quad \text{and} \quad B_+ B_- = B_0 \left(1 - B_0\right). \tag{2.6}$$

The inclusion $\mathcal{O}(S_q^2) \subset \mathcal{O}(SL_q(2))$ is a noncommutative principal bundle.[5] The total space algebra decomposes as $\mathcal{O}(SL_q(2)) = \bigoplus_{n \in \mathbb{Z}} A_n$, where

$$A_n := \left\{ x \in \mathcal{O}(SL_q(2)) \mid \delta(x) = x \otimes z^{-n} \right\}.$$

In the classical case, $q = 1$ of commutative algebras, one gets the $U(1)$ bundle over the sphere S^2 and the above is the decomposition of $SL(2)$ in monopole harmonics, with the integer n, the monopole charge of the line bundle A_n over S^2. As B-modules, A_{-1} is generated by a, c, while A_1 is generated by d, b. Any gauge transformation is determined by the images

$$F(a) = Xa + Yc, \quad F(c) = Za + Wc,$$
$$F(d) = \tilde{X}d + \tilde{Y}b, \quad F(b) = \tilde{Z}d + \tilde{W}b, \tag{2.7}$$

with coefficient elements in the algebra B and extended as an algebra map. In the classical case, asking for the coinvariant generators in (2.5) to be left unchanged by F in (2.7) reduces the coefficient to a single one X, any non-vanishing function from $S^2 \to \mathbb{C}$,

$$F(a) = Xa, \quad F(c) = Xc, \quad F(d) = X^{-1}d, \quad F(b) = X^{-1}b. \tag{2.8}$$

One gets $\mathrm{Aut}_H(\mathcal{O}(SL(2))) = \mathrm{Map}(S^2 \to \mathbb{C}^*)$. In contrast, when $q \neq 1$, requiring that F be an algebra map and so to respect the commutation relations in (2.3), one gets $X \in \mathbb{C}^*$ since the center of $\mathcal{O}(S_q^2)$ is just the algebra $\mathbb{C}$. Thus, $\mathrm{Aut}_H(\mathcal{O}(\mathrm{SL}_q(2))) = \mathbb{C}^*$ are the non-vanishing complex numbers.

3. Hopf Algebroids

To any Hopf–Galois extension $B = A^{co\,H} \subseteq A$, one associates a bialgebroid[16] (see Ref. [6, S34.13; 34.14]). As mentioned, this can be viewed as a quantization of the gauge groupoid that is associated with a principal fiber bundle (see Ref. 10). It can be given in a few equivalent ways. Let $B = A^{co\,H} \subseteq A$ be a Hopf–Galois extension with right coaction $\delta^A : A \to A \otimes H$. Consider the diagonal coaction, given for all $a, a' \in A$, by

$$\delta^{A \otimes A} : A \otimes A \to A \otimes A \otimes H, \quad a \otimes a' \mapsto a_{(0)} \otimes a'_{(0)} \otimes a_{(1)} a'_{(1)}.$$

Let τ be the translation map of the Hopf–Galois extension. Then, we have the following.

Lemma 3.1. *The B-bimodule of coinvariant elements,*

$$(A \otimes A)^{coH} = \{a \otimes \tilde{a} \in A \otimes A;\ a_{(0)} \otimes \tilde{a}_{(0)} \otimes a_{(1)} \tilde{a}_{(1)} = a \otimes \tilde{a} \otimes 1_H\}$$

for the diagonal coaction, is the same as the B-bimodule

$$\mathcal{C} := \{a \otimes \tilde{a} \in A \otimes A;\ a_{(0)} \otimes \tau(a_{(1)})\tilde{a} = a \otimes \tilde{a} \otimes_B 1_A\}. \tag{3.1}$$

Then, the B-bimodule $\mathcal{C}$ in (3.1) has natural coproduct and counit:

$$\Delta(a \otimes \tilde{a}) = a_{(0)} \otimes \tau(a_{(1)}) \otimes \tilde{a} = a_{(0)} \otimes a_{(1)}{}^{<1>} \otimes_B a_{(1)}{}^{<2>} \otimes \tilde{a}, \tag{3.2}$$

$$\varepsilon(a \otimes \tilde{a}) = a\tilde{a}. \tag{3.3}$$

The B-bimodule $\mathcal{C} = \mathcal{C}(A, H)$ is in fact a bialgebroid[16] called the *Ehresmann–Schauenburg bialgebroid* (see Ref. [6, 34.14, 34.14]). One sees that $\mathcal{C}(A, H) = (A \otimes A)^{coH}$ is a subalgebra of $A \otimes A^{op}$ with a product

$$(x \otimes \tilde{x}) \bullet_{\mathcal{C}(A,H)} (y \otimes \tilde{y}) = xy \otimes \tilde{y}\tilde{x},$$

for all $x \otimes \tilde{x}, y \otimes \tilde{y} \in \mathcal{C}(A, H)$. The target and the source maps t and s, algebra maps with commuting range from B to $\mathcal{C}(A, H)$ are

$$t(b) = 1_A \otimes b \quad \text{and} \quad s(b) = b \otimes 1_A.$$

With some caveats on the existence of an antipode, the bialgebroid of a Hopf–Galois extension is viewed as a quantization of the classical gauge groupoid of a (classical) principal bundle.[10] Dually to the notion of a bisection on the classical gauge groupoid, there is the notion of a bisection on the bialgebroid. These bisections correspond to gauge transformations.

Definition 3.2. Let $\mathcal{C}(A, H)$ be the Ehresmann–Schauenburg bialgebroid of a Hopf–Galois extension $B = A^{coH} \subseteq A$. A bisection of $\mathcal{C}(A, H)$ is a B-bilinear unital left character. That is, a map $\sigma : \mathcal{C}(A, H) \to B$ such that

(1) $\sigma(1_A \otimes 1_A) = 1_B$,   unitality,

(2) $\sigma\big(s(b)t(\tilde{b})(x \otimes \tilde{x})\big) = b\sigma(x \otimes \tilde{x})\tilde{b}$, B-bilinearity,

(3) $\sigma\big((x \otimes \tilde{x}) s(\sigma(y \otimes \tilde{y}))\big) = \sigma\big((x \otimes \tilde{x})(y \otimes \tilde{y})\big)$,   associativity,

for all $b, \tilde{b} \in B$ and $x \otimes \tilde{x}, y \otimes \tilde{y} \in \mathcal{C}(A, H)$.

The collection $\mathcal{B}(\mathcal{C}(A, H))$ of bisections of the bialgebroid $\mathcal{C}(A, H)$ is made a group by the convolution product of any two σ_1 bisections σ_2:

$$\sigma_1 * \sigma_2(x \otimes \tilde{x}) := \sigma_1((x \otimes \tilde{x})_{(1)}) \, \sigma_2((x \otimes \tilde{x})_{(2)})$$
$$= \sigma_1(x_{(0)} \otimes x_{(1)}{}^{<1>}) \, \sigma_2(x_{(1)}{}^{<2>} \otimes \tilde{x}) \qquad (3.4)$$

for any element $x \otimes \tilde{x} \in \mathcal{C}(A, H)$, recalling the coproduct (3.2), with notation $\Delta(x \otimes \tilde{x}) = (x \otimes \tilde{x})_{(1)} \otimes_B (x \otimes \tilde{x})_{(2)}$. One shows that this product is associative. The counit of the bialgebroid is a bisection by definition and one checks that $\varepsilon * \sigma = \sigma = \sigma * \varepsilon$ for any bisection σ and thus ε is the unit element. The inverse of the bisection σ is

$$\sigma^{-1}(x \otimes \tilde{x}) = x \, \sigma(\tilde{x}_{(0)} \otimes \tilde{x}_{(1)}{}^{<1>}) \, \tilde{x}_{(1)}{}^{<2>}, \qquad (3.5)$$

for $x \otimes \tilde{x} \in \mathcal{C}(A, H)$. Indeed, all group properties of $\mathcal{B}(\mathcal{C}(A, H))$ with the product (3.4) follow from the following proposition that parallels Prop. 1.

Prop 2. *Let $B = A^{coH} \subseteq A$ be a Hopf–Galois extension, and let $\mathcal{C}(A, H)$ be the corresponding bialgebroid. There is a group isomorphism $\alpha : \mathrm{Aut}_H(A) \to \mathcal{B}(\mathcal{C}(A, H))$ between gauge transformations and bisections.*

First, given a bisection $\sigma \in \mathcal{B}(\mathcal{C}(A, H))$, we define a map $F_\sigma : A \to A$ by

$$F_\sigma(a) := \sigma(a_{(0)} \otimes a_{(1)}{}^{<1>}) \, a_{(1)}{}^{<2>}, \quad \text{for } a \in A. \qquad (3.6)$$

This is well defined and it is shown to be in $\text{Aut}_H(A)$. Conversely, let $F \in \text{Aut}_H(A)$ be a gauge transformation and define $\sigma_F \in \mathcal{B}(\mathcal{C}(A, H))$ by

$$\sigma_F(a \otimes \tilde{a}) := F(a)\tilde{a}, \quad \text{for } a \otimes \tilde{a} \in \mathcal{C}(A, H). \tag{3.7}$$

This is well defined and it is shown to be in $\mathcal{B}(\mathcal{C}(A, H))$.

3.1. *The monopole bundle over the quantum sphere*

We illustrate the bialgebroid for the $U(1)$ principal bundle over the quantum sphere in Section 2.1. There is also a suitable invertible antipode that upgrades the bialgebroid to a Hopf algebroid. From Section 2.1, let us denote $H = \mathcal{O}(U(1))$ coacting as in (2.4) on the generators in (2.3) of $A = \mathcal{O}(\text{SL}_q(2))$ with algebra of coinvariants $B = \mathcal{O}(S_q^2)$ with generators in (2.6). From its definition, the bialgebroid $\mathcal{C}(A, H) = (A \otimes A)^{coH}$ is generated by elements

$$\alpha = a \otimes d, \quad \gamma = -q^{-1}c \otimes b, \quad \tilde{\alpha} = -q^{-1}a \otimes b, \quad \tilde{\gamma} = c \otimes d \tag{3.8}$$

and their 'conjugated'

$$\delta = d \otimes a, \quad \beta = -q^{-1}b \otimes c, \quad \tilde{\beta} = d \otimes c, \quad \tilde{\delta} = -q^{-1}b \otimes a. \tag{3.9}$$

A direct computation leads to

$$\beta\gamma + \tilde{\delta}\tilde{\gamma} = B_0 \otimes 1 = s(B_0), \quad \beta\gamma + \tilde{\beta}\tilde{\alpha} = 1 \otimes B_0 = t(B_0),$$

$$\alpha\tilde{\delta} + q^2\tilde{\alpha}\beta = B_- \otimes 1 = s(B_-), \quad \tilde{\alpha}\delta + q^2\gamma\tilde{\delta} = 1 \otimes B_- = t(B_-),$$

$$\tilde{\gamma}\delta + q^2\gamma\tilde{\beta} = B_+ \otimes 1 = s(B_+), \quad \alpha\tilde{\beta} + q^2\tilde{\gamma}\beta = 1 \otimes B_+ = t(B_+),$$

with source $s : B \to \mathcal{C}(A, H)$ and target $t : B \to \mathcal{C}(A, H)$ maps, respectively.

The eight generators in (3.8) and (3.9) are not independent. Indeed, define

$$A = a \otimes d - q\,b \otimes c, \quad B = b \otimes a - q^{-1}a \otimes b,$$
$$C = c \otimes d - q\,d \otimes c, \quad D = d \otimes a - q^{-1}c \otimes b.$$

Then, a direct computation shows that

$$DA - q^{-1}CB = 1 \otimes 1 = AD - qBC.$$

Also, the sphere relations in (2.6) translate into

$$\bigl(B_+ B_- - B_0\,(1 - B_0)\bigr) \otimes 1 = 0 = 1 \otimes \bigl(B_+ B_- - B_0\,(1 - B_0)\bigr).$$

Note that the above relations which survive the classical limit $q = 1$ are constraints among the generators and not commutation relations.

The bialgebroid $\mathcal{C}(A, H)$ has a structure of a Hopf algebroid with coproduct (3.2) which results into

$$\Delta(\alpha) = \alpha \otimes_B \alpha + \tilde{\alpha} \otimes_B \tilde{\gamma}, \quad \Delta(\tilde{\alpha}) = \alpha \otimes_B \tilde{\alpha} + \tilde{\alpha} \otimes_B \gamma,$$

$$\Delta(\gamma) = \tilde{\gamma} \otimes_B \tilde{\alpha} + \gamma \otimes_B \gamma, \quad \Delta(\tilde{\gamma}) = \tilde{\gamma} \otimes_B \alpha + \gamma \otimes_B \tilde{\gamma},$$

$$\Delta(\delta) = \delta \otimes_B \delta + q^2\tilde{\beta} \otimes_B \tilde{\delta}, \quad \Delta(\tilde{\beta}) = \delta \otimes_B \tilde{\beta} + q^2\tilde{\beta} \otimes_B \beta,$$

$$\Delta(\beta) = \tilde{\delta} \otimes_B \tilde{\beta} + q^2\beta \otimes_B \beta, \quad \Delta(\tilde{\delta}) = \tilde{\delta} \otimes_B \delta + q^2\beta \otimes_B \tilde{\delta}$$

and counit (3.3) which results into

$$\varepsilon(\alpha) = 1 - q^2 B_0, \quad \varepsilon(\gamma) = B_0, \quad \varepsilon(\tilde{\alpha}) = B_-, \quad \varepsilon(\tilde{\gamma}) = B_+$$

$$\varepsilon(\delta) = 1 - B_0, \quad \varepsilon(\beta) = B_0, \quad \varepsilon(\tilde{\beta}) = q^{-1}B_+, \quad \varepsilon(\tilde{\delta}) = q^{-1}B_-.$$

As mentioned, now there is antipode $S = S^{-1}$ given by

$$S(\alpha) = \delta, \quad S(\gamma) = \beta \quad S(\tilde{\alpha}) = \tilde{\delta}, \quad S(\tilde{\gamma}) = \tilde{\beta}. \tag{3.10}$$

Again, Prop. 2 determines the group of bisections $\mathcal{B}(\mathcal{C}(A, H))$ out of gauge transformations worked out in Section 2.1. Both Eqs. (3.6) and (3.7) are well defined. In particular, for Eq. (3.6), the map

$$A \ni h \mapsto h_{(0)} \otimes h_{(1)}{}^{<1>} \otimes_B h_{(1)}{}^{<2>} \in \mathcal{C}(A, H) \otimes_B A$$

is well defined. Given the generic gauge transformation in (2.8), formula (3.7) determines a generic bisection. This is then given on generators by

$$\sigma(\alpha) = Xad, \quad \sigma(\gamma) = -q^{-1}Xcb, \quad \sigma(\tilde{\alpha}) = -q^{-1}Xab, \quad \sigma(\tilde{\gamma}) = Xcd$$

$$\sigma(\delta) = X^{-1}da, \quad \sigma(\beta) = -q^{-1}X^{-1}bc, \quad \sigma(\tilde{\beta}) = X^{-1}dc, \quad \sigma(\tilde{\delta}) = -q^{-1}X^{-1}ba.$$

As before, in Section 2.1, X is any map from the sphere $S^2 \to \mathbb{C}^*$ when $q = 1$, while is any non-vanishing element, $X \in \mathbb{C}^*$ when $q \neq 1$.

In Ref. 9, there are two classes of examples of Hopf algebroids associated with noncommutative principal bundles. The first comes from deforming the principal bundle while leaving unchanged the structure Hopf algebra. The second is related to deforming a quantum homogeneous space; this needs a careful deformation of the structure Hopf algebra in order to preserve the compatibilities between the Hopf algebra operations.

4. Noncommutative Bundles over a Point

We now consider noncommutative principal bundles over a point, also called *Galois objects*. In contrast to the classical result, the set $\text{Gal}_H(\mathbb{C})$ of isomorphic classes of H-Galois objects need not be trivial (see Ref. 7). We shall illustrate this non-triviality with examples coming from Taft algebras.

Definition 4.1. Let H be a Hopf algebra. A *Galois object* of H is an H-Hopf–Galois extension A of the ground field $\mathbb{C}$.

Thus, for a Galois object, the coinvariant subalgebra is the ground field $\mathbb{C} = A^{co\,H}$. With coaction $\delta^A : A \to A \otimes H$, $\delta^A(a) = a_{(0)} \otimes a_{(1)}$ and translation map $\tau : H \to A \otimes A$, $\tau(h) = h^{<1>} \otimes h^{<2>}$, for the bialgebroid of a Galois object, being $B = \mathbb{C}$, one has (see also Ref. [15, Def. 3.1])

$$\mathcal{C}(A, H) = \{a \otimes \tilde{a} \in A \otimes A : a_{(0)} \otimes \tilde{a}_{(0)} \otimes a_{(1)}\tilde{a}_{(1)} = a \otimes \tilde{a} \otimes 1_H\}$$

$$= \{a \otimes \tilde{a} \in A \otimes A : a_{(0)} \otimes a_{(1)}{}^{<1>} \otimes a_{(1)}{}^{<2>}\tilde{a} = a \otimes \tilde{a} \otimes 1_A\}.$$

The coproduct (3.2) becomes $\Delta_\mathcal{C}(a \otimes \tilde{a}) = a_{(0)} \otimes a_{(1)}{}^{<1>} \otimes a_{(1)}{}^{<2>} \otimes \tilde{a}$ and the counit (3.3) is $\varepsilon_\mathcal{C}(a \otimes \tilde{a}) = a\tilde{a} \in \mathbb{C}$, for any $a \otimes \tilde{a} \in \mathcal{C}(A, H)$. But now there is also an antipode [15, Thm. 3.5] given, for any $a \otimes \tilde{a} \in \mathcal{C}(A, H)$, by

$$S_\mathcal{C}(a \otimes \tilde{a}) := \tilde{a}_{(0)} \otimes \tilde{a}_{(1)}{}^{<1>} a\tilde{a}_{(1)}{}^{<2>}. \tag{4.1}$$

Thus, the bialgebroid $\mathcal{C}(A, H)$ of a Galois object is a Hopf algebra. Being algebra maps, now bisections are characters of $\mathcal{C}(A, H)$ with product in (3.4) and inverse in (3.5), written as $\sigma^{-1} = \sigma \circ S_\mathcal{C}$ with antipode (4.1). From Prop. 2, we have for the group of gauge transformations the isomorphism

$$\text{Aut}_H(A) \simeq \mathcal{B}(\mathcal{C}(A, H)) = \text{Char}(\mathcal{C}(A, H)). \tag{4.2}$$

Taft algebras

Let $N \geq 2$ be an integer and let q be a primitive N-th root of unity. The *Taft algebra* T_N, introduced in Ref. 18, is a Hopf algebra which is neither commutative nor cocommutative. First, T_N is the N^2-dimensional unital algebra generated by generators x, g subject to the relations

$$x^N = 0, \quad g^N = 1, \quad xg - q\,gx = 0.$$

It is a Hopf algebra with coproduct

$$\Delta(x) := 1 \otimes x + x \otimes g, \quad \Delta(g) := g \otimes g,$$

counit $\varepsilon(x) := 0, \varepsilon(g) := 1$ and antipode $S(x) := -xg^{-1}, S(g) := g^{-1}$. The four-dimensional algebra T_2 is also known as the *Sweedler algebra*.

G. Landi

For any $s \in \mathbb{C}$, let A_s be the unital algebra generated by elements X, G with relations

$$X^N = s, \quad G^N = 1, \quad XG - qGX = 0.$$

The algebra A_s has a right T_N coaction, $\delta^A : A_s \to A_s \otimes T_N$, defined by

$$\delta^A(X) := 1 \otimes x + X \otimes g, \quad \delta^A(G) := G \otimes g. \tag{4.3}$$

The algebra of corresponding coinvariants is just the ground field $\mathbb{C}$ and so A_s is a T_N-Galois object. It is known (see Ref. 11, Prop. 2.17 and Prop. 2.22) that any T_N-Galois object is isomorphic to A_s for some $s \in \mathbb{C}$ and that any two such Galois objects A_s and A_t are isomorphic if and only if $s = t$. Thus, the equivalence classes of T_N-Galois objects are in bijective correspondence with the abelian group $\mathbb{C}$. It is easy to see that the translation map of the coaction (4.3) is given on generators by

$$\tau(g) = G^{-1} \otimes G, \quad \tau(x) = 1 \otimes X - XG^{-1} \otimes G. \tag{4.4}$$

For the bialgebroid $\mathcal{C}(A_s, T_N)$, we then have the following.

Prop 3. *For any $s \in \mathbb{C}$, there is a Hopf algebra isomorphism*

$$\Phi : \mathcal{C}(A_s, T_N) \simeq T_N.$$

Indeed, it is easy to see that the elements

$$\Xi = X \otimes G^{-1} - 1 \otimes XG^{-1}, \quad \Gamma = G \otimes G^{-1}$$

are coinvariants for the right diagonal coaction of T_N on $A_s \otimes A_s$ and that they generate $\mathcal{C}(A_s, T_N) = (A_s \otimes A_s)^{co\,T_N}$ as an algebra. These elements satisfy the relations

$$\Xi^N = 0, \quad \Gamma^N = 1, \quad \Xi \bullet_c \Gamma = q \Xi \bullet_c \Gamma.$$

Thus, Ξ and Γ generate a copy of the algebra T_N and the isomorphism Φ maps Ξ to x and Γ to g. One shows that the map Φ is a coalgebra map.

The group of characters of the algebra T_N (and then of $\mathcal{C}(A_s, T_N)$) is the cyclic group $\mathbb{Z}_N$: indeed any character ϕ is such that $\phi(x) = 0$, while $\phi(g)^N = \phi(g^N) = \phi(1) = 1$. Then, for the group of gauge transformations of the Galois object A_s — the same as the group of bisections of the bialgebroid $\mathcal{C}(A_s, T_N)$ — due to Prop. 3, we have

$$\mathrm{Aut}_{T_N}(A_s) \simeq \mathcal{B}(\mathcal{C}(A_s, T_N)) = \mathrm{Char}(T_N) = \mathbb{Z}_N.$$

References

1. P. Aschieri, G. Landi and C. Pagani, The gauge group of a noncommutative principal bundle and twist deformations, *J. Noncommut. Geom.* **14** (2020) 1501–1559.
2. P. Aschieri, G. Landi and C. Pagani, Braided Hopf algebras and gauge transformations. arXiv:2203.13811.
3. G. Böhm, *Hopf Algebroids.* Handbook of Algebra, Vol. 6, pp. 173–235. North-Holland (2009).
4. T. Brzeziński, Translation map in quantum principal bundles, *J. Geom. Phys.* **20** (1996) 349–370.
5. T. Brzeziński and S. Majid, Quantum group gauge theory on quantum spaces, *Commun. Math. Phys.* **157** (1993) 591–638; Erratum **167** (1995) 235.
6. T. Brzeziński and R. Wisbauer, *Corings and Comodules.* London Mathematical Society Lecture Notes 309, CUP (2003).
7. C. Kassel, Principal fiber bundles in noncommutative geometry, in *Quantization, Geometry and Noncommutative Structures in Mathematics and Physics, Mathematical Physics Studies*, pp. 75–133. Springer (2017).
8. X. Han and G. Landi, Gauge groups and bialgebroids, *Lett. Math. Phys.* (2021) 111:140 (43 pages).
9. X. Han, G. Landi and Y. Liu, Hopf algebroids from noncommutative bundles, arXiv:2201.01612.
10. K. Mackenzie, *General Theory of Lie Groupoids and Lie Algebroids.* London Mathematical Society Lecture Notes 213, CUP (2005).
11. A. Masuoka, Cleft extensions for a Hopf algebra generated by a nearly primitive element, *Commun. Algebra* **22** (1994) 4537–4559.
12. S. Montgomery, *Hopf Algebras and their Actions on Rings*, AMS (1993).
13. C. Nastasescu and F. Van Oystaeyen, *Graded Ring Theory*, Elsevier (1982).
14. P. Podleś, Quantum spheres, *Lett. Math. Phys.* **14** (1987) 193–202.
15. P. Schauenburg, Hopf bi-Galois extensions, *Commun. Algebra* **24** (1996) 3797–3825.
16. P. Schauenburg, Bialgebras over noncommutative rings and a structure theorem for Hopf bimodules, *Appl. Categ. Struct.* **6** (1998) 193–222.
17. H.-J. Schneider, Principal homogeneous spaces for arbitrary Hopf algebras, *Isr. J. Math.* **72** (1990) 167–195.
18. E. J. Taft, The order of the antipode of finite-dimensional Hopf algebra, *Proc. Natl. Acad. Sci. USA* **68** (1971) 2631–2633.

Chapter 11

316

Fedele Lizzi

*Dipartimento di Fisica "Ettore Pancini", Università di Napoli Federico II,
INFN, Sezione di Napoli, Napoli, Italy
Departament de Física Quàntica i Astrofísica and Institut de Ciencies del
Cosmos (ICCUB), Universitat de Barcelona, Barcelona, Spain
fedele.lizzi@unina.it*

fedele.lizzi@na.infn.it

A room, a teacher and many friends.

Let us begin with a celebrated anecdote involving an Indian scientist from Tamil Nadu: Ramanujan. The great number theorist G. Hardy recalls[1]: *I remember once going to see him when he was ill at Putney. I had ridden in taxi cab number 1729 and remarked that the number seemed to me rather a dull one, and that I hoped it was not an unfavourable omen. "No," he replied, "it is a very interesting number; it is the smallest number expressible as the sum of two cubes in two different ways."* It is not clear what would Ramanujan had answered if Hardy had taken cab 316. The Penguin Dictionary of Curious and Interesting Numbers[2] jumps from 306 to 319 (which cannot be represented as the sum of fewer than 19 4th powers). Even a perusal of *Wikipedia* does not illuminate; it is a centered triangular number and a centered heptagonal number, a psalm of John (3:16), the area code of Wichita (Texas) and the year of the consulship of Sabinus and Rufinus.

But for Balachandran's student and collaborators, 316 is very important; it may be said it is a room, but this would be very reductive!

Fig. 1.   Room 316 in 2022.

The physical locations is on the third floor of the Syracuse University
Physics Building. It is a common room (in the British sense); it has a
kitchenette in a corner, a sofa, a table with ensuing chairs and a blackboard. If bibliometrical indices were valid for places rather than people, it
would have an h-parameter of at least 40. I will mainly talk of 316 during
the eighties, but things did not change in the following decades. The room
is still there, and there is a photographic proof of it. The blackboard is
white, and now there is a projector, and probably the chairs have been
reupholstered.

For generations of students and collaborators of Balachandran, 316 is
not a number, it a locus of the soul. She has nurtured us, in my case,
gave me some of the best years of my life. There is no doubt in our mind
that in that place we became the physicists we are now. More, it shaped
us in a more profound way as persons. We met daily in 316, having coffee,
discussing physics, solving the problems of the world, gossiping, doing more
physics, politics and, you will forgive the vocabulary, but I cannot think
of a better word, bullshitting. The room was certainly alive from about
twelve noon until about six or seven in the evening; often it was active in

the middle of the night. I personally have no idea of its uses before twelve; I was rarely up before the crack of noon.

Like everything in the Universe, 316 is a quantum object, and it has a ground state. Between about 1 p.m. and 5 p.m., the ground state of the room was with Balachandran sitting on the sofa, a student at the board, maybe another more senior person also on the sofa and other students (or post-doc) at the table, possibly taking notes. The empty room was an excited state, something must have happened (like traveling or a downtown march against Reagan's politics in Latin America).

Bal would arrive shortly after lunch. His modus operandi was to wake up early, work in a room under the roof in his house,[a] then walk along Euclid Avenue to the physics building, attend a few bureaucratic chores, fill a mug with instant coffee and walk to 316. Along the route, he would encounter someone or drop to one of the offices and peremptorily say: "Get the others!" And we, the "al", of the many papers "A.P. Balachandran *et al.*", would soon gather in 316. At first, we would all be sitting, starting to tell what we had done, a calculation performed, or planned, a paper we had read or a question. Soon, we would be explaining to each other all the things we did not understand. Then, Bal would address whoever was talking an undescribable gesture with his hand, and the ground state would be reached. The board would become the center of attention, and the discussion would flow. At times, these could be very animated; newcomers had the impression that we were at each other's throat. Other moments, there would be long silences, Bal with his chin in his hand, all staring at a blank blackboard, where a wrong idea had been canceled. Bal would rarely stand up and go himself to the board, and when he did, it meant that we (but not him) were stuck or that he had come up with a new important idea.

This little note is too short to write down all of the people who, at one time or another, passed for 316. It suffices to say that in the years I was Bal's student in Syracuse (1980–1985), he has written papers with 21 different collaborators. There were remote branches as well, wherever Bal goes, there is a 316, in various continents. The very room in which I am presently writing, in my flat, has a whiteboard and an armchair.

---

[a]Come to think of it, I have been to Bal's house countless times while I was a student there (1980–1985), enjoyed Indra cuisine, played with Vinod as a child (he is now a surgeon at Sloane-Kettle hospital in New York city) and even slept there when I first arrived in Syracuse, but I have no recollection of this other room. The "room" for me is only 316.

When I refurbished it, I made sure that Bal could have his space during his visits to Napoli. Not far from Napoli, the countryside branch, the *Policeta*, near Beppe Marmo's birthplace, has an outdoor room 316, just to mention those nearby. Napoli has several connections with Bal, many of his students were Neapolitans, he visited often and we celebrated his 65th birthday with a conference in nearby Vietri sul Mare, on the Amalfi Coast.

316 survived unscathed Bal's retirement, if you wish to call Bal a retired scientist; but I have personally not noticed any difference. The COVID pandemic forced all of us to find new ways to do our work, and 316 found a new life as a virtual seminar room. Bal is running a series of Zoom seminars called 316. So that now 316 can be found also in the cloud!

References

1. G. Hardy and Srinivasa Ramanujan, *Proc. London Math. Soc.* **s2-19**(1) (1921) xl–lviii.
2. D. Wells, *The Penguin Dictionary of Curious and Interesting Numbers*, Penguin, London (1986).

https://doi.org/10.1142/9789811270437_0012

Chapter 12

Eductions of Edge Mode Effects

V. P. Nair

Physics Department, City College of the CUNY
New York, NY 10031, USA

vpnair@ccny.cuny.edu

Edge modes in gauge theories, whose *raison d'être* is in the nature of the test functions used for imposing the Gauss law, have implications in many physical contexts. I discuss two such cases: (1) how edge modes are related to the interface term in the BFK formula and how they generate the so-called contact term for entanglement entropy in gauge theories and (2) how they describe the dynamics of particles in generalizing the Einstein–Infeld–Hoffmann approach to particle dynamics in theories of gravity.

1. Introduction

For many students of high energy physics in the early 1980s in Syracuse, afternoons repeated a familiar pattern. Balachandran, or Bal as everyone referred to him, would appear at the door of the tiny office I shared with a fellow graduate student, in his signature attire, the greenish-grey sweater with elbow patches, coffee cup in hand, saying "Let us discuss". We would walk down the hallway, collecting more students, sometimes visitors, to gather in Room 316, made famous by Fedele Lizzi, in his contribution to this volume. All matters great and small would come up for discussion. And some days we would then repair to his home where Indra's tolerance and the excellent food would let us talk late into the evening. Bal was always fascinated by the mathematical side of things in physics. He, and his students and collaborators, had just been working on applying the coadjoint orbit actions to various problems when I arrived in Syracuse. This was also a time of an effervescence of many ideas in particle physics: the role of topology in physics was beginning to be appreciated, monopoles, solitons and instantons made frequent appearances in papers, anomalies were still intriguing, Bal talked of why he felt the effective action for anomalies constructed by Wess and Zumino should be important (prescient comments as it turned out) and the long dormant idea of skyrmions was just about to be revived. Over time, all of us, his students and collaborators, have branched out in different ways, but a fascination and engagement with the mathematical side of things, from the Syracuse days with Bal, have remained the leitmotif of our work.

In appreciation, in this chapter dedicated to Bal's 85th birthday, I would like to highlight a couple of effects related to one of his many favorite topics, namely edge modes in gauge theories and gravity.

2. Entanglement in a Gauge Theory and the "Contact Term"

As is well known, in a gauge theory, physical states are annihilated by the Gauss law operator $G(\theta)$ defined with test functions $\theta(x)$ which vanish at the boundary of the space under consideration. In this case, $e^{iG(\theta)}$ generates gauge transformations $e^{i\theta}$ which become the identity on the boundary. The same operator $e^{iG(\tilde{\theta})}$, with $\tilde{\theta}$ which do not vanish on the boundary, generates physical states; these are the edge modes. In the 1990s, Bal and collaborators explored the properties of such states for the Maxwell–Chern–Simons theory, as well as in more general contexts.[1,2] More recently, as interest in entanglement has increased, it has become clear that edge modes do make a contribution, the so-called contact term, to the entanglement entropy.[3] This is the facet of edge modes that I would first like to highlight.

We will consider the Maxwell theory in 2+1 dimensions as this suffices to illustrate the main point. In a Hamiltonian framework, we can take $A_0 = 0$. The spatial components of the gauge potential A_i and the electric field E_i, ($i = 1, 2$) can be parametrized as

$$A_i = \partial_i \theta + \epsilon_{ij}\partial_j \phi, \quad E_i = \dot{A}_i = \partial_i \sigma + \epsilon_{ij}\partial_j \Pi. \qquad (2.1)$$

We consider the spatial manifold $\mathcal{M}$ to be a square which is separated, by a straight line interface, into two rectangular regions which we will label as L and R. The idea is to split the fields into three terms each: a field in L which vanishes on the interface, a field in R which also vanishes on the interface and a field on the interface itself. Focusing on the region L first, the fields can be split as

$$\chi_L(x) = \tilde{\chi}_L(x) + \int_{\partial L} \chi_0(y)\, n \cdot \partial_y G_L(y, x), \qquad (2.2)$$

where $\chi = \theta,\ \phi,\ \sigma,\ \Pi$. $\tilde{\chi}_L$ vanishes on the boundary of $\mathcal{M}$ as well as on the interface between L and R. χ_0 denotes the value of χ on the interface and $G_L(y, x)$ is Green's function for the Laplacian, again obeying Dirichlet (vanishing) conditions on the boundary of $\mathcal{M}$ and on the interface ∂L. In (2.2), the value of the field on the interface, namely χ_0, is continued into the interior of L by Laplace's equation and Green's theorem so that

$$\nabla_x^2 \int_{\partial L} \chi_0(y)\, n \cdot G_L(y, x) = 0. \qquad (2.3)$$

This does not introduce any functional degrees of freedom in addition to χ_0, and the freedom of choosing arbitrary values of the field in L is contained in $\tilde{\chi}_L$. Equation (2.2) gives a general parametrization of the fields in L.

Consider now the phase space path integral where, for the constraint $\nabla \cdot E \approx 0$, we choose the conjugate constraint $\nabla \cdot A \approx 0$ (Coulomb gauge). The path integral is given by

$$Z = \int d\mu\, \delta(\nabla \cdot E)\, \delta(\nabla \cdot A)\, \det(-\nabla^2)\, e^{iS},$$
$$S = \int d^3x \left[E_i \dot{A}_i - \mathcal{H} \right], \qquad (2.4)$$

where $\mathcal{H}$ is the Hamiltonian density. The canonical two-form $\int \delta E_i\, \delta A_i$ serves to define the phase space measure $d\mu$ in (2.4). The canonical one-form $\mathcal{A} = \int E_i\, \delta A_i$ is given in terms of the parametrization (2.1) as

$$\mathcal{A} = \int_L \left[(-\nabla^2 \tilde{\sigma}_L)\, \delta\tilde{\theta}_L + \tilde{\Pi}_L\, \delta B_L \right] + \int_{\partial L} \mathcal{E}_0\, \delta\theta_0(x) + \int_{\partial L} Q_0\, \delta\phi_0(x), \quad (2.5)$$

where $B = -\nabla^2 \phi$ is the magnetic field and

$$\mathcal{E}_0(x) = \int_y \sigma_0(y) M_L(y, x) + \partial_\tau \Pi_0(x),$$
$$Q_0(x) = \int_y \Pi_0(y) M_L(y, x) - \partial_\tau \sigma_0(x), \qquad (2.6)$$

where ∂_τ signifies the tangential derivative and $M_L(x, y) = n \cdot \partial_x n \cdot \partial_y G_L(x, y)$ (with x, y on ∂L) is the Dirichlet-to-Neumann operator for the geometry we have. It is easily verified that this is essentially

$(\sqrt{-\nabla^2})_{x,y}$. As a result, we can see that $\mathcal{E}_0$ and Q_0 are related by $\mathcal{C} = \partial_y \int \mathcal{E}_I(x) M^{-1}(x,y) + Q_I(y) = 0$. This is another constraint in the problem; it is due to the freedom in defining B. Note that $B = -\nabla^2(\phi + f) = -\nabla^2\phi$, if $\nabla^2 f = 0$. Non-trivial choices of such functions exist, for example, as

$$f(x) = \int_{\partial L} f_0(y)\, n \cdot \partial_y G_L(y, x). \tag{2.7}$$

The constraint $\mathcal{C}$ encodes this additional "gauge freedom", which is really an ambiguity of the parametrization (2.1). We can use the freedom of f to set ϕ_0 to zero, i.e., choose $\phi_0 \approx 0$ as the constraint conjugate to $\mathcal{C}$. Eliminating $\mathcal{C}$ and ϕ_0 by standard symplectic reduction, we get

$$\mathcal{A} = \int_L \left[(-\nabla^2\tilde{\sigma}_L)\, \delta\tilde{\theta}_L + \tilde{\Pi}_L\, \delta B_L \right] + \int_{\partial L} \mathcal{E}_0\, \delta\theta_0. \tag{2.8}$$

The corresponding phase space volume element is

$$d\mu_I = [d\tilde{\sigma}d\tilde{\theta}]_L\, [d\mathcal{E}_0\, d\theta_0]\, [d\tilde{\Pi}dB]_L \det(-\nabla^2)_L \tag{2.9}$$

where the determinant is understood to be evaluated with Dirichlet conditions on its eigenfunctions. The action, the Hamiltonian and the constraints in (2.4) are given by

$$\mathcal{S}_L = \int_L \left[(-\nabla^2\tilde{\sigma}_L)\, \dot{\tilde{\theta}}_L + \tilde{\Pi}_L\, \dot{B}_L \right] + \int_{\partial L} \mathcal{E}_0\, \dot{\theta}_0(x) - \int dt\, \mathcal{H}, \tag{2.10}$$

$$\mathcal{H} = \int \frac{1}{2} \left[(\nabla\tilde{\sigma}_L)^2 + (\nabla\tilde{\Pi}_L)^2 + B_L^2 \right] + \frac{1}{2} \int_{\partial L} \mathcal{E}_0(x) M_L^{-1}(x, y)\, \mathcal{E}_0(y),$$

$$\delta(\nabla \cdot E) = (\det(-\nabla^2)_L)^{-1}\, \delta(\tilde{\sigma}_L) \quad \delta(\nabla \cdot A) = (\det(-\nabla^2)_L)^{-1}\, \delta(\tilde{\theta}_L). \tag{2.11}$$

Using these results and carrying out integrations, including over B_L, we find

$$Z_I = \int [d\mathcal{E}_0 d\theta_0]\, [d\tilde{\Pi}_L]\, \exp(i\tilde{S}),$$

$$\tilde{S}_L = \int \frac{1}{2} \left[\dot{\tilde{\Pi}}_L^2 - (\nabla\tilde{\Pi}_L)^2 \right] + \int_{\partial L} \left[\mathcal{E}_0\, \dot{\theta}_0 - \frac{1}{2}\mathcal{E}_0\, M_L^{-1}\, \mathcal{E}_0 \right]. \tag{2.12}$$

We see that the dynamics has been reduced to that of a scalar field $\tilde{\Pi}_L$ (which obeys Dirichlet conditions on all of ∂L) and "edge modes" with dynamics given by the second term in $\tilde{S}_L$.

We now carry out exactly the same analysis for the full space $\mathcal{M}$. This will give an expression similar to (2.12), but for the whole space, and with no edge modes since we chose Dirichlet conditions on $\partial\mathcal{M}$. However, we can also choose to split any field into $\tilde{\chi}_L$, $\tilde{\chi}_R$ and χ_0 with

$$\chi = \begin{cases} \tilde{\chi}_L(x) + \int_{\partial L} \chi_0(y)\, n \cdot \partial G_L(y,x) & \text{in L and on } \partial L \\ \tilde{\chi}_R(x) + \int_{\partial R} \chi_0(y)\, n \cdot \partial G_R(y,x) & \text{in R and on } \partial R = \partial L. \end{cases} \tag{2.13}$$

With this parametrization of the fields, the action takes the form

$$\mathcal{S}_{\text{split}} = \int_L \left[(-\nabla^2 \tilde{\sigma}_L)\, \dot{\tilde{\theta}}_L - \frac{1}{2}(\nabla \tilde{\sigma}_L)^2 \right] + \int_R \left[(-\nabla^2 \tilde{\sigma}_R)\, \dot{\tilde{\theta}}_R - \frac{1}{2}(\nabla \tilde{\sigma}_R)^2 \right]$$
$$+ \int \Pi \dot{B} - \frac{1}{2}\left[(\nabla \Pi)^2 + B^2 \right] + \int_{\partial L} \mathcal{E}_0\, \dot{\theta}_0 - \frac{1}{2}\mathcal{E}_0(M_L + M_R)^{-1}\mathcal{E}_0. \tag{2.14}$$

From this, we can read off the canonical one-form

$$\mathcal{A} = \int_{L \cup R} \Pi\, \delta B + \int_L (-\nabla^2 \tilde{\sigma}_L)\, \delta\tilde{\theta}_L + \int_R (-\nabla^2 \tilde{\sigma}_R)\delta\tilde{\theta}_R + \int_{\partial L} \mathcal{E}_0\, \delta\theta_0. \tag{2.15}$$

The phase volume associated with this is

$$d\mu_{\text{split}} = [d\tilde{\sigma} d\tilde{\theta}]_L\, [d\tilde{\sigma} d\tilde{\theta}]_R\, \det(-\nabla^2)_L\, \det(-\nabla^2)_R\, [d\mathcal{E}_0 d\theta_0]\, [d\Pi dB]. \tag{2.16}$$

Note that $d\mu$ involves $\det(-\nabla^2)$ calculated separately for L and R with Dirichlet conditions. The determinant $\det(-\nabla^2)$ for the full space appearing in the path integral (as in (2.4)) can also be displayed in the split form using the BFK gluing formula as[4]

$$\det(-\nabla^2) = \det(-\nabla^2)_L\, \det(-\nabla^2)_R\, \det(M_L + M_R). \tag{2.17}$$

As regards the constraints, we can write them as

$$\int \partial_i f E_i = \int_L \tilde{f}_L(-\nabla^2 \tilde{\sigma}_L) + \int_R \tilde{f}_R(-\nabla^2 \tilde{\sigma}_R) + \int f_0 \mathcal{E} \approx 0,$$
$$\int \partial_i h A_i = \int_L \tilde{h}_L(-\nabla^2 \tilde{\theta}_L) + \int_R \tilde{h}_R(-\nabla^2 \tilde{\theta}_R) + \int h_0\, (M_L + M_R)\, \theta_0 \approx 0. \tag{2.18}$$

If we plan to integrate over the full space, the constraints eliminate θ_0-dependence everywhere, which is equivalent to writing

$$
\delta(\nabla \cdot E)\,\delta(\nabla \cdot A) = \delta[-\nabla^2 \tilde{\sigma}_{\mathrm{L}}]\,\delta[-\nabla^2 \tilde{\sigma}_{\mathrm{R}}]\;\delta[-\nabla^2 \tilde{\theta}_{\mathrm{L}}]\,\delta[-\nabla^2 \tilde{\theta}_{\mathrm{R}}]
$$
$$
\times\,\delta[\mathcal{E}_0]\delta[(M_{\mathrm{L}} + M_{\mathrm{R}})\theta_0]
$$
$$
= \delta\left[\cdots\right]\det(-\nabla^2)_{\mathrm{L}}^{-1}\det(-\nabla^2)_{\mathrm{R}}^{-1}\det(M_{\mathrm{L}} + M_{\mathrm{R}})^{-1},
$$

$$(2.19)$$

where $\delta\left[\cdots\right]$ indicates the product of δ-functions for fields $\tilde{\sigma}_{\mathrm{L}}$, $\tilde{\sigma}_{\mathrm{R}}$, $\tilde{\theta}_{\mathrm{L}}$, $\tilde{\theta}_{\mathrm{R}}$, $\mathcal{E}_0$, θ_0. Note again the appearance of determinants separately for L and R, but there is also one factor of $\det(M_{\mathrm{L}} + M_{\mathrm{R}})^{-1}$. This arises from the last term in the second of the constraints (2.18). It is then easy to check that the path integral for the full space is reproduced.

However, if we now consider integrating over fields in R, the edge modes $\mathcal{E}_0$ and θ_0 on the interface are physical degrees of freedom from the point of view of region R. The test functions f_0, h_0 in (2.18) are to be taken to be zero, so we do not have the corresponding δ-functions or the factor $\det(M_{\mathrm{L}} + M_{\mathrm{R}})^{-1}$ in (2.19). We then find, after integration over fields in R,

$$
Z = \det(M_{\mathrm{L}} + M_{\mathrm{R})}\int[d\mathcal{E}d\theta_0][d\Pi dB]\,e^{i\mathcal{S}},
$$
$$
\mathcal{S} = \int\left[\mathcal{E}\,\dot{\theta}_0 - \frac{1}{2}\mathcal{E}(M_{\mathrm{L}} + M_{\mathrm{R}})^{-1}\,\mathcal{E}\right] + \mathcal{S}_{\Pi,B}.
$$

$$(2.20)$$

The Π, B part, which we have not displayed in split form or elaborated on, behaves as a scalar field and gives what is expected for a scalar field as regards entanglement. The new ingredient is that, while the result reduces to the expected one in region L, *there is an extra factor* $\det(M_{\mathrm{L}} + M_{\mathrm{R}})$ *from the phase volume, i.e., an extra degeneracy factor due to the edge modes.* This contributes to the entanglement entropy, which is now of the form

$$
S_E = \log\det(M_{\mathrm{L}} + M_{\mathrm{R}}) + S_{E\,\Pi},
$$

$$(2.21)$$

where $S_{E\,\Pi}$ is due to the (Π, B)-sector. The extra contribution from the edge modes

1. is the so-called contact term, calculated many years ago via the replica trick by Kabat,[5]
2. is the interface term needed for the BFK gluing formula.

This is one of the key effects of the edge modes we wanted to emphasize. For more details and extensions to the non-abelian, as well as Maxwell–Chern–Simons theories, see Ref. [3].

3. The EIH Method for Chern–Simons + Einstein Gravity

We now turn to another effect of edge modes, namely how they play a role in the context of the Einstein–Infeld–Hoffmann (EIH) method.[6] Recall that EIH argued that the Einstein field equations for gravity in the vacuum are sufficient to determine the dynamics of point particles (i.e., matter dynamics) interacting gravitationally. They defined point particles as singularities in the gravitational field, excised small spheres (or tubes when we include time) around the singularities to keep fields well defined and imposed the field equations on the resulting configurations. This led to a set of equations which are not only conceptually interesting but actually can also be applied in some astrophysical contexts. The question we pose here is as follows: How does this work in more general theories of gravity, say, Chern–Simons (CS) gravity or in CS gravity with an Einstein–Hilbert term added?[7] To begin with, let us consider the 2+1-dimensional CS action, with connections in the algebra of some Lie group G, given as

$$S = \frac{k}{4\pi} \int \mathrm{Tr}(A\,dA + \tfrac{2}{3}A^3) + S_b(A, \psi). \tag{3.1}$$

The space-time manifold is taken as $M \times \mathbb{R}$, where M has the topology of the disk. $S_b(A, \psi)$ is a boundary action (which may depend on some other fields ψ) which ensures the full gauge invariance of the action (3.1), including transformations on the boundary ∂M.

The bulk equation of motion is $F = 0$, i.e., A is a pure gauge in the bulk. Consider now singular classical solutions on the disk M of the form

$$A_i = a_i, \quad da + a^2 = \sum_{s=1}^{N} q_s\,\delta^{(2)}(x - x_s), \; . \; A_0 = a_0 = 0. \tag{3.2}$$

For simplicity, we take all q_s to be in the Cartan subalgebra of G so that $da + a^2 = da$. We still have $F = 0$ on $\tilde{M} = M - \{C_s\}$, where C_s denote small disks around the singularities; thus, A is still a pure gauge on $\tilde{M}$. The general solution to the field equation is then a gauge transform of (3.2),

$$A_i = g^{-1}a_i\,g + g^{-1}\partial_i g, \quad A_0 = g^{-1}\partial_0 g. \tag{3.3}$$

The evaluation of the action on this configuration gives

$$S = S[a] - \frac{k}{4\pi} \sum_s \oint_{\partial C_s} \left[\text{Tr}(a\, dg\, g^{-1}) \right]. \tag{3.4}$$

We see that the dynamics is given in terms of the group elements g on ∂C_α; these represent the "edge modes". Consider shrinking the size of the disks to almost zero radius so that g can be taken to be $g_s = g(\vec{x}_s) \equiv h_s^{-1}$. The g-dependent part of the action, after integrating over the spatial boundaries, is thus

$$S = \frac{k}{4\pi} \sum_s \int dt \left[\text{Tr}(q_s\, h_s^{-1} \partial_0 h_s) \right]. \tag{3.5}$$

This is a sum of coadjoint orbit actions.[a] We see that there is no real dynamical evolution, which is not surprising for the CS theory, but, in the usual manner of quantizing such actions, the states in the quantum theory will carry unitary representations of G, of highest weights specified by the q_s.

With this observation in mind, consider Einstein gravity in 2+1 dimensions (with a cosmological constant $\sim l^{-2}$) which may be described by the CS action[8]

$$S = -\frac{k}{4\pi} \left[\int \text{Tr} \left(AdA + \tfrac{2}{3}A^3 \right)_L - \int \text{Tr} \left(AdA + \tfrac{2}{3}A^3 \right)_R \right]$$

$$= -\frac{k}{4\pi l} \int d^3x \det e \left[R - \frac{2}{l^2} \right] + \text{total derivative} \tag{3.6}$$

where the gauge connections are given in terms of the spin connection ω^{ab} and the frame field (dreibein in this case) e^a by

$$A_L = (-iM_a)\, A_L^a = (-iM_a) \left(-\tfrac{1}{2}\eta^{ak}\epsilon_{kbc}\omega^{bc} + \tfrac{e^a}{l} \right),$$
$$A_R = (-iN_a)\, A_R^a = (-iN_a) \left(-\tfrac{1}{2}\eta^{ak}\epsilon_{kbc}\omega^{bc} - \tfrac{e^a}{l} \right), \tag{3.7}$$

where M_a, N_a are the generators of two independent $SO(2,1)$ Lie algebras, with parity exchanging them. Newton's constant for this problem can be identified as $G = l/(4k)$.

Since there is a non-zero cosmological constant, it is clear from (3.6) that the solution for the vacuum state, i.e., the solution of the bulk equation of motion, is the anti-de Sitter (AdS) space in 2+1 dimensions. As with (3.1),

---

[a] Another of Bal's favorite topics.

we can now add a boundary term (on ∂M) to obtain full gauge invariance for (3.6) and then introduce "point particles" via ansätze of the form (3.2), (3.3). The action on these configurations becomes

$$S = -\frac{k}{4\pi} \sum_s \int dt \; \left[\mathrm{Tr}(q_s \, h_s^{-1} \, \partial_0 h_s)_L - \mathrm{Tr}(q_s \, h_s^{-1} \, \partial_0 h_s)_R \right]. \qquad (3.8)$$

We get two sets of coadjoint orbit actions, leading to unitary representations of $SO(2,1) \times SO(2,1)$ upon quantization. The full isometry group for the AdS space is $SO(2,1) \times SO(2,1)$, with the diagonal $SO(2,1)$ as the Lorentz group, while the coset directions correspond to translations. Thus, a point particle in AdS space must be defined in the quantum theory as a unitary irreducible representation of $SO(2,1) \times SO(2,1)$, and this is exactly what is obtained, realizing the EIH strategy. The mass and spin, read off from the identification of translations and Lorentz transformations, are

$$m = \frac{q_R + q_L}{32\pi G}, \qquad s = \frac{l}{16G}(q_L - q_R). \qquad (3.9)$$

In 2+1-dimensional gravity, curvatures are localized at the positions of the particles so that the result (3.8) with no interactions between particles is indeed what we expect. So, it is interesting to consider a similar analysis in 4+1 dimensions, where the Einstein–Hilbert action is distinct from a combination of CS terms[9] and can be included as an additional term in the action. We can construct an $SO(4,2)$ algebra using 4×4 Dirac γ-matrices, γ_a, $\Sigma_{ab} = (i/4)[\gamma_a, \gamma_b]$, $a,b = 0,1,2,3,5$. We then define the connections

$$A_L = -\frac{i}{2}(\omega^{ab}\Sigma_{ab} + e^a\gamma_a), \qquad A_R = -\frac{i}{2}(\omega^{ab}\Sigma_{ab} - e^a\gamma_a). \qquad (3.10)$$

As for the action, we will consider the parity invariant combination

$$S = CS(A_L) - CS(A_R) + S_b(A, \psi) + S_{\mathrm{EH}},$$

$$CS(A) = -\frac{ik}{24\pi^2} \int \mathrm{Tr} \left[AdAdA + \frac{3}{2}A^3 dA + \frac{3}{5}A^5 \right]. \qquad (3.11)$$

$S_b(A, \psi)$ is, as before, included to cancel any boundary term from gauge transformations, and S_{EH} is the Einstein–Hilbert action with a cosmological constant $\sim l^{-2}$. In terms of the curvature $R^{ab} = d\omega^{ab} + (\omega\omega)^{ab}$ and torsion $T^a = de^a + \omega^{ab}e^b$, we find

$$F_{L,R} = (-i\Sigma_{ab}/2)(R^{ab} + e^a e^b) \pm (-i\gamma_a)T^a. \qquad (3.12)$$

The bulk equation of motion for the CS-part in (3.11) is thus satisfied by $R^{ab} = -e^a e^b$ and $T^a = 0$. With the scaling, $e^a \to \sqrt{12}\, e^a/l$, we see that this is AdS space-time. For simplicity, we take S_{EH} to have the same value for the cosmological constant so that the term from S_{EH} in the bulk equation of motion is also zero for the same AdS space-time.

We now introduce solutions with point-like singularities on the spatial manifold. These will be taken as the point-like limit of 4d-instantons in $SO(4) \in SO(4,2)$. We excise small balls around each point, and on $M - \{C_s\}$, we take the solution to be of the form

$$a = t_1\, U^{-1}dU, \quad t_1 = \begin{pmatrix} 1 & 0 \\ 0 & 0 \end{pmatrix}, \quad \text{or} \quad \begin{pmatrix} 0 & 0 \\ 0 & 1 \end{pmatrix}$$

$$U = \phi^0 + i\sigma_i \phi^i, \quad \phi^0 \phi^0 + \phi^i \phi^i = 1.$$

(3.13)

t_1 is a projector to either of the two $SU(2)$s in $SO(4) \sim SU(2) \times SU(2)$. The instanton number is the integral of $(1/12\pi^2)\epsilon_{\mu\nu\alpha\beta}\phi^\mu d\phi^\nu d\phi^\alpha d\phi^\beta$. We then consider the more general configurations with the same singularity structure:

$$A_L = g^{-1}(a_L + \delta a_L)g + g^{-1}dg, \quad A_R = g^1(a_R + \delta a_R)dg + g^{-1}dg, \quad (3.14)$$

where a_L and a_R are terms of the form (3.13) and are related by parity. δa_L and δa_R correspond to perturbations of the metric as $g_{\mu\nu} \to g_{\mu\nu} + h_{\mu\nu}$, i.e.,

$$\delta a_{L,R} = (-i/2) \left[\pm \gamma_a\, e^{-1\nu a} h_{\mu\nu} - \Sigma_{ab}\, e^{-1\alpha a} e^{-1\beta b} \nabla_\alpha h_{\beta\mu} \right] dx^\mu. \quad (3.15)$$

Also, g is an element of $SO(4,2)$ parametrized as

$$g = S^{-1}\Lambda V, \quad V = \frac{1}{\sqrt{z}} \begin{pmatrix} z & iX \\ 0 & 1 \end{pmatrix}, \quad S = \frac{1}{\sqrt{2}} \begin{pmatrix} 1 & 1 \\ 1 & -1 \end{pmatrix}, \quad (3.16)$$

where $X = x^0 - i\sigma_i x^i$, and $z = x^5$ is the radial coordinate of AdS and $\Lambda \in SO(4,1)$ denotes a Lorentz transformation. Evaluating the action (3.11) on the configurations (3.14), we find

$$S = \frac{k}{2} \sum_s \int (Q_s^1 - Q_s^2)\eta_{ab}(\Lambda_s)_0^a \left[e_\mu^b + e^{-1\nu b} h_{\mu\nu} \right] dx^\mu + \frac{1}{2} \int h_{\mu\nu} \mathbb{L}^{\mu\nu\alpha\beta} h_{\alpha\beta},$$

(3.17)

where Q_s^1 and Q_s^2 are the instanton numbers for the two $SU(2)$s in $SO(4)$ and $\mathbb{L}^{\mu\nu\alpha\beta}$ is the Lichnerowicz operator (i.e., the quadratic fluctuation operator) for S_{EH}. Without the $h_{\mu\nu}$ terms, this is a sum of coadjoint orbit actions. The mass may be identified as $m_s = -\frac{1}{2}k(Q_s^1 - Q_s^2)$. (Spin can be included by considering different orientations of the $SU(2)$s in $SO(4)$ for different instantons.) Further, if we write $\Lambda_0^a = e_\nu^a \frac{dx^\nu}{ds}$, which is consistent with its properties, the first term in S becomes $-m \int ds$, corresponding to free particle motion. Solving for $h_{\mu\nu}$ from its equation of motion *à la* (3.17), and using it back again in (3.17), we get interactions between the particles. In the nonrelativistic limit, the result is

$$S = -\sum_s \int m_s \int ds_s + \frac{1}{6} \sum_{s \neq s'} \int dx_s^0 \frac{m_s m_{s'}}{4\pi^2 r_{ss'}^2}, \quad x_{s'}^0 = x_s^0 + |\vec{x}_s - \vec{x}_{s'}|$$

(3.18)

showing, correctly, the 4d Coulomb-like potential, if we neglect retardation effects. In principle, one can include higher corrections systematically, but this suffices to prove our main point: The edge modes of the CS+Einstein gravity allow us to realize the Einstein–Infeld–Hoffmann method of defining point particles as singularities of the gravitational field and then obtaining their equations of motion from the field equations.

This work was supported in part by U.S. National Science Foundation research grants PHY-2112729 and PHY-1820271.

References

1. Entanglement is a topic with an enormous literature by now, it is impossible to make even a vaguely comprehensive list in the short space available. The papers cited below can be used to trace most of the relevant articles.
2. A. P. Balachandran *et al.*, *Int. J. Mod. Phys.* **A9** (1994) 3417; A. P. Balachandran, L. Chandar and E. Ercolessi, *ibid.* **A10** (1995) 1969; A. P. Balachandran, L. Chandar and A. Momen, *ibid.* **A12** (1997) 625 + other articles.
3. A. Agarwal, D. Karabali and V. P. Nair, *Phys. Rev.* **D 96** (2017) 125008.
4. D. Burghelea, L. Friedlander and T. Kappeler, *J. Funct. Anal.* **107** (1992) 34.
5. D. N. Kabat, *Nucl. Phys. B* **453** (1995) 281.
6. A. Einstein, L. Infeld and B. Hoffmann, *Ann. Math.* **39** (1938) 65; L. Infeld, *Rev. Mod. Phys.* **29** (1957) 398.
7. What is discussed here is based on some joint work with Lei Jiusi, some of it published in L. Jiusi and V. P. Nair, *Phys. Rev.* **D 96** (2017) 065019. I also use some unpublished material from his PhD dissertation, CUNY, 2020, with more details and generalization to be published.
8. A. Achúcarro and P. K. Townsend, *Phys. Lett.* **B180** (1986) 89; E. Witten, *Nucl. Phys.* **B311** (1988) 46.
9. For a review of CS gravity, see J. Zanelli, *Lecture notes on Chern-Simons (super-)gravities*, Second edition (February 2008), arXiv:hep-th/0502193.

Chapter 13

Loop Braid Groups and Integrable Models

Pramod Padmanabhan* and Abhishek Chowdhury[†]

*School of Basic Sciences Indian Institute of Technology,
Bhubaneswar, India*
*ppadmana@iitbbs.ac.in
[†] achowdhury@iitbbs.ac.in

Loop braid groups characterize the exchange of extended objects, namely loops, in three-dimensional space generalizing the notion of braid groups that describe the exchange of point particles in two-dimensional space. Their interest in physics stems from the fact that they capture anyonic statistics in three dimensions which is otherwise known to only exist for point particles on the plane. Here, we explore another direction where the algebraic relations of the loop braid groups can play a role: quantum integrable models. We show that the *symmetric loop braid group* can naturally give rise to solutions of the Yang–Baxter equation, proving the integrability of certain models through the RTT relation. For certain representations of the symmetric loop braid group, we obtain integrable deformations of the XXX-, XXZ- and XYZ-spin chains.

1. Introduction

Two-dimensional spaces offer more possibilities for point particles — *anyons* — that obey exotic statistics, unlike their three-dimensional counterparts that appear as either bosons or fermions. However, this narrative fails for *extended objects* in higher dimensions giving rise to novel statistical properties and spin–statistics relations.[1–5] While the two-dimensional anyons are seen as representations of the braid group, the statistics groups of extended objects in higher-dimensional spaces are the *motion groups*.[6] Anyons have received renewed attention in recent years due to their potential applications in *topological quantum computation*.[7]

Though these questions address some important aspects of fundamental physics, the physicist's interest in the braid group also extends to the study of *quantum integrable models* which form the basis of the method of *algebraic Bethe ansatz*.[8-10] The method of *Baxterization*[11] helps us obtain R-matrices that solve the *spectral parameter*-dependent *Yang–Baxter equation* (YBE) from braid group generators. In this chapter, we show the construction of R-matrices using the generators of a quotient of the *loop braid group* — the *symmetric loop braid group*,[12-14] that study the exchange of loops (S^1) in three dimensions. We will see that they naturally give rise to solutions of the YBE and hence open the possibility for obtaining new integrable models.

The contents of this chapter are as follows: We review the construction of local integrable models using the R-matrix method in Section 2. We then move on to describing the algebraic relations of the loop braid group in Section 3 and use these to construct YBE solutions in Section 4. By choosing appropriate representations of the symmetric loop braid group, we find associated integrable models in Section 5. We end with an outlook in Section 6.

This chapter is to honor the immense contribution of A. P. Balachandran to theoretical physics and its community on the occasion of his 85th birthday. Bal, as he is fondly known among his friends, has been a continuous source of inspiration for several generations of physicists including us and this work is one such example coming from his school of thought.

2. Integrability *via* R-matrices

In 1931, Hans Bethe proposed an original method (*the coordinate Bethe ansatz*) to arrive at the spectrum of the quantum Heisenberg spin chain which later on in 1970–1980s gave rise to a new algebraic approach (*the algebraic Bethe ansatz*) to study a wide class of quantum systems in various dimensions. It was found that the same algebra of operators had different representations which solved many quantum models with completely different physical interpretations.

The algebraic Bethe ansatz (ABA) provides a mathematical reasoning for the integrability of these myriad quantum systems by arriving at the existence of an infinite-dimensional symmetry. Here, we will focus on closed spin chains with N sites and periodic boundary conditions, whose Hilbert space has the form

$$\mathcal{H} = \bigotimes_{n=1}^{N} \mathcal{H}_n, \tag{2.1}$$

where $\mathcal{H}_n$ is the Hilbert space at the nth site of the chain.

The fundamental object of ABA is a generating object called the *Lax operator* or the *monodromy matrix*, $T(u)$. In order to define it, we have to introduce an auxiliary space V (which we take as $\mathbb{C}^2$) in addition to the Hilbert space $\mathcal{H}$ of the quantum system. We also have to introduce a continuous complex parameter u, the *spectral parameter* which allow us to recover the integrals of motions as coefficients of a series expansion in u. The auxiliary space V on the other hand is essential to show that the integrals of motion commute. The Lax operator acts on $\mathcal{H} \otimes V$, i.e., the matrix elements T^{ij} are operators acting on the Hilbert space $\mathcal{H}$:

$$T(u) = \begin{pmatrix} A(u) & B(u) \\ C(u) & D(u) \end{pmatrix}. \tag{2.2}$$

They are assumed to follow the commutation relations (RTT) in the space $\mathcal{H} \otimes V_1 \otimes V_2$:

$$R_{12}(u - v)T_1(u)T_2(v) = T_2(v)T_1(u)R_{12}(u - v), \tag{2.3}$$

where the *R-matrix* $R_{12}(u)$ acts on $V_1 \otimes V_2$ and are solutions to the YBE acting on the tensor product of auxiliary spaces $V_1 \otimes V_2 \otimes V_3$:

$$R_{12}(u) R_{13}(u + v) R_{23}(v) = R_{23}(v) R_{13}(u + v) R_{12}(u). \tag{2.4}$$

It is straightforward to verify that $R(u) = u\,\mathbb{I} + c\,s$ (we have the freedom to multiply by an arbitrary function), where c is a constant and s (defined in Section 3) is the permutation matrix, solves the above YBE. For this R-matrix, Eq. (2.3) leads to the following commutation relations:

$$\left[T^{ij}(u), T^{kl}(v)\right] = \frac{c}{u - v} \left(T^{kj}(v)T^{il}(u) - T^{kj}(u)T^{il}(v)\right), \tag{2.5}$$

which gives the *ABCD* algebra crucial to generating the spectrum of the given quantum system.

From the RTT relation Eq. (2.3), multiplying by $R_{12}^{-1}(u - v)$ from the left, taking trace w.r.t. the space $V_1 \otimes V_2$ and using the cyclic permutations in the trace, we get $[\mathcal{T}(u), \mathcal{T}(v)] = 0$. It is only natural to work with the *transfer matrix*, $\mathcal{T}(u) = \text{tr}\, T(u) = A(u) + D(u)$, as it satisfies the periodic boundary condition of one-dimensional quantum chains. Expanding $\mathcal{T}(u)$ into a power series over u centered around some point u_0 gives

$$\mathcal{T}(u) = \sum_k (u - u_0)^k I_k, \tag{2.6}$$

which leads to a set of commuting operators in the Hilbert space $\mathcal{H}$

$$[I_k, I_n] = 0, \quad \forall \, k, n. \tag{2.7}$$

If we now set one of these I_k to be the Hamiltonian of some quantum system, then we have an infinite set of integrals of motion (for a finite chain of size N, the expansion will get truncated at $k = N$), i.e., an integrable system. Before discussing the Hamiltonians, we note two important points:

- In order to ensure locality of the operators I_k, the correct choice(s) of decomposition point u_0 is crucial.
- Instead of expanding the operator $\mathcal{T}(u)$ into a series, we could have expanded a function of this operator (almost always the case).

In the following section, we will demonstrate this with the concrete example of the XXX-spin chain.

2.1. Local Hamiltonians

Now, we show one way of obtaining a local operator from the transfer matrix, $\mathcal{T}(u)$, through the expression

$$H(u_0) = \frac{d\mathcal{T}(u)}{du} \mathcal{T}^{-1}(u) \Bigg|_{u=u_0}, \tag{2.8}$$

where the derivative is taken at some point $u = u_0$. We give the final answer in terms of a general invertible R-matrix and show that the XXX-spin chain can be obtained for a particular choice of the R-matrix.

A particular solution of the RTT relation, Eq. (2.3), is the R-matrix itself, and hence, using this, we obtain the transfer matrix, $\mathcal{T}(u)$, as

$$\mathcal{T}(u) = \mathrm{tr}_0 \left[R_{0N}(u) \cdots R_{01}(u) \right], \tag{2.9}$$

where the entries of the R-matrix act on the Hilbert space, $\mathcal{H} = \otimes_{n=1}^{N} \mathbb{C}_n^2$, the N-site spin chain. The auxiliary space is denoted by V_0. The derivative of $\mathcal{T}(u)$ is evaluated as follows:

$$\frac{d\mathcal{T}(u)}{du}\bigg|_{u_0} = \sum_{k=1}^{N} \text{tr}_0\left[R_{0N}(u_0)\cdots R'_{0k}(u_0)\cdots R_{01}(u_0)\right]$$

$$= \sum_{k=1}^{N} \text{tr}_0\left[R'_{0k}\cdots R_{01}R_{0N}\cdots R_{0k+1}s^2_{0,k+1}\right]$$

$$= \sum_{k=1}^{N} \text{tr}_0\left[R'_{k+1,k}\cdots R_{k+1,1}R_{k+1,N}\cdots R_{k+1,k+2}R_{k+1,0}\right]$$

$$= \sum_{k=1}^{N}\left[R'_{k+1,k}\cdots R_{k+1,1}R_{k+1,N}\cdots R_{k+1,k+2}\right], \qquad (2.10)$$

where we have repeatedly used the cyclicity properties of the trace operation and we also take $\text{tr}_0(R_{k+1,0}) \propto 1$. This is indeed the case in the examples we consider; nevertheless, relaxing this assumption slightly modifies the computations and can still be handled in this procedure. In a similar fashion, we find

$$\mathcal{T}(u)^{-1}\bigg|_{u=u_0} = R^{-1}_{k+1,k+2}\cdots R^{-1}_{k+1,N}R^{-1}_{k+1,1}\cdots R^{-1}_{k+1,k}, \qquad (2.11)$$

and multiplying this with the derivative of $\mathcal{T}(u)$, we obtain the Hamiltonian as

$$H(u_0) = \sum_{k=1}^{N} R'_{k+1,k}(u_0)R^{-1}_{k+1,k}(u_0). \qquad (2.12)$$

Note that though this operator is a sum of *local* terms, for a general R-matrix, it need not be *self-adjoint*.

On choosing $R(u) = u\,\mathbb{I} + c\,s$, we obtain the XXX-spin chain with the local Hamiltonian,

$$H\left(\frac{c}{2}\right) = \frac{2}{3c} \sum_{k=1}^{N} \left[X_{k+1}X_k + Y_{k+1}Y_k, +Z_{k+1}Z_k\right], \qquad (2.13)$$

where we have used the representation of the permutation operator, s (see Eq. (4.5)).

3. Loop Braid Group

The loop braid group, LB_N, is defined on N sites just as the braid group. A convenient presentation is given by the relations of the generators, σ_i, s_i with $i \in \{1, \ldots, N-1\}$. The braid generators, σ_i, satisfy the braid group relations,

$$\sigma_i \sigma_{i+1} \sigma_i = \sigma_{i+1} \sigma_i \sigma_{i+1}, \tag{B1}$$

$$\sigma_i \sigma_j = \sigma_j \sigma_i ; \quad |i - j| > 1, \tag{B2}$$

where $(B2)$ is the far commutativity of the braid group generators. The s_i operators generate the permutation group,

$$s_i s_{i+1} s_i = s_{i+1} s_i s_{i+1}, \tag{S1}$$

$$s_i s_j = s_j s_i; \quad |i - j| > 1, \tag{S2}$$

$$s_i^2 = \mathbb{I}. \tag{S3}$$

Each of the generators, σ_i and s_i, act non-trivially on sites labeled by i and $i+1$. Furthermore, we have the *mixed* relations between the generators, σ_i and s_i,

$$\sigma_i s_j = s_j \sigma_i; \quad |i - j| > 1 \tag{M1}$$

$$s_i s_{i+1} \sigma_i = \sigma_{i+1} s_i s_{i+1}, \tag{M2}$$

$$\sigma_i \sigma_{i+1} s_i = s_{i+1} \sigma_i \sigma_{i+1}. \tag{M3}$$

When all the above relations except $(M3)$ are satisfied, the relations realize the virtual braid group, VB_N.[15] The relation $(M2)$ read backwards, $s_{i+1} s_i \sigma_{i+1} = \sigma_i s_{i+1} s_i$, is always satisfied, however the relation $(M3)$ read backwards,

$$\sigma_{i+1} \sigma_i s_{i+1} = s_i \sigma_{i+1} \sigma_i, \tag{M3$'$}$$

is not always satisfied. If $(M3)'$ is satisfied instead of $(M3)$, we have the opposite loop braid group, OLB_N, and if both $(M3)$ and $(M3)'$ are satisfied, we obtain the *symmetric loop braid group*, SLB_N. We will use the relations of SLB_N along with

$$\sigma_i s_{i+1} \sigma_i = \sigma_{i+1} s_i \sigma_{i+1} \tag{M4}$$

in constructing the solutions of the YBE.

Though the above algebraic relations of SLB_N will suffice for the purpose of constructing R-matrices, we also mention the possibility of understanding the relations of the loop braid group pictorially. This is especially useful to visualize the exchange of loops realized by the generators of SLB_N.[16]

4. R-matrices from the Loop Braid Group

We now develop the technique to find R-matrices using SLB_N. The following *ansatz*,

$$R_{i,i+1}(u) \equiv R_i(u) = s_i + a(u)\,\sigma_i, \tag{4.1}$$

is easily seen to satisfy the braided form of the YBE,

$$R_{i,}(u-v)R_{i+1}(u)R_i(v) = R_{i+1}(v)R_i(u)R_{i+1}(u-v), \tag{4.2}$$

for an arbitrary function, $a(u)$, provided s_i and σ_i satisfy the relations of SLB_N along with the relation in $(M4)$. Multiplying the R-matrix in Eq. (4.1) with s_i, we obtain another R-matrix,

$$s_i R_{i,i+1}(u) = \mathbb{I} + a(u)\,s_i\sigma_i, \tag{4.3}$$

that satisfies

$$R_{i,i+1}(u-v)R_{i,i+2}(u)R_{i+1,i+2}(v) = R_{i+1,i+2}(v)R_{i,i+2}(u)R_{i,i+1}(u-v) \tag{4.4}$$

which resembles the RTT relation with the R-matrices assuming the role of the monodromy matrices. As seen earlier, this is the starting point for proving the integrability of various quantum models and the ABA.

Having constructed the R-matrix out of the generators of SLB_N, we are left with finding appropriate representations. To this end, we work on a closed chain comprising N sites, with periodic boundary conditions. On each site, we place a qubit with Hilbert space $\mathcal{H}_n = \mathbb{C}^2$ and the total Hilbert space is $\otimes_{n=1}^{N} \mathbb{C}_n^2$. On this space, the local representation for the permutation group generators, s_i is well known,

$$s_i = \frac{\mathbb{I} + X_i X_{i+1} + Y_i Y_{i+1} + Z_i Z_{i+1}}{2}, \tag{4.5}$$

where X, Y, Z are the Pauli matrices,

$$X = \begin{pmatrix} 0 & 1 \\ 1 & 0 \end{pmatrix}, \quad Y = \begin{pmatrix} 0 & -i \\ i & 0 \end{pmatrix}, \quad Z = \begin{pmatrix} 1 & 0 \\ 0 & -1 \end{pmatrix}. \tag{4.6}$$

We are left with finding the representations of the braid group generators, σ_i, which we obtain from projectors that commute among themselves and are left invariant by the permutation operator. Such an operator can be written as

$$\sigma_i = s_i + \alpha\, B_{i,i+1}. \tag{4.7}$$

The operator $B_{i,i+1}$ satisfies

$$B_{i,i+1}^2 = B_{i,i+1}, \quad B_{i,i+1}B_{i+1,i+2} = B_{i+1,i+2}B_{i,i+1}, \quad s_iB_{i,i+1} = B_{i,i+1}. \tag{4.8}$$

While there are plenty of projectors on $\mathbb{C}^2 \otimes \mathbb{C}^2$, only a few among them are left invariant by an action of the permutation operator, s_i. One natural choice is when $B_{i,i+1} = P_iP_{i+1}$, that is, an operator that is not 'entangled' on the neighboring sites. As it is symmetric in the indices i and $i+1$, it is left invariant under a left and right s_i action. With these properties satisfied, a small computation shows that the operator in Eq. (4.7) satisfies the braid relations, with the inverse given by $\sigma_i^{-1} = s_i - \dfrac{\alpha}{1+\alpha}\, B_{i,i+1}$ and thus generates the braid group. A further calculation shows that the braid generator in Eq. (4.7) along with the permutation operator satisfies all the relations of SLB_N, especially the relations $(M3)$, $(M3)'$ and $(M4)$ in Section 3.

Next, we will use the permutation operator and the braid group generator in Eq. (4.7) to construct the R-matrix using the ansatz in Eq. (4.1).

5. Integrable Spin Chains

The Hamiltonian of the integrable system is taken to be a local operator following the discussion in Section 2.1. Using the formula for the local Hamiltonian in Eq. (2.12) with the R-matrix in Eq. (4.3), we obtain

$$H = \sum_{i=1}^{N} \frac{a'(u)}{1+a(u)} \left[1 + \frac{\alpha}{1+(\alpha+1)\,a(u)}\, B_{i,i+1} \right], \tag{5.1}$$

on a closed chain. According to the representation of the B operators satisfying Eq. (4.8), we obtain both trivial and non-trivial spin chains. By trivial we mean those models which does not require the machinery of the ABA for a solution. The scope of the ansatz for the R-matrix in Eq. (4.1) is limited to this type of model, but we would like to go further.

In order to obtain a more non-trivial model, consider the ansatz

$$R_{i,i+1}(u) = s_i + a(u)\,\sigma_i + b(u)\,\sigma_i^2, \tag{5.2}$$

which is equivalent to working with

$$R_{i,i+1}(u) = \mathbb{I} + a(u)\,s_i + b(u)\,B_{i,i+1}, \tag{5.3}$$

as we have the identities

$$\sigma_i^{2k+1} = s_i + \left[(\alpha+1)^{2k+1} - 1\right] B_{i,i+1}, \quad \sigma_i^{2k} = \mathbb{I} + \left[(\alpha+1)^{2k} - 1\right] B_{i,i+1}. \tag{5.4}$$

The ansatz in Eq. (5.3) satisfies the YBE for $a(u) = \alpha u$, $b(u) = -2\alpha u$. Using Eq. (2.12), the resulting nearest neighbor term in the Hamiltonian is given by

$$H(u) = \sum_{i=i}^{N} \frac{1}{1-\alpha^2 u^2}\left[(1-\alpha^2 u)\mathbf{1} + \alpha(1-u)s_i + (2\alpha(u-1))\,B_{i,i+1}\right]. \tag{5.5}$$

A naive reading of this local operator suggests that it is a deformation of the XXX-chain. While this is true for certain representations of $B_{i,i+1}$, there are many other non-trivial ones as we shall see in the following.

Model 1 Consider the case when $B_{i,i+1} = P_i P_{i+1}$. A general projector on $\mathbb{C}^2$ is given by

$$P = \frac{1}{2}\mathbf{1} + lX + mY + nZ, \tag{5.6}$$

provided $(l^2 + m^2 + n^2) = \frac{1}{4}$, where l, m, n are complex numbers. This reduces Eq. (5.5) to a deformation of the XYZ-spin chain. We include the cumbersome expression

$$P_i P_{i+1} = \frac{1}{4}\mathbf{1} + l^2 X_i X_{i+1} + m^2 Y_i Y_{i+1} + n^2 Z_i Z_{i+1}$$

$$+ \frac{l}{2}(X_i + X_{i+1}) + \frac{m}{2}(Y_i + Y_{i+1}) + \frac{n}{2}(Z_i + Z_{i+1})$$

$$+ lm(X_i Y_{i+1} + Y_i X_{i+1}) + mn(Y_i Z_{i+1} + Z_i Y_{i+1})$$

$$+ ln(Z_i X_{i+1} + X_i Z_{i+1}), \tag{5.7}$$

to give a glimpse of the models available for this choice of $B_{i,i+1}$. When $l = m = 0$, we obtain a deformation of the XXZ-spin chain.

Model 2 The XXZ-spin chain is obtained by choosing $B_{i,i+1} = \dfrac{\mathbb{I} + Z_i Z_{i+1}}{2}$ and the corresponding Hamiltonian is given by

$$H_{XXZ} \sim \frac{\alpha(1-u)}{2(1-\alpha^2 u^2)}[X_i X_{i+1} + Y_i Y_{i+1} - Z_i Z_{i+1}]. \tag{5.8}$$

We note that the XXZ-spin chain is usually obtained from a *trigonometric* R-matrix, but here we have shown a *rational* R-matrix (Eq. (5.3)) to obtain the same model.

6. Outlook

The usual approaches to obtaining integrable deformations of the XXX and XXZ chains include Drinfeld twisting of the associated quantum groups.[17] Here, we have shown how the loop braid group can be used to obtain solutions of the YBE opening up the possibility of finding new integrable models. For a simple choice of the representation of the symmetric loop braid group, we ended up with integrable deformations of the XYZ-spin chain. Using these R-matrices in the RTT relation, we can find new $ABCD$ algebras. We intend to develop the algebraic Bethe ansatz method for such algebras in future publications. As a final remark, the fact that we obtain the R-matrices algebraically renders itself to an easy extension to higher spin chains.

References

1. C. Aneziris, A. P. Balachandran, L. Kauffman and A. M. Srivastava, Novel statistics for strings and string "Chern-Simon" terms, *Int. J. Mod. Phys. A* **06**(14) (1991) 2519–2558.

2. J. L. Friedman and R. D. Sorkin, Spin $\frac{1}{2}$ from gravity, *Phys. Rev. Lett.* **44** (1980) 1100; **45** (1980) 148.

3. R. D. Sorkin, Introduction to topological geons, in Bergmann, P. G., De Sabbata, V. (eds.) *Topological Properties and Global Structure of Space-Time.* NATO ASI Series. Springer, Boston, MA (1986).

4. A. P. Balachandran, G. Marmo, B. S. Skagerstam and A. Stern, *Classical Topology and Quantum States,* World Scientific (1991).

5. J. L. Friedman and D. M. Witt, Internal symmetry groups of quantum geons, *Phys. Lett. B* **120** (1983) 324.

6. D. L. Goldsmith, The theory of motion groups, *Michigan Math. J.* **28**(1) (1981) 3–17.

7. M. H. Freedman, A. Kitaev, M. J. Larsen and Z. Wang, Topological quantum computation, arXiv:quant-ph/0101025.

8. N. A. Slavnov, Algebriac Bethe Ansatz, arXiv:1804.07350 [math-ph].

9. V. E. Korepin, N. M. Bogoliubov and A. G. Izergin, *Quantum Inverse Scattering Method and Correlation Functions,* Cambridge University Press (1993).

10. M. Jimbo, Introduction to the Yang Baxter equation, *Int. J. Mod. Phys. A* **04**(15) (1989) 3759–3777.

11. V. F. R. Jones, Baxterization, *Int. J. Mod. Phys. B* **04**(05) (1990) 701–713.

12. C. Damiani, A journey through loop braid groups, arXiv:1605.02323 [math.GT].

13. J. C. Baez, D. K. Wise and A. S. Crans, Exotic statistics for strings in 4d BF theory, *Adv. Theor. Math. Phys.* **11** (2007) 707–749.

14. D. M. Dahm, A Generalisation of Braid Theory, PhD Thesis, Princeton University (1962).

15. L. H. Kauffman and S. Lambropoulou, Virtual Braids, arXiv:math/0407349.

16. L. Chang, Representations of the loop braid groups from braided tensor categories, *J. Math. Phys.* **61** (2020) 051702.

17. P. Kulish, Twist deformations of quantum integrable spin chains, in *Noncommutative Spacetimes.* Lecture Notes in Physics, Vol. 774. Springer, Berlin, Heidelberg (2009).

Chapter 14

On Uhlmann's Proof of the Monotonicity of the Relative Entropy

Juan Manuel Pérez-Pardo*

Universidad Carlos III de Madrid
Avda. de la Universidad 30, 28911 Leganés (Madrid), Spain
Instituto de Ciencias Matemáticas (CSIC - UAM - UC3M - UCM)
Nicolás Cabrera, 13-15, 28049 Cantoblanco (Madrid), Spain

jmppardo@math.uc3m.es

This chapter presents in a self-contained way A. Uhlmann's celebrated theorem of monotonicity of the relative entropy under completely positive and trace preserving maps. The theorem is presented in its more general form and meaningful examples are given.

1. Previous Remarks

The main results presented in this chapter are the construction of the functional calculus of positive Hermitian quadratic forms developed by W. Pusz and S.L. Woronowicz and A. Uhlmann's celebrated Theorem of monotonicity of the relative entropy on a C^*-algebra $\mathcal{A}$ under the action of a positive and unital map, i.e., with trace preserving dual, $\Phi : \tilde{\mathcal{A}} \to \mathcal{A}$.

My interest in this topic originated during discussions with Balachandran around the years 2018/2019 on the properties of monotonicity of the

---

*This work was partially supported by the "Ministerio de Ciencia e Innovación" Research Project PID2020-117477GB-I00, by the Madrid Government (Comunidad de Madrid-Spain) under the Multiannual Agreement with UC3M in the line of "Research Funds for Beatriz Galindo Fellowships" (C&QIG-BG-CM-UC3M), and in the context of the V PRICIT (Regional Programme of Research and Technological Innovation), by the QUITEMAD Project P2018/TCS-4342 funded by Madrid Government (Comunidad de Madrid-Spain) and by the Severo Ochoa Programme for Centers of Excellence in R&D" (CEX2019-000904-S).

relative entropy when considering the subalgebras generated by observables of a quantum subsystem and in connection with previous results obtained by him and coworkers.[1–3]

The results presented in this chapter can be found in the original papers by W. Pusz and S.L. Woronowicz[4] and by A. Uhlmann.[5] A. Uhlmann's original proof is often perceived as difficult to follow. In my opinion, this is so because most of the technical results and constructions depend on the construction of the functional calculus developed by Pusz and Woronowicz. This is the reason why I originally prepared the set of notes that have served as the core of this chapter and that I have shared many times since. On this occasion, I decided to revise the notes and give them a more polished format in the hope that they can be helpful for researchers on these topics in the future.

2. Functional Calculus of Positive Quadratic Forms

In order to proof the monotonicity of the relative entropy for completely positive and trace preserving maps, we are going to take Uhlmann's definition of Relative Entropy.[5] This is based on interpolations of quadratic forms and uses their properties intensively. The interpolation of quadratic forms relies on a remarkable construction due to W. Pusz and S.L. Woronowicz[4] that allows for the definition of a functional calculus of quadratic forms defined on any vector space $\mathcal{V}$. For our purposes, this vector space will be in the following sections the C^*-algebra itself.

We shall now give a brief account on Pusz and Woronicz's construction. Let $\mathcal{V}$ be a complex vector space and consider two positive, Hermitian quadratic forms defined on it

$$q, p \colon \mathcal{V} \times \mathcal{V} \to \mathbb{C}.$$

The null space associated with these quadratic forms is defined as

$$\mathcal{N} := \{a \in \mathcal{V} \mid p(a, a) + q(a, a) = 0\}.$$

The set $\mathcal{N}$ is a linear subspace of $\mathcal{V}$. Using an approach analogous to the Gelfand–Naimark–Segal construction, one can obtain a representation of the quadratic forms p and q on a Hilbert space. Indeed, the quotient $\mathcal{V}/\mathcal{N}$ defines a pre-Hilbert space with inner product

$$\langle [a], [b] \rangle = p(a, b) + q(a, b), \quad [a], [b] \in \mathcal{V}/\mathcal{N},$$

and one can define $\mathcal{H}$ to be the closure of $\mathcal{V}/\mathcal{N}$ with respect to the norm induced by the inner product. Both quadratic forms on $\mathcal{V}$ define canonically quadratic forms on $\mathcal{H}$ which we will keep denoting with the same symbols, i.e.,

$$p : \mathcal{H} \times \mathcal{H} \to \mathbb{C}$$
$$[a], [b] \mapsto p(a, b),$$

and analogously for q. Moreover, both quadratic forms are continuous with respect to the norm induced by $\langle \cdot , \cdot \rangle$, and by Riesz representation theorem, they can be represented by two positive and bounded operators on $\mathcal{H}$. That is, there exist $P, Q \in \mathcal{B}_+(\mathcal{H})$ such that for all $a, b \in \mathcal{H}$ one has that

$$p(a, b) = \langle a , Pb \rangle \quad \text{and} \quad q(a, b) = \langle a , Qb \rangle. \tag{2.1}$$

Remarkably, these two operators commute. Note that by the definition of the scalar product and of the operators P and Q, one has that $P + Q = \mathbb{I}_{\mathcal{H}}$. As a pair of commuting, self-adjoint operators, one can use them to define a functional calculus of quadratic forms. On their original paper, Pusz and Woronowicz consider the set of homogeneous measurable and locally bounded functions, but we will not pursue such generality here. Let $f : \mathbb{R}_+^2 \to \mathbb{R}$ be a homogeneous continuous function and let $f(P, Q)$ be the positive operator defined by the functional calculus on self-adjoint operators. Define a new quadratic form on $\mathcal{V}$ by the formula

$$f_{[p,q]}(a, b) := \langle [a] , f(P, Q)[b] \rangle. \tag{2.2}$$

The quadratic form so defined does not depend on the representation chosen. That is, consider that there is another Hilbert space $\mathcal{H}'$ and a pair of positive commuting operators on it P' an Q' representing the forms p and q. The quadratic form defined by the functional calculus of the self-adjoint operators P' and Q' gives rise to the same quadratic form on $\mathcal{V}$. We refer to [4, Theorem 1.2] for the details of this proof. The particular representation obtained here by means of the GNS-like construction is just a convenient way of providing such representation. A triple $(\mathcal{H}, P, Q)$ of a Hilbert space and two positive, bounded, commuting operators $P, Q \in \mathcal{B}_+(\mathcal{H})$ that satisfy (2.1) will be called a **compatible representation for the quadratic forms** p and q.

To exemplify how this construction works, we are going to consider a particular example. Let $\hat{P}$ and $\hat{Q}$ be the positive, Hermitian operators on

$\mathbb{C}^{\not\equiv}$ given by

$$\hat{P} = \begin{pmatrix} 2 & 1 \\ 1 & 2 \end{pmatrix} \quad \hat{Q} = \begin{pmatrix} 2 & i \\ -i & 2 \end{pmatrix} \quad [\hat{P},\hat{Q}] = 2i \begin{pmatrix} -1 & 0 \\ 0 & 1 \end{pmatrix}.$$

These two operators do not commute. They define two Hermitian, positive quadratic forms on $\mathbb{C}^2$. That is, for $a, b \in \mathbb{C}^2$,

$$q(a,b) = a^+\hat{Q}b \; ; \quad p(a,b) = a^+\hat{P}b,$$

where a^+ is the Hermitian transpose vector of a. The null space is in this case $\mathcal{N} = \{0\}$. This is so since the eigenvalues of the matrix $\hat{Q} + \hat{P}$ are $4 + \sqrt{2} > 0$. The Hilbert space in which one can define the representation of the forms is $\mathcal{H} = (\mathbb{C}^2, \langle \cdot, \cdot \rangle)$ where the scalar product is defined as $\langle a, b \rangle = a^+(\hat{P} + \hat{Q})b$. Since there is no null space, we dropped the notation for the equivalence classes.

The quadratic forms q and p can be represented in this Hilbert space,

$$p(a,b) = a^+\hat{P}b = \langle a, Pb \rangle = a^+(\hat{P} + \hat{Q})Pb,$$

and similarly for q. Therefore, the matrices P and Q representing the forms q and p become respectively $P = (\hat{P} + \hat{Q})^{-1}\hat{P}$ and $Q = (\hat{P} + \hat{Q})^{-1}\hat{Q}$. Clearly, one has that $P + Q = (\hat{P} + \hat{Q})^{-1}(\hat{P} + \hat{Q}) = \mathbb{I}$. Note also that although the matrices $\hat{P}$ and $\hat{Q}$ do not commute, the matrices P and Q do commute,

$$[P,Q] = PQ + P^2 - P^2 - QP = P(P+Q) - (P+Q)P = 0.$$

3. Interpolation of Quadratic Forms and Relative Entropy

In what follows, we will consider a particular family of quadratic forms constructed using the functional calculus of quadratic forms.

Definition 1. Let p and q be positive, Hermitian quadratic forms on a vector pace $\mathcal{V}$ and let $(\mathcal{H}, P, Q)$ be a compatible representation for them. Consider the family of functions $f^t \colon \mathbb{R}_+^2 \to \mathbb{R}$ defined by $f^t(x,y) = x^{1-t}y^t$ for $t \in [0,1]$. The **interpolation of the quadratic forms** p and q is the family $\{\gamma_{[p,q]}^t\}_{t \in [0,1]}$ of quadratic forms on $\mathcal{V}$ defined by

$$\gamma_{[p,q]}^t(a,b) := \langle [a], P^{1-t}Q^t[b] \rangle, \quad a, b \in \mathcal{V}.$$

Note that one has that $\gamma_{[p,q]}^0 = p$ and $\gamma_{[p,q]}^1 = q$ and the following interpolation property:

$$\gamma_{[\gamma_{[p,q]}^{t_1}, \gamma_{[p,q]}^{t_2}]}^t = \gamma_{[p,q]}^{t'}, \quad t' = t_1(1-t) + t_2 t. \tag{3.1}$$

This implies that successive interpolations of quadratic forms give rise to quadratic forms on the previous interpolation. We shall be interested in the following element of the interpolation.

Definition 2. Let $\{\gamma^t_{[p,q]}\}_{t\in[0,1]}$ be the interpolation of the quadratic forms p and q on $\mathcal{V}$. The **geometric mean** of the quadratic forms p and q is the quadratic form $\gamma^{1/2}_{[p,q]}$.

Definition 3. Let p and q and r be positive, Hermitian quadratic forms on $\mathcal{V}$. We will say that r is **dominated by** p and q if for all $a, b \in \mathcal{V}$ one has that

$$|r(a,b)|^2 \le p(a,a)q(b,b).$$

Theorem 3.1 ([4, Theorem 2.1]). *Let p and q be quadratic forms on $\mathcal{V}$ and let S be the space of quadratic forms dominated by p and q. The geometric mean of the forms p and q satisfies*

$$\gamma^{1/2}_{[p,q]}(a,a) = \sup_{r\in S} r(a,a), \quad a \in \mathcal{V}.$$

Proposition 1. *Let, p, p', q and q' be quadratic forms on $\mathcal{V}$ such that for all $a \in \mathcal{V}$ one has that $p(a,a) \ge p'(a,a)$ and $q(a,a) \ge q'(a,a)$. Then, their respective geometric means satisfy*

$$\gamma^{1/2}_{[p,q]}(a,a) \ge \gamma^{1/2}_{[p',q']}(a,a), \quad a \in \mathcal{V}.$$

Proof. From the definition of geometric mean and the Cauchy–Schwarz inequality, it follows that the geometric mean of the quadratic forms p' and q' is dominated by p' and q' and therefore we have for all $a, b \in \mathcal{V}$ that

$$\gamma^{1/2}_{[p',q']}(a,b) \le p'(a,a)q'(b,b) \le p(a,a)q(b,b).$$

This shows that $\gamma^{1/2}_{[p',q']}$ is dominated by p and q and therefore Theorem 3.1 implies that

$$\gamma^{1/2}_{[p',q']}(a,a) \le \gamma^{1/2}_{[p,q]}(a,a), \quad a \in \mathcal{V},$$

as we wanted to show. □

We will show now that this property extends to the full interpolation.

Proposition 2. *Let $\gamma^t_{[p,q]}$ and $\gamma^t_{[p',q']}$ be quadratic interpolations such that for all $a \in \mathcal{V}$ one has that $p(a,a) \ge p'(a,a)$ and $q(a,a) \ge q'(a,a)$. Then,*

$$\gamma^t_{[p,q]}(a,a) \ge \gamma^t_{[p',q']}(a,a), \quad t \in [0,1], \ a \in \mathcal{V}.$$

Proof. Having into account that $\gamma^0_{[p,q]} = p$ and $\gamma^1_{[p,q]} = q$ and respectively for the quadratic interpolation of p' and q', Proposition 1 shows the result for $t = 1/2$. From the interpolation property, see Eq. (3.1), and the definition of geometric mean, we have that

$$\gamma^{1/2}_{[\gamma^{t_1}_{[p,q]}, \gamma^{t_2}_{[p,q]}]} = \gamma^{(t_1+t_2)/2}_{[p,q]}, \quad t_1, t_2 \in [0,1].$$

Therefore, we can iteratively prove the result for the middle point of all the successive bisections of the interval, which is a dense subset. Note that the functions $t \mapsto \gamma^t_{[p,q]}(a,a)$ and $t \mapsto \gamma^t_{[p',q']}(a,a)$, $a \in \mathcal{V}$, are continuous. Let $t_0 \in [0,1]$. For every $\epsilon > 0$, there exists an interval $[s_1, s_2]$ containing t_0 such that $\gamma^{s_1}_{[p,q]}(a,a) \geq \gamma^{t_0}_{[p',q']}(a,a)$, $|\gamma^{s_1}_{[p,q]}(a,a) - \gamma^{t_0}_{[p,q]}(a,a)| < \epsilon$ and $|\gamma^{s_1}_{[p',q']}(a,a) - \gamma^{t_0}_{[p',q']}(a,a)| < \epsilon$. Hence, we have

$$\gamma^{t_0}_{[p,q]}(a,a) + \epsilon \geq \gamma^{s_1}_{[p,q]}(a,a) \geq \gamma^{s_1}_{[p',q']}(a,a) \geq \gamma^{t_0}_{[p',q']}(a,a) - \epsilon.$$

Since this is true for every epsilon, the proof is complete. □

Consider now that $\mathcal{V}$ and $\mathcal{V}'$ are two vector spaces and let $\Phi \colon \mathcal{V}' \to \mathcal{V}$ be a linear mapping. Given a quadratic form $p : \mathcal{V} \times \mathcal{V} \to \mathbb{C}$, the linear map Φ induces a new quadratic form $\Phi^\dagger p$ on $\mathcal{V}'$ by pull-back:

$$\Phi^\dagger p \colon \mathcal{V}' \times \mathcal{V}' \to \mathbb{C}$$
$$a', b' \mapsto p(\phi(a'), \phi(b'))^{\cdot}$$

Proposition 3. *Let $\mathcal{V}$ and $\mathcal{V}'$ be two linear vector spaces and let $\Phi \colon \mathcal{V}' \to \mathcal{V}$ be a liner map. Let p and q be positive, Hermitian quadratic forms over $\mathcal{V}$. Then, the pull-backs of the quadratic interpolations satisfy*

$$\Phi^\dagger \gamma^t_{[p,q]}(a',a') \leq \gamma^t_{[\Phi^\dagger p, \Phi^\dagger q]}(a',a'), \quad a' \in \mathcal{V}'.$$

Proof. Let r be a positive, Hermitian quadratic form dominated by p and q and note that the inequality $|r(a,b)|^2 \leq p(a,a)q(b,b)$, $a,b \in \mathcal{V}$ implies that

$$|\Phi^\dagger r(a',b')|^2 \leq \Phi^\dagger p(a',a')\Phi^\dagger q(b',b'), \quad a',b' \in \mathcal{V}'.$$

By the maximality property of the geometric mean, Theorem 3.1, one has that for every such r

$$\Phi^\dagger r(a',a') \leq \gamma^{1/2}_{[\Phi^\dagger p, \Phi^\dagger q]}(a',a').$$

In particular, this holds for $r = \gamma_{[p,q]}^{1/2}$ and this proves the statement for $t = 1/2$. The cases $t = 0$ and $t = 1$ are trivial. Repeating the final part of the argument in the proof of Proposition 2 finishes the proof. $\qquad\square$

We are now ready to define the relative entropy between two states on a unital C^*-algebra. This is the definition originally introduced by A. Uhlmann.[5] In this section and for the rest of this chapter, the generic vector spaces $\mathcal{V}$ of the previous section are going to be provided by the vector space structure of the C^*-algebra. A state ω on a unital C^*-algebra $\mathcal{A}$ is a real, normalized, positive, linear functional $\omega \colon \mathcal{A} \to \mathbb{C}$, i.e., for $\lambda, \mu \in \mathbb{C}$, $a, b \in \mathcal{A}$ and $e \in \mathcal{A}$, the identity element in the algebra

   (i) $\omega(\lambda a + \mu b) = \lambda\omega(a) + \mu\omega(b)$,
  (ii) $\omega(a^*a) \geq 0$,
 (iii) $\omega(a^*) = \overline{\omega(a)}$,
 (iv) $\omega(e) = 1$.

In terms of any state on a C^*-algebra, one can define two Hermitian, positive quadratic forms as follows:

$$\omega^R(a, b) = \omega(ba^*), \quad a, b \in \mathcal{A},$$

$$\omega^L(a, b) = \omega(a^*b), \quad a, b \in \mathcal{A}.$$

Definition 4. Let ω, ν be two states on a C^*-algebra $\mathcal{A}$. And let $\gamma_{[\omega^R, \nu^L]}^t$ be the quadratic interpolation of the forms ω^R and ν^L. The **relative entropy functional** between the states ω, ν is defined by

$$S_{[\omega,\nu]}(a, b) = -\liminf_{t \to 0^+} \frac{1}{t}\left(\gamma_{[\omega^R, \nu^L]}^t(a, b) - \omega^R(a, b)\right), \quad a, b \in \mathcal{A}.$$

Definition 5. Let ω, ν be two states on a C^*-algebra $\mathcal{A}$. The **relative entropy** between the states ω and ν is the evaluation on the identity of the relative entropy functional:

$$S[\omega, \nu] = S_{[\omega,\nu]}(e, e).$$

Next, we are going to provide an example with the connection of this definition of relative entropy with von Neumann's. Suppose that we are given two density matrices $\hat{\omega}, \hat{\nu} \colon \mathbb{C}^n \to \mathbb{C}^n$, that is, positive, Hermitian matrices with trace one. For simplicity, we are going to assume that they are strictly positive definite. They define respectively, by means of the trace, two linear functionals ω, ν on $\mathcal{A} = M(\mathbb{C}^n)$. That is, for $a \in M(\mathbb{C}^n)$,

$$\omega(a) = \mathrm{Tr}(\hat{\omega}a), \quad \nu(a) = \mathrm{Tr}(\hat{\nu}a).$$

The quadratic forms associated with them are therefore

$$\omega^R(a,b) = \mathrm{Tr}(\hat{\omega}ba^*), \quad \nu^L(a,b) = \mathrm{Tr}(\hat{\nu}a^*b), \quad a,b \in M(\mathbb{C}^n).$$

We are going to obtain a compatible representation, cf. Section 2, for these quadratic forms on $\mathcal{H} = M(\mathbb{C}^n)$ with Hilbert–Schmidt scalar product,

$$\langle a, b \rangle = \mathrm{Tr}(a^*b).$$

We have that

$$\omega^R(a,b) = \mathrm{Tr}(\hat{\omega}ba^*) = \mathrm{Tr}(a^*\hat{\omega}b) = \langle a, \hat{\omega}b \rangle = \langle a, L_\omega b \rangle,$$

where L_ω is the operator of left-multiplication on the algebra. Equivalently, we have

$$\nu^L(a,b) = \mathrm{Tr}(\hat{\nu}a^*b) = \mathrm{Tr}(a^*b\hat{\nu}) = \langle a, b\hat{\nu} \rangle = \langle a, R_\nu b \rangle,$$

where R_ν is the operator of right-multiplication on the algebra. Note that L_ω and R_ν are self-adjoint, commuting operators and therefore define a compatible representation for the forms ω and ν. The interpolation of quadratic forms becomes in this case

$$\gamma^t_{[\omega^R,\nu^L]}(a,b) = \mathrm{Tr}(a^* L_\omega^{1-t} R_\nu^t b) = \mathrm{Tr}(a^*\hat{\omega}^{1-t}b\hat{\nu}^t).$$

Applying now the definition of the relative entropy, we get

$$S[\omega,\nu] = -\liminf_{t\to 0} \frac{1}{t}\left(\mathrm{Tr}(\hat{\omega}^{1-t}\hat{\nu}^t - \mathrm{Tr}(\hat{\omega}))\right).$$

This limit is minus the derivative of the first summand evaluated at zero, if it exists. A straightforward calculation shows that

$$S[\omega,\nu] = -\mathrm{Tr}(\hat{\omega}\ln\hat{\nu}) + \mathrm{Tr}(\hat{\omega}\ln\hat{\omega}),$$

which is the expression of von Neumann's relative entropy.

4. Proof of the Monotonicity of the Relative Entropy

We are going to prove in this section the monotonicity of the relative entropy under completely positive and unital (with trace preserving dual) maps. In fact, the prove is slightly more general and proves monotonicity under Schwarz maps. Let $\Phi : \tilde{\mathcal{A}} \to \mathcal{A}$ be a linear map between $*$-algebras with the following properties:

- $\Phi(a^*) = \Phi(a)^*$,
- $\Phi(a^*)\Phi(a) \leq \Phi(a^*a)$.

Given a state defined on $\mathcal{A}$, $\omega : \mathcal{A} \to \mathbb{C}$, the map Φ induces a state in $\tilde{\mathcal{A}}$ in the following way:

$$\omega_\Phi : \tilde{A} \to \mathbb{C}$$
$$\tilde{a} \mapsto \omega(\Phi(\tilde{a})).$$

This state satisfies

$$\omega_\Phi(\tilde{a}^*\tilde{a}) = \omega(\Phi(\tilde{a}^*\tilde{a})) \geq \omega(\Phi(\tilde{a})^*\Phi(\tilde{a})) \geq 0,$$

which proves that a Schwarz map is a positive map. It is a known result that a completely positive map that is trace non-increasing, i.e., $e_\mathcal{A} \geq \Phi(e_{\tilde{\mathcal{A}}})$, is Schwarz. We will prove the monotonicity of the relative entropy under Schwarz maps that are trace preserving.

Theorem 4.1 (Monotonicity of the relative entropy). *Let $\tilde{\mathcal{A}}$ and $\mathcal{A}$ be unital C^*-algebras and $\Phi : \tilde{\mathcal{A}} \to \mathcal{A}$ be a unital (with trace preserving dual) Schwarz map. Let ω, ν be positive, linear functionals on $\mathcal{A}$. Then,*

$$S[\omega, \nu] \geq S[\omega_\Phi, \nu_\Phi].$$

Before we proceed with the proof, it is important to remark that $\Phi^\dagger \omega^R \neq \omega_\Phi^R$. Indeed, on one hand, we have for $\tilde{a} \in \tilde{\mathcal{A}}$ that

$$\Phi^\dagger \omega^R(\tilde{a}, \tilde{a}) = \omega^R(\Phi(\tilde{a}), \Phi(\tilde{a})) = \omega(\Phi(a)\Phi(a)^*).$$

On the other hand,

$$\omega_\Phi^R(\tilde{a}, \tilde{a}) = \omega_\Phi(\tilde{a}\tilde{a}^*) = \omega(\Phi(aa^*)).$$

Hence, only in the case that Φ is a $*$-homomorphism, we get the equality.

Proof. Let e and $\tilde{e}$ be the identity elements in $\mathcal{A}$ and $\tilde{\mathcal{A}}$, respectively. By definition, we have that

$$S[\omega, \nu] = -\liminf_{t \to 0^+} \frac{1}{t} \left(\gamma^t_{[\omega^R, \nu^L]}(e, e) - \omega^R(e, e) \right),$$

$$S[\omega_\Phi, \nu_\Phi] = -\liminf_{t \to 0^+} \frac{1}{t} \left(\gamma^t_{[\omega^R_\Phi, \nu^L_\Phi]}(\tilde{e}, \tilde{e}) - \omega^R_\Phi(\tilde{e}, \tilde{e}) \right).$$

Since the map is unital, we have the following:

$$\omega^R(e, e) = \omega(ee^*) = \omega(e) = \omega(\Phi(\tilde{e})) = \omega(\Phi(\tilde{e}\tilde{e}^*)) = \omega_\Phi(\tilde{e}\tilde{e}^*) = \omega^R_\Phi(\tilde{e}, \tilde{e}).$$

Therefore, it is enough to show that

$$\gamma^t_{[\omega^R, \nu^L]}(e, e) = \gamma^t_{[\omega^R, \nu^L]}(\Phi(\tilde{e}), \Phi(\tilde{e})) \leq \gamma^t_{[\omega^R_\Phi, \nu^L_\Phi]}(\tilde{e}, \tilde{e}).$$

In particular, this will hold if

$$\gamma^t_{[\omega^R, \nu^L]}(\Phi(\tilde{a}), \Phi(\tilde{a})) \leq \gamma^t_{[\omega^R_\Phi, \nu^L_\Phi]}(\tilde{a}, \tilde{a}), \quad \forall \tilde{a} \in \tilde{\mathcal{A}}.$$

Using the Schwarz property of the map Φ, we get that

$$\begin{aligned}
\Phi^\dagger \omega^R(\tilde{a}, \tilde{a}) &= \omega^R(\Phi(\tilde{a}), \Phi(\tilde{a})) \\
&= \omega(\Phi(\tilde{a})\Phi(\tilde{a})^*) \\
&\leq \omega(\Phi(\tilde{a}\tilde{a}^*)) \\
&= \omega_\Phi(\tilde{a}\tilde{a}^*) \\
&= \omega^R_\Phi(\tilde{a}, \tilde{a}),
\end{aligned}$$

and equivalently for ν^L.

Now, one can use Proposition 2 and Proposition 3 to get

$$\gamma^t_{[\omega^R, \nu^L]}(\Phi(\tilde{a}), \Phi(\tilde{a})) = \Phi^\dagger \gamma^t_{[\omega^R, \nu^L]}(\tilde{a}, \tilde{a}) \leq \gamma^t_{[\Phi^\dagger \omega^R, \Phi^\dagger \nu^L]}(\tilde{a}, \tilde{a}) \leq \gamma^t_{[\omega^R_\Phi, \nu^L_\Phi]}(\tilde{a}, \tilde{a}),$$

which completes the proof. $\qquad\square$

The conditions of the theorem apply in the particular case that $\tilde{\mathcal{A}} \subset \mathcal{A}$ is a unital $*$-subalgebra and $\Phi : \tilde{\mathcal{A}} \to \mathcal{A}$ is the injection map. Since the injection is a $*$-homomorphism and unital, it is a completely positive and trace preserving map. This means that the restriction of the observables to a subalgebra decreases the relative entropy.

References

1. A. P. Balachandran, T. R. Govindarajan, A. R. de Queiroz and A. F. Reyes-Lega, Algebraic approach to entanglement and entropy, *Phys. Rev. A* **88** (2) (2013) 022301.
2. A. P. Balachandran, T. R. Govindarajan, A. R. de Queiroz and A. F. Reyes-Lega, Entanglement and particle identity: A unifying approach, *Phys. Rev. Lett.* **110** (8) (2013) 080503.
3. A. P. Balachandran, A. R. de Queiroz and S. Vaidya, Quantum entropic ambiguities: Ethylene, *Phys. Rev. D* **88** (2) (2013) 025001.
4. W. Pusz and S. L. Woronowicz, Functional Calculus for sesquilinear forms and the purification map, *Rep. Math. Phys.* **8** (1975) 159–170.
5. A. Uhlmann, Relative entropy and the Wigner-Yanase-Dyson-Lieb concavity in an interpolation theory, *Comm. Math. Phys.* **54** (1977) 21–32.

Chapter 15

Noncommutative AdS_2 I: Exact Solutions

Aleksandr Pinzul[*,‡] and Allen Stern[†,§]

*Universidade de Brasília, Instituto de Física
and International Center of Physics
70910-900, Brasília, DF, Brasil
†Department of Physics, University of Alabama,
Tuscaloosa, Alabama 35487, USA
‡apinzul@unb.br
§astern@ua.edu

We study the exact solutions of both, massless and massive, scalar field theories on the noncommutative AdS_2. We also discuss some important limits in order to compare with known results.

1. Introduction

The study of noncommutative spaces and field theories on them has a great number of motivations and applications, ranging from quantum Hall effect[1] to superstring theory.[2] This, of course, is in addition to the purely mathematical interest in noncommutative geometry.[3] Of special interest is the study of noncommutative spaces that preserve all the isometries of their commutative counterparts, one of the most famous examples being the so-called fuzzy sphere.[4] Such spaces serve as good models to simulate the ultraviolet properties of field theories due to universal quantum gravity effects (see Refs. 5, 6 for a general model independent argument for the appearance of the space-time noncommutativity as a quasi-classical quantum gravitational effect and, e.g., Ref. 7 for a review of some recent studies of a scalar field theory on a fuzzy sphere).

In this chapter, we study the exact solutions to the dynamics of a free scalar field on noncommutative (Euclidean) AdS_2 ($ncAdS$). This space is the non-compact cousin of the fuzzy sphere. One of the main motivations to study such a model, aside from those briefly mentioned above, is its importance to the celebrated AdS/CFT correspondence,[8] specifically to the 2-dimensional version, AdS_2/CFT_1. If one believes that the correspondence holds in the strong sense, i.e., including at the quantum gravity scale, then, from the observation that noncommutativity models quasi-classical quantum gravitational effects, it follows that there should be some form of correspondence between a field theory on $ncAdS$ and some conformal theory on the one-dimensional boundary of the AdS space. This strategy was adopted in Refs. 9, 10 where strong perturbative evidence in favor of such a correspondence was found. The results of Ref. 11 suggest that this also should be true in higher dimensions. The work Ref. 12 was the first attempt to go beyond the perturbative (in the noncommutative parameter) analysis. For the case of a free massless scalar field, the exact solutions for the dynamics were found, and it was shown that the exact 2-point correlator of boundary fields obtained via the AdS/CFT correspondence prescription is conformally invariant.

In this chapter, we develop another approach for the construction of the exact solutions of a free scalar theory on $ncAdS$, which allows for an extension of the results[12] to the massive case. This, in turn, allows for a more general test of the correspondence. This development is reported in another contribution to the volume.[13]

The plan of this chapter is as follows. In Section 2, we introduce our model, as well as give some technical details. The general construction of the exact solutions is done in Section 3, while Section 4 discusses some important limits, as well as reproduces the results of Ref. 12 for the massless case. In Section 5, we discuss possible future developments of the project.

2. The Model

In this section, we briefly introduce our model. For more details, see Refs. 9, 10, 12.

Noncommutative Euclidean AdS_2 is obtained by quantizing the $SU(1,1)$ invariant Poisson structure of commutative AdS_2. Commutative Euclidean AdS_2 ($EAdS_2$) can be defined in terms of imbedding coordinates X^a, $a = 0, 1, 2$, spanning $\mathbb{R}^{1,2}$ and satisfying

$$\eta_{ab}X^a X^b = -1, \tag{2.1}$$

where the metric on $\mathbb{R}^{1,2}$ is $[\eta_{ab}] = \mathrm{diag}(-1, +1, +1)$. The corresponding $SU(1,1)$-invariant Poisson structure is given by

$$\{X^a, X^b\} = \epsilon^{abc} X_c, \tag{2.2}$$

where raising/lowering of indices is done with the help of η_{ab}, and ϵ^{abc} is totally antisymmetric with $\epsilon^{012} = 1$. The field equations for a free massive scalar field Φ on $EAdS$ can be compactly written in terms of this structure:

$$\{X^a, \{X_a, \Phi\}\} = m^2 \Phi. \tag{2.3}$$

To study the AdS/CFT correspondence, it is convenient to also introduce local coordinates. The so-called Fefferman–Graham coordinates, (t, z), can be defined in terms of the imbedding coordinates as[a]

$$z = \frac{1}{X^2 - X^0}, \quad t = -\frac{X^1}{X^2 - X^0} \equiv -zX^1, \tag{2.4}$$

and the Poisson structure (2.2) takes the form

$$\{t, z\} = z^2.$$

The quantization of the model, preserving the constraint (2.1), is rather straightforward (see Refs. 9, 10, 12 for details). Essentially, it amounts to promoting the imbedding coordinates X^a to operators $\hat{X}^a$ satisfying the quantized version of the Poisson structure

$$[\hat{X}^a, \hat{X}^b] = i\alpha \epsilon^{abc} \hat{X}_c,$$

where α is the parameter of noncommutativity, which is related to the quantum gravitational scale. The noncommutative version of the scalar field equation (2.3) becomes[12]

$$\frac{\alpha^2}{2} \hat{\Delta} \hat{\Phi} = \hat{X}_a \hat{\Phi} \hat{X}^a + \hat{\Phi} = \frac{\alpha^2 m^2}{2} \hat{\Phi}, \tag{2.5}$$

where $\hat{\Delta} := [\hat{X}^a, [\hat{X}_a, \cdot\,]]$ is the noncommutative Laplacian.

---

[a]Here, we restrict to the single-sheeted hyperboloid with $X^0 \leq -1$, which implies $z > 0$ and so (t, z) span the half-plane.

3. Exact Solutions

To find the solutions of (2.5), we will first re-write (2.5) in terms of the quantized Fefferman–Graham coordinates, $(\hat{t}, \hat{z})$, which are given by the quantization of (2.4)[b]

$$\hat{z} = \frac{1}{\hat{X}^2 - \hat{X}^0}, \quad \hat{t} = -\frac{1}{2}(\hat{z}\hat{X}^1 + \hat{X}^1\hat{z}).$$

The only non-trivial step is the inversion of these relations to obtain $\hat{X}^a$ as a function of $\hat{z}$ and $\hat{t}$. It is more convenient to formally introduce the quantized "radial" coordinate, $\hat{r} = \hat{z}^{-1}$, satisfying a simple, canonical, commutation relation

$$[\hat{t}, \hat{r}] = -i\alpha. \tag{3.1}$$

Then, the noncommutative imbedding coordinates can be written in terms of the local ones as[12]

$$\hat{X}^0 = -\frac{1}{2}(\kappa^2\hat{r}^{-1} + \hat{t}\hat{r}\hat{t} + \hat{r}), \hat{X}^1 = -\frac{1}{2}(\hat{r}\hat{t} + \hat{t}\hat{r}),$$

$$\hat{X}^2 = -\frac{1}{2}(\kappa^2\hat{r}^{-1} + \hat{t}\hat{r}\hat{t} - \hat{r}), \tag{3.2}$$

where the non-trivial deformation parameter, $\kappa = \sqrt{1 + \frac{\alpha^2}{4}}$, is necessary to preserve the quantum version of (2.1).

Using (3.2) in (2.5), one gets

$$\begin{aligned}
2\alpha^2\hat{\Delta}\hat{\Phi} &= -(\kappa^2\hat{r}^{-1} + \hat{t}\hat{r}\hat{t} + \hat{r})\hat{\Phi}(\kappa^2\hat{r}^{-1} + \hat{t}\hat{r}\hat{t} + \hat{r}) \\
&\quad + (\hat{r}\hat{t} + \hat{t}\hat{r})\hat{\Phi}(\hat{r}\hat{t} + \hat{t}\hat{r}) \\
&\quad + (\kappa^2\hat{r}^{-1} + \hat{t}\hat{r}\hat{t} - \hat{r})\hat{\Phi}(\kappa^2\hat{r}^{-1} + \hat{t}\hat{r}\hat{t} - \hat{r}) + 4\hat{\Phi} \\
&= (\hat{r}\hat{t} + \hat{t}\hat{r})\hat{\Phi}(\hat{r}\hat{t} + \hat{t}\hat{r}) - 2\left(\hat{t}\hat{r}\hat{t}\hat{\Phi}\hat{r} + \hat{r}\hat{\Phi}\hat{t}\hat{r}\hat{t}\right) \\
&\quad - 2\kappa^2\left(\hat{r}^{-1}[\hat{\Phi}, \hat{r}] + [\hat{r}, \hat{\Phi}]\hat{r}^{-1}\right) - \alpha^2\hat{\Phi}.
\end{aligned}$$

This can be simplified further by using (3.1)

$$\alpha^2\hat{\Delta}\hat{\Phi} = -[\hat{t}, \hat{r}[\hat{t}, \hat{\Phi}]\hat{r}] - \kappa^2\hat{r}^{-1}[\hat{r}, [\hat{r}, \hat{\Phi}]]\hat{r}^{-1},$$

---

[b]In, Ref. 9 we showed that $\hat{X}^2 - \hat{X}^0$ has a positive spectrum for the discrete series irreducible representation of $su(1,1)$, which exactly corresponds to the noncommutative Euclidean *AdS*.

so the full noncommutative dynamics will be given by

$$-[\hat{t}, \hat{r}[\hat{t}, \hat{\Phi}]\hat{r}] - \kappa^2 \hat{r}^{-1}[\hat{r}, [\hat{r}, \hat{\Phi}]]\hat{r}^{-1} = \alpha^2 m^2 \hat{\Phi}$$

or

$$-\hat{r}[\hat{t}, \hat{r}[\hat{t}, \hat{\Phi}]\hat{r}]\hat{r} - \kappa^2[\hat{r}, [\hat{r}, \hat{\Phi}]] = \alpha^2 m^2 \hat{r}\hat{\Phi}\hat{r}. \tag{3.3}$$

To further simplify the equation of motion, we note that a trivial consequence of the commutation relation (3.1) is

$$[\hat{r}, T] = i\alpha\dot{T}, \quad [\hat{t}, R] = -i\alpha R', \tag{3.4}$$

where $T = T(\hat{t})$, $R = R(\hat{r})$, $\dot{T} = \frac{d}{dt}T(t)|_{t=\hat{t}}$ and $R' = \frac{d}{dr}R(r)|_{r=\hat{r}}$. Then, it is easy to see that the following ansatz for $\hat{\Phi}$

$$\hat{\Phi} = T(\hat{t})R(\hat{r})T(\hat{t}) \quad \text{with} \quad \dot{T} = i\beta T, \tag{3.5}$$

for any β (it will be necessary to restrict to real β, see the following), separates variables in (3.3) in the following sense. Plugging (3.5) into (3.3) and using (3.4), one immediately gets

$$T(\hat{r} - \alpha\beta)\frac{d}{dr}\left((\hat{r} - \alpha\beta)R'(\hat{r} + \alpha\beta)\right)(\hat{r} + \alpha\beta)T - 4\kappa^2\beta^2 TRT$$

$$= m^2 T(\hat{r} - \alpha\beta)R(\hat{r} + \alpha\beta)T. \tag{3.6}$$

Using that T is invertible (actually from (3.5), $T = \exp(i\beta\hat{t})$) and any two functions of $\hat{r}$ commute, (3.6) can be written as[c]

$$[(\alpha^2\beta^2 - \hat{r}^2)R']' + \left[m^2 - \frac{4\kappa^2\beta^2}{\alpha^2\beta^2 - \hat{r}^2}\right]R = 0. \tag{3.7}$$

For $\alpha\beta \neq 0$, (3.7) is essentially the general Legendre equation

$$[(1 - x^2)L']' + \left[\nu(\nu + 1) - \frac{\mu^2}{1 - x^2}\right]L = 0$$

---

[c]We also have to assume that, at least formally, $\alpha^2\beta^2 - \hat{r}^2$ is invertible too. This could be achieved by adding an infinitesimal imaginary part, which will amount to choosing the correct analytic continuation, see eq. (3.8).

that has eight standard solutions (of course, only two of them being linearly independent). For the case of (3.7), the solutions can be written as[14]

$$R_{1\pm} = P_{\nu-\frac{1}{2}}^{\pm\frac{2}{\alpha}\kappa}\left(\frac{\hat{r}}{\alpha\beta}\right), \quad R_{2\pm} = Q_{\nu-\frac{1}{2}}^{\pm\frac{2}{\alpha}\kappa}\left(\frac{\hat{r}}{\alpha\beta}\right),$$

$$R_{3\pm} = P_{-\nu-\frac{1}{2}}^{\pm\frac{2}{\alpha}\kappa}\left(\frac{\hat{r}}{\alpha\beta}\right), \quad R_{4\pm} = Q_{-\nu-\frac{1}{2}}^{\pm\frac{2}{\alpha}\kappa}\left(\frac{\hat{r}}{\alpha\beta}\right),$$

$$\tag{3.8}$$

where $\nu = \sqrt{m^2 + \frac{1}{4}}$.

For $\alpha = 0$, after a change of the variable (which now can be thought as the commutative one), $r = 1/z$, and substituting $R = \sqrt{z}\tilde{R}$, Eq. (3.7) becomes the modified Bessel's equation

$$z^2\frac{d^2}{dz^2}\tilde{R} + z\frac{d}{dz}\tilde{R} - (m^2 + \frac{1}{4} + 4\beta^2 z^2)\tilde{R} = 0,$$

which leads to the well-known commutative solution[15]

$$R = \sqrt{z}K_\nu(2\beta z), \quad \nu = \sqrt{m^2 + \frac{1}{4}}, \tag{3.9}$$

where K_ν is the modified Bessel function of the second type (the other solution given in terms of the modified Bessel function of the first type, I_ν, exponentially blows up when $z \to \infty$, so it does not satisfy the asymptotic regularity condition).

In order to better understand the above results, we next examine some important limits.

4. Special Cases

Here, we would like to study two important limits of the general result (3.8): the commutative limit and the massless one.

4.1. Commutative limit

A more careful study of the commutative limit is necessary to understand which of the solutions in (3.8) should be taken in the fully noncommutative case. Formally, the limit corresponds to sending α to zero. This limit is somewhat tricky because it corresponds to simultaneously sending both, the upper index and the argument, to infinity in either of the solutions (3.8). It is more convenient to do this using the integral representations. By doing so, we will show that the correct choice, i.e., the one that has the correct commutative limit (3.9), is R_{1-}.

The well-known integral representation of $P_\lambda^{-\mu}(x)$, valid for $\Re\mu > \Re\lambda > -1$ and $x > 1$, is given by[14]

$$P_\lambda^{-\mu}(x) = \frac{(x^2-1)^{\mu/2}}{2^\lambda \Gamma(\mu-\lambda)\Gamma(\lambda+1)} \int_0^\infty \frac{(\sinh t)^{2\lambda+1}}{(x+\cosh t)^{\mu+\lambda+1}} dt. \qquad (4.1)$$

Applying this to $P_\lambda^{-\mu}(\mu x)$, we get

$$P_\lambda^{-\mu}(\mu x) = \frac{\mu^{-\lambda-1}x^{-\lambda-1}(1-\frac{1}{\mu^2 x^2})^{\mu/2}}{2^\lambda \Gamma(\mu-\lambda)\Gamma(\lambda+1)} \int_0^\infty \frac{(\sinh t)^{2\lambda+1}}{(1+\frac{\cosh t}{\mu x})^{\mu+\lambda+1}} dt.$$

We can obtain the $\mu \to \infty$ limit with the help of the known asymptotics

$$(1+y\epsilon)^{1/\epsilon} = e^y + \mathcal{O}(\epsilon),$$

$$\Gamma(z) = \sqrt{\frac{2\pi}{z}} \left(\frac{z}{e}\right)^z \left(1 + \mathcal{O}\left(\frac{1}{z}\right)\right). \qquad (4.2)$$

It is straightforward to get

$$\Gamma(\mu+1)P_\lambda^{-\mu}(\mu x) = \frac{x^{-\lambda-1}}{2^\lambda \Gamma(\lambda+1)} \int_0^\infty e^{-\frac{\cosh t}{x}} (\sinh t)^{2\lambda+1} dt + \mathcal{O}\left(\frac{1}{\mu}\right), \qquad (4.3)$$

where we used $\frac{\Gamma(\mu+1)}{\Gamma(\mu-\lambda)} = \mu^{\lambda+1}\left(1+\mathcal{O}\left(\frac{1}{\mu}\right)\right)$, which follows from (4.2).

The asymptotic result (4.3) should be compared to the integral representation for $K_\nu(x)$ (valid for the same range of the parameter and argument)[16]

$$K_\nu(x) = \frac{\sqrt{\pi}x^\nu}{2^\nu \Gamma(\nu+\frac{1}{2})} \int_0^\infty e^{-x\cosh t}(\sinh t)^{2\nu} dt.$$

It is immediately seen that by taking $\lambda = \nu - \frac{1}{2}$, we have

$$\Gamma(\mu+1)P_{\nu-\frac{1}{2}}^{-\mu}(\mu x) = \sqrt{\frac{2}{\pi}}\sqrt{\frac{1}{x}}K_\nu\left(\frac{1}{x}\right) + \mathcal{O}\left(\frac{1}{\mu}\right).$$

Upon identifying x with $\frac{1}{2\beta z}$ and μ with $\frac{2}{\alpha}\kappa$, this result shows that the R_{1-} solution in (3.8) has the correct commutative limit (3.9). Incidentally, this also shows that to get the correct limit, R_{1-} should be normalized by a factor of $\Gamma\left(\frac{2}{\alpha}\kappa + 1\right)$. As a result, the full exact noncommutative solution in the free massive case takes the form

$$\hat{\Phi} \sim \Gamma\left(\frac{2}{\alpha}\kappa + 1\right) e^{i\beta\hat{t}} \, P_{\nu-\frac{1}{2}}^{-\frac{2}{\alpha}\kappa}\left(\frac{\hat{r}}{\alpha\beta}\right) e^{i\beta\hat{t}}. \qquad (4.4)$$

Another consistency check of (4.4) can be done by observing that near the boundary ($r \to \infty$, where now r should be thought of as a symbol of $\hat{r}$) of noncommutative AdS_2, the space effectively becomes commutative.[9, 10] Then, the near-boundary behavior of (4.4) should reproduce the known result for the commutative field.[15]

From Ref. 17 the asymptotic expansion for $P_{\nu-1/2}^{-\mu}(x)$ as $x \to \infty$ is

$$P_{\nu-1/2}^{-\mu}(x) = \left\{ \frac{2^{\nu-1/2}\Gamma(\nu)}{\sqrt{\pi}\Gamma(\mu + \nu + 1/2)} x^{\nu-1/2} \right.$$
$$\left. + \frac{\Gamma(-\nu)}{2^{\nu+1/2}\sqrt{\pi}\Gamma(\mu - \nu + 1/2)} x^{-\nu-1/2} \right\} (1 + \mathcal{O}(x^{-2})),$$
$$(\nu - 1/2) \neq \pm\frac{1}{2}, \pm\frac{3}{2}, \ldots.$$

The two terms in the braces are the dominant ones when $\nu - 1/2$ and $-\nu-1/2$ are greater than both $\nu-5/2$ and $-\nu-5/2$. Then, for $-1 < \nu < 1$ and $\nu \neq 0$ (the Breitenlohner–Freedman (BF) bound further restricts to $0 < \nu < 1$), defining $\Delta_\pm = \frac{1}{2} \pm \nu$, we obtain

$$P_{-\Delta_-}^{-\frac{2}{\alpha}\kappa}\left(\frac{r}{\alpha\beta}\right) \to \frac{2^{-\Delta_-}\Gamma(\frac{1}{2} - \Delta_-)}{\sqrt{\pi}\Gamma(\Delta_+ + \frac{2}{\alpha}\kappa)} (\alpha\beta z)^{\Delta_-} + \frac{2^{-\Delta_+}\Gamma(\Delta_- - \frac{1}{2})}{\sqrt{\pi}\Gamma(\Delta_- + \frac{2}{\alpha}\kappa)} (\alpha\beta z)^{\Delta_+},$$

as $z = \frac{1}{r} \to 0$ (here, z should be thought of as the symbol of $\hat{z}$), which agrees with the asymptotic behavior of the commutative scalar field.[15]

4.2. *Massless case*

The massless case corresponds to $m = 0$, $\alpha \neq 0$, i.e., $\nu = \frac{1}{2}$. We can easily evaluate the integral (4.1) for $\lambda = \nu - \frac{1}{2} = 0$

$$P_0^{-\mu}(x) = \frac{(x^2 - 1)^{\mu/2}}{\Gamma(\mu)} \int_0^\infty \frac{\sinh t}{(x + \cosh t)^{\mu+1}} dt = \frac{1}{\Gamma(\mu + 1)} \left(\frac{x - 1}{x + 1}\right)^{\mu/2}. \qquad (4.5)$$

(The branch cut is taken in the usual way as to exclude the points $x = \pm 1$.) Actually, it is easily seen that in the massless case, two of the solutions in

(3.8) are allowed, $R_{1\pm}$, since passing from one to another is equivalent to replacing β by $-\beta$. Using (4.5), we see that the general massless solution is given by an arbitrary combination of (as before $\hat{z} = \hat{r}^{-1}$)

$$e^{i\beta\hat{t}} \left(\frac{1 \pm \alpha\beta\hat{z}}{1 \mp \alpha\beta\hat{z}} \right)^{\frac{\kappa}{\alpha}} e^{i\beta\hat{t}}. \tag{4.6}$$

Note that the factor of $\Gamma\left(\frac{2}{\alpha}\kappa + 1\right)$ in (4.4) exactly cancels out.

The result (4.6) should be compared with Ref. 12 where exact solutions for the massless case were studied directly. There it was found that $\hat{\Phi}$ is an arbitrary combination (if one does not require reality) of operator functions $F_\pm$

$$\hat{\Phi} = aF_+(\Xi_+) + bF_-(\Xi_-),$$

where $\Xi_\pm := \pm\kappa\hat{z} + i\hat{t}$ and a and b are complex coefficients. We should show that this is consistent with (4.6).

Let us choose the upper sign in (4.6) (the other case is treated identically). Then, using (3.1), we get

$$e^{i\beta\hat{t}} \left(\frac{1 + \alpha\beta\hat{z}}{1 - \alpha\beta\hat{z}} \right)^{\frac{\kappa}{\alpha}} e^{i\beta\hat{t}} = (1 + 2\alpha\beta\hat{z})^{\frac{\kappa}{\alpha}} e^{2i\beta\hat{t}}.$$

We will show that

$$(1 + 2\alpha\beta\hat{z})^{\frac{\kappa}{\alpha}} e^{2i\beta\hat{t}} = e^{2\beta(\kappa\hat{z} + i\hat{t})} \equiv e^{2\beta\Xi_+}, \tag{4.7}$$

which would prove that the solutions are equivalent.

From (3.1) follows the commutator $[\hat{t}, \hat{z}] = i\alpha\hat{z}^2$. Then, the Baker–Campbell–Hausdorff formula (or rather, the related Zassenhaus formula) gives

$$e^{2\beta(\kappa\hat{z} + i\hat{t})} e^{-2i\beta\hat{t}} = f(\hat{z}) \tag{4.8}$$

for some function f of only $\hat{z}$, which depends on β as a parameter. Taking the derivative of (4.8) with respect to β, we find

$$-2\alpha\hat{z}^2 f'(\hat{z}) + 2\kappa\hat{z}f(\hat{z}) = \frac{\partial}{\partial\beta}f(\hat{z}), \tag{4.9}$$

where $f'(\hat{z}) = \frac{\partial}{\partial z}f(z)|_{z=\hat{z}}$ as usual. The solution should satisfy the condition $f(\hat{z})|_{\beta=0} = 1$. It is easy to see that

$$f(\hat{z}) = (1 + 2\alpha\beta\hat{z})^{\frac{\kappa}{\alpha}}$$

satisfies (4.9) as well as the condition for $\beta = 0$. In this way, (4.7) is established uniquely, and the proof that the exact solutions from Ref. 12 are identical to the massless limit of (4.4) is complete.

5. Discussion

The fact that $ncAdS$ leaves all the isometries of AdS intact has always suggested that the free theory is exactly solvable. Yet, in the earlier works on $ncAdS$,[9, 10] only perturbative solutions of free massless[9] and massive[10] scalar field theories were constructed. In a recent paper,[12] we obtained the general form of the solutions for free massless scalar and fermionic theories. In the work presented here, using a completely different approach, we extended this result to the case of a massive scalar field. We believe that the extension to massive fermionic fields can also be done with the help of a map between the solutions in scalar and fermionic theories, which was found in Ref. 12.

The results obtained here are very important because knowledge of the exact (in the noncommutative parameter) free solutions is a necessary starting point for the perturbative treatment of interactions. In the presence of an interaction, there are two parameters: the noncommutativity parameter, α, and the coupling constant, λ. Without the exact free solution, one has to resort to a perturbative analysis in *both* parameters. This was the approach adopted in Ref. 10. Since only limited use of the isometry is made in this approach, the calculations are very involved and not intuitive. As a consequence, all the results were obtained only up to α^2 order. Now, having the exact noncommutative solution, we hope to significantly improve the construction of the perturbation theory in a manner that will be exact in the noncommutative parameter. To make progress in this direction, one has to generalize the notion of Green's function to the fully noncommutative setting. This problem is quite difficult both technically and conceptually and is the subject of our current research.

Another importance of the result of this work is its relation to nonperturbative checks of the AdS_2/CFT_1 correspondence. More details on this are discussed in Ref. 13.

References

1. L. Susskind, The Quantum Hall fluid and noncommutative Chern-Simons theory, [arXiv:hep-th/0101029 [hep-th]].

2. N. Seiberg and E. Witten, String theory and noncommutative geometry, *JHEP* **09** (1999) 032. doi: 10.1088/1126-6708/1999/09/032 [arXiv:hep-th/9908142 [hep-th]].

3. A. Connes, *Noncommutative Geometry*, Academic Press (1994).

4. J. Madore, The Fuzzy sphere, *Class. Quant. Grav.* **9** (1992) 69–88. doi: 10.1088/0264-9381/9/1/008.

5. S. Doplicher, K. Fredenhagen and J. E. Roberts, Space-time quantization induced by classical gravity, *Phys. Lett.* B **331** (1994) 39–44. doi: 10.1016/0370-2693(94)90940-7.

6. S. Doplicher, K. Fredenhagen and J. E. Roberts, The Quantum structure of space-time at the Planck scale and quantum fields, *Commun. Math. Phys.* **172** (1995) 187–220. doi: 10.1007/BF02104515 [arXiv:hep-th/0303037 [hep-th]].

7. M. Šubjaková and J. Tekel, Multitrace matrix models of fuzzy field theories, *PoS* **CORFU2019** (2020) 234. doi: 10.22323/1.376.0234 [arXiv:2006.13577 [hep-th]].

8. J. M. Maldacena, The Large N limit of superconformal field theories and supergravity, *Adv. Theor. Math. Phys.* **2** (1998) 231–252. doi: 10.1023/A:1026654312961 [arXiv:hep-th/9711200 [hep-th]].

9. A. Pinzul and A. Stern, Noncommutative AdS_2/CFT_1 duality: The case of massless scalar fields, *Phys. Rev.* D **96**(6) (2017) 066019. doi: 10.1103/PhysRevD.96.066019 [arXiv:1707.04816 [hep-th]].

10. F. R. de Almeida, A. Pinzul and A. Stern, Noncommutative AdS_2/CFT_1 duality: The case of massive and interacting scalar fields, *Phys. Rev.* D **100**(8) (2019) 086005. doi: 10.1103/PhysRevD.100.086005 [arXiv: 1907.07298 [hep-th]].

11. F. Lizzi, A. Pinzul, A. Stern and C. Xu, Asymptotic commutativity of quantized spaces: The case of $\mathbb{CP}^{p,q}$, *Phys. Rev.* D **102**(6) (2020) 065012. doi: 10.1103/PhysRevD.102.065012 [arXiv:2006.13204 [hep-th]].

12. A. Pinzul and A. Stern, Exact solutions for scalars and spinors on quantized Euclidean AdS2 space and the correspondence principle, *Phys. Rev.* D **104**(12) (2021) 126034. doi: 10.1103/PhysRevD.104.126034 [arXiv: 2106.13376 [hep-th]].

13. A. Pinzul and A. Stern, Noncommutative AdS_2 II: The correspondence principle, see the contribution to this volume (2023).

14. H. Bateman and A. Erdélyi, *Higher Transcendental Functions*, Vol. 1, California Institute of technology. Bateman Manuscript project, McGraw-Hill, New York, NY (1955).

15. D. Z. Freedman, S. D. Mathur, A. Matusis and L. Rastelli, Correlation functions in the CFT(d)/AdS(d+1) correspondence, *Nucl. Phys.* B **546** (1999) 96–118. doi: 10.1016/S0550-3213(99)00053-X [arXiv:hep-th/9804058 [hep-th]].

16. H. Bateman and A. Erdélyi, *Higher Transcendental Functions*, Vol. 2, California Institute of technology. Bateman Manuscript project, McGraw-Hill, New York, NY (1955).

17. I. S. Gradshteyn and I. M. Ryzhik, *Table of Integrals, Series, and Products*, 7th edition. Elsevier/Academic Press, Amsterdam (2007).

https://doi.org/10.1142/9789811270437_0016

Chapter 16

The Mass Hyperboloid
as a Poisson–Lie Group

S. G. Rajeev[*] and Patrizia Vitale[†]

Department of Physics and Astronomy
Department of Mathematics
University of Rochester, Rochester, NY 14627, USA
Dipartimento di Fisica "Ettore Pancini",
Università di Napoli Federico II and INFN, Napoli, Italy
[*]*s.g.rajeev@rochester.edu*
[†]*patrizia.vitale@na.infn.it*

The light cone formalism of a massive scalar field has been shown by Dirac to have many advantages. But it is not manifestly Lorentz invariant. We will show that this is a feature not a bug: Lorentz invariance is indeed a symmetry but in a different sense as defined by Drinfel'd. The key idea is that the mass shell (mass hyperboloid) is a Poisson–Lie group: there is a non-abelian group multiplication and non-zero Poisson brackets between components of four-momentum. Rotations form the dual group of the hyperboloid in the sense of Drinfel'd. Infinitesimal Lorentz transformations form a Lie bi-algebra.

1. Introduction

Quantum groups, originally introduced in relation with quantum integrability more than 30 years ago (see, for example, Ref. 1), have acquired a central role in modern literature as appropriate symmetries of noncommutative space-time, noncommutative dynamics and noncommutative models of matter and gauge fields. All this activity is relevant in view of a consistent theory of quantum gravity which would imply noncommutative models of space-time as effective theories, at scales which are large in comparison with Planck length. Poisson–Lie groups are semiclassical approximations of quantum groups, where noncommutativity of the algebra of functions on the groups manifold is replaced by non-zero Poisson brackets. As such, they are mainly studied in the same perspective as above, namely to grasp hints on the deviation from the standard pseudo-Riemannian framework, which is foreseen by most of the QG models on the market.

In this contribution, we wish to show that these structures do not necessarily need to depart from standard physics in order to be detected but may play a role in conventional, undeformed dynamical systems. We will discuss a specific example, the Poisson–Lie group $SU(2)$ and its Drinfel'd double[2] $SL(2, \mathbb{C})$, namely the (double covering) Lorentz group, and show that they are intimately related with the light-front formalism of particle physics, being indeed the most natural framework where to describe the mass hyperboloid of massive scalar fields. The example will give us the opportunity to make explicit the geometric structures of Poisson–Lie groups, such as Poisson–Lie brackets, dressing actions and Poisson–Lie duality, and to place them into a familiar context.

2. The Hyperboloid

The four-momentum of a particle with unit mass belongs to one-sheet of a hyperboloid. There are equivalent descriptions of this manifold:

$$p_0^2 - p_1^2 - p_2^2 - p_3^2 = 1, \quad p_0 > 0 \iff p_+ p_- - (p_2^2 + p_3^2) = 1, \quad p_+ > 0,$$

where $p_+ = p_0 + p_1, p_- = p_0 - p_1$. We can solve for

$$p_- = \frac{1 + p_2^2 + p_3^2}{p_+}, \quad p_0 = \frac{1 + p_+^2 + p_2^2 + p_3^2}{2p_+}, \quad p_1 = \frac{-(1 + p_2^2 + p_3^2) + p_+^2}{2p_+}$$

in terms of p_+, p_2, p_3. Let us refer to the latter as $\mathcal{P}_+$. Thus, the hyperboloid $\mathcal{P}_+$ can be identified with the half-space $\mathbb{R}^+ \times \mathbb{R}^2 \ni (p_+, p_2, p_3)$. Note that, if we were to use p_1, p_2, p_3 as independent variables, solving for p_0 would involve a square root. This is a disadvantage: we would have to somehow project out the negative energy solutions. Dirac was the first to point the convenience of the light-cone (more accurately light-front) framework for field theory.

Let $\mathcal{H} = \{A \mid A^\dagger = A, A > 0, \det A = 1\}$ be the set of 2×2 positive Hermitian matrices with unit determinant. There is a one-to-one correspondence between $\mathcal{P}_+$ and $\mathcal{H}$. In light-front coordinates, the correspondence goes as follows:

$$\mathcal{P}_+ \ni (p_+, p_2, p_3) \to \mathcal{P}(p) = \begin{pmatrix} p_+ & p_2 + ip_3 \\ p_2 - ip_3 & p_- \end{pmatrix} \in \mathcal{H} \qquad (2.1)$$

with $p_+ p_- - (p_2^2 + p_3^2) = 1$. Moreover, a positive Hermitian matrix with unit determinant has a unique decomposition in terms of upper triangular matrices with real diagonal and unit determinant (Cholesky factorization), $\mathcal{P}(p) = \ell^\dagger \ell$ with

$$\ell = \frac{1}{\sqrt{p_+}} \begin{pmatrix} p_+ & p_2 + ip_3 \\ 0 & 1 \end{pmatrix}. \qquad (2.2)$$

The latter form a group, which we shall refer to as $SB(2, C)$, it being the Borel subgroup of $SL(2, \mathbb{C})$, as we shall explain in a moment. Therefore, the Cholesky "square root" of a positive Hermitian matrix with unit determinant is a Lie group.

Let us remark the following: the Lorentz group acts on $\mathcal{P}(p) \in \mathcal{H}$ by $\mathcal{P}(p) \mapsto \Lambda \mathcal{P}(p) \Lambda^\dagger, \Lambda \in SL(2, C)$; the subgroup $SU(2)$ acts with 1 as fixed point, i.e., rotations; and the trace of $\mathcal{P}(p)$ is invariant under $SU(2)$: $\mathrm{Tr}\mathcal{P}(p) = 2p_0$, twice the energy.

Besides the standard realization of the Lie algebra of the Lorentz group, which we shall identify with $\mathfrak{sl}(2, C)$ from now on, in terms of rotations and boosts, it is possible to choose a different basis, which emphasizes its bi-algebra structure. Let us indicate with e_a, f^a dually related generators with respect to the inner product $\mathrm{Im Tr}$. There are two maximally isotropic, non-commuting subalgebras, $\mathfrak{su}(2)$ (spanned by e_a) and the Borel algebra, $\mathfrak{sb}(2, C)$ (spanned by f^a). The latter is the Lie algebra of the group $SB(2, C)$. As a vector space, $\mathfrak{sl}(2, \mathbb{C}) = \mathfrak{su}(2) \oplus \mathfrak{sb}(2, \mathbb{C})$. All together, they form a Manin triple. As a basis, we may choose

$$e_1 = \begin{pmatrix} \frac{i}{2} & 0 \\ 0 & -\frac{i}{2} \end{pmatrix}, \quad e_2 = \begin{pmatrix} 0 & \frac{i}{2} \\ \frac{i}{2} & 0 \end{pmatrix}, \quad e_3 = \begin{pmatrix} 0 & -\frac{1}{2} \\ \frac{1}{2} & 0 \end{pmatrix}$$

$$f^1 = \begin{pmatrix} 1 & 0 \\ 0 & -1 \end{pmatrix}, \quad f^2 = \begin{pmatrix} 0 & 2 \\ 0 & 0 \end{pmatrix}, \quad f^3 = \begin{pmatrix} 0 & 2i \\ 0 & 0 \end{pmatrix} \qquad (2.3)$$

which verify $\mathrm{Im\, Tr}\, e_a f^b = \delta_b^a$ with Lie brackets

$$[e_i, e_j] = \epsilon_{ij}{}^k e_k, \quad [f^i, f^j] = f^{ij}{}_k f^k, \quad [e_i, f^j] = \epsilon_{ki}{}^j f^k + f_i{}^{jk} e_k \qquad (2.4)$$

and $f_i{}^{jk} = 2\epsilon_{1i\ell}\epsilon^{\ell jk}$ the structure constants of the Lie algebra $\mathfrak{sb}(2,\mathbb{C})$. Note the mutual adjoint action of one subalgebra onto the other in the last Lie bracket.

The corresponding dual subgroups of $SL(2,\mathbb{C})$, which realize the latter as a Drinfel'd double, are $SU(2) = \{u \mid u^\dagger u = 1, \ \det u = 1\}$ and $SB(2,\mathbb{C}) = \left\{ \lambda = \begin{pmatrix} a & b+ic \\ 0 & a^{-1} \end{pmatrix}, a > 0, b+ic \in \mathbb{C} \right\}$. Therefore, an element of the Lorentz group $\gamma \in SL(2,\mathbb{C})$ may be locally parametrized as a product $\gamma = u \cdot \lambda$ or alternatively $\gamma = \lambda' \cdot u'$. Note that this is nothing but the familiar Iwasawa decomposition of $SL(2,\mathbb{C})$, $g = k \cdot a \cdot n$, with $k \in SU(2)$, $a \cdot n = b \in SB(2,\mathbb{C})$ (see Ref. 4 for a review).

3. The Hyperboloid Group

As we have seen, the hyperboloid $\mathcal{P}_+$ is itself a group, it being $\ell \in SB(2,\mathbb{C})$. Therefore, $SB(2,\mathbb{C})$ plays multiple roles: on one hand, it is a factor of the Lorentz group in the Iwasawa decomposition, and, as such, it acts with the appropriate representation; on the other hand, it models the manifold of momenta for particles of unit mass, therefore, it is acted upon by the Lorentz group and its factors; moreover, it being the dual group of $SU(2)$, its group elements replace standard angular momentum which is usually associated with the abelian dual Lie algebra of $\mathfrak{su}(2)$ through momentum map. Let us see these structures in more detail.

The left-invariant Maurer–Cartan form can be easily calculated:

$$\alpha = \lambda^{-1}d\lambda = \frac{da}{a}f^1 + \frac{1}{2}\left(\frac{db}{a} + \frac{b}{a^2}da\right)f^2 + \frac{1}{2}\left(\frac{dc}{a} + \frac{c}{a^2}da\right)f^3. \quad (3.1)$$

This yields, under the identification of $\lambda \to \ell$, namely $a \to \sqrt{p_+}$, $b \to \frac{p_2}{\sqrt{p_+}}$, $c \to \frac{p_3}{\sqrt{p_+}}$, a basis of left-invariant one-forms

$$\alpha_1 = \frac{1}{2}\frac{dp_+}{p_+}, \quad \alpha_2 = \frac{1}{2}\frac{dp_2}{p_+}, \quad \alpha_3 = \frac{1}{2}\frac{dp_3}{p_+} \quad (3.2)$$

with left-invariant vector fields

$$Y^1 = 2p_+\frac{\partial}{\partial p_+}, \quad Y^2 = 2p_+\frac{\partial}{\partial p_2}, \quad Y^3 = 2p_+\frac{\partial}{\partial p_3} \quad (3.3)$$

satisfying $[Y^i, Y^j] = f^{ij}{}_k Y^k$. The latter generate the right action of $SB(2, \mathbb{C})$ on the momentum hyperboloid. The left action is obtained by repeating the above for the right-invariant Maurer–Cartan form. We get the right-invariant forms

$$\theta_1 = \frac{1}{2}\frac{dp_+}{p_+}, \quad \theta_2 = \frac{1}{2}\left(dp_2 - \frac{p_2}{p_+}dp_+\right), \quad \theta_3 = \frac{1}{2}\left(dp_3 - \frac{p_3}{p_+}dp_+\right) \quad (3.4)$$

with right-invariant vector fields

$$Z^1 = 2\left(p_+ \frac{\partial}{\partial p_+} + p_2 \frac{\partial}{\partial p_2} + p_3 \frac{\partial}{\partial p_3}\right), \quad Z^2 = 2\frac{\partial}{\partial p_2}, \quad Z^3 = 2\frac{\partial}{\partial p_3} \quad (3.5)$$

satisfying $[Z^i, Z^j] = -f^{ij}_k Z^k$.

3.1. *Poisson–Lie brackets*

In terms of a generalization of the Kirillov–Souriau–Konstant bracket (KSK), which is naturally defined on the dual of any Lie algebra, dually related Lie groups possess a natural Poisson bracket (that reduces to the KSK bracket when evaluated at the identity), which is compatible with the group action, namely the group multiplication results to be a Poisson morphism. It has been shown (see Ref. 3 for a review) that Poisson–Lie brackets for a given Lie group G may always be written as

$$\{\gamma_1, \gamma_2\} = [r, \gamma_1 \gamma_2], \quad (3.6)$$

with $\gamma \in G$ and $r \in \mathfrak{g} \otimes \mathfrak{g}$ the so-called classical r-matrix, solution of the classical (possibly modified) Yang–Baxter equation (CYB).[3] It is standard notation to denote $\gamma_1 = \gamma \otimes 1, \gamma_2 = 1 \otimes \gamma$ with $\gamma_1 \gamma_2 = \gamma \otimes \gamma$. The Lorentz group itself is a Poisson–Lie group with

$$r = e_i \otimes f^i, \quad [r_{12}, r_{13} + r_{23}] + [r_{13}, r_{23}] = 0 \quad \text{(CYB)}. \quad (3.7)$$

When specialized to $\lambda \in SB(2, \mathbb{C})$, Eq. (3.6) yields

$$\{\lambda \overset{\otimes}{,} \lambda\} = e_i \lambda \otimes f^i \lambda - \lambda e_i \otimes \lambda f^i \quad (3.8)$$

$$\{\lambda \overset{\otimes}{,} \lambda^\dagger\} = e_i \lambda \otimes (f^i \lambda)^\dagger - \lambda e_i \otimes (\lambda f^i)^\dagger, \quad (3.9)$$

namely in terms of the hyperboloid coordinates,

$$\{p_+, p_2\} = 2p_+ p_3, \quad \{p_3, p_+\} = 2p_+ p_2, \quad \{p_2, p_3\} = p_+^2 - p_2^2 - p_3^2 - 1. \quad (3.10)$$

The Poisson algebra (3.10) admits a Casimir function which corresponds to the energy p_0, $C = \frac{1}{2}\left(p_+ + \frac{1}{p_+} + \frac{p_2^2 + p_3^2}{p_+}\right)$.

3.2. *Dressing action of SU(2) on the hyperboloid*

The Drinfel'd double $SL(2, C)$ is a deformation of the semidirect product $SU(2) \ltimes \mathbb{R}^3$, with $\mathbb{R}^3$ the dual algebra of $SU(2)$. In standard Poisson geometry, the momentum map associated with the Hamiltonian action of $SU(2)$ on $\mathbb{R}^3$ gives rise to Hamiltonian functions which close the algebra of $SU(2)$ with respect to the KSK Poisson bracket. In classical mechanics, these functions are the three components of the angular momentum, ℓ_i, according to $\Lambda(d\ell_i) = -V_i$, where Λ is the KSK Poisson tensor and V_i is the vector fields realizing the infinitesimal action of $SU(2)$ on $\mathbb{R}^3$.

The generalization of the above construction to the Drinfel'd double $SL(2, \mathbb{C})$ yields the infinitesimal dressing action of $SU(2)$ on $SB(2, \mathbb{C})$. The Maurer–Cartan forms α^i derived in Eq. (3.2) play now the role of the exact one-forms $d\ell_i$, whereas Λ is replaced by the Poisson–Lie tensor computed in Sec. 3.1,

$$
P = 2p_+ p_3 \frac{\partial}{\partial p_+} \wedge \frac{\partial}{\partial p_2} - 2p_+ p_2 \frac{\partial}{\partial p_+} \wedge \frac{\partial}{\partial p_3} - (1 - p_+^2 + p_2^2 + p_3^2) \frac{\partial}{\partial p_2} \wedge \frac{\partial}{\partial p_3}.
$$
$$(3.11)$$

This yields $L_i = -P(\alpha_i)$, that is, in the light-front coordinates of the hyperboloid

$$
L_1 = -p_2 \frac{\partial}{\partial p_3} + p_3 \frac{\partial}{\partial p_2},
$$

$$
L_2 = -p_3 \frac{\partial}{\partial p_+} - \frac{1}{2p_+}(1 - p_+^2 + p_2^2 + p_3^2)\frac{\partial}{\partial p_3}, \qquad (3.12)
$$

$$
L_3 = p_2 \frac{\partial}{\partial p_+} + \frac{1}{2p_+}(1 - p_+^2 + p_2^2 + p_3^2)\frac{\partial}{\partial p_2}.
$$

They satisfy the commutation relations of the Lie algebra $\mathfrak{su}(2)$, $[L_i, L_j] = \varepsilon_{ij}{}^k L_k$. That is, they generate the dressing action of $SU(2)$ on the hyperboloid. The latter is the dressing action generated through left-invariant one-forms. Analogously, we have another one, generated by right-invariant forms (3.4). We have $\widehat{L}_\ell = -P(\theta_\ell)$, that is, in the light-front coordinates of the hyperboloid

$$
\widehat{L}_1 = p_3 \frac{\partial}{\partial p_2} - p_2 \frac{\partial}{\partial p_3},
$$

$$
\widehat{L}_2 = -p_3 \left(p_+ \frac{\partial}{\partial p_+} + p_2 \frac{\partial}{\partial p_2} \right) - \frac{1}{2}(1 - p_+^2 - p_2^2 + p_3^2) \frac{\partial}{\partial p_3}, \qquad (3.13)
$$

$$
\widehat{L}_3 = p_2 \left(p_+ \frac{\partial}{\partial p_+} + p_3 \frac{\partial}{\partial p_3} \right) + \frac{1}{2}(1 - p_+^2 + p_2^2 - p_3^2) \frac{\partial}{\partial p_2}.
$$

They close the same $SU(2)$ algebra, $[\widehat{L}_i, \widehat{L}_j] = \varepsilon_{ij}{}^k \widehat{L}_k$.

4. The Group Manifold of $SU(2)$

As well as the group manifold of $SB(2,\mathbb{C})$, which we have regarded as the mass hyperboloid, the group manifold $SU(2)$, namely the three-sphere S^3, is an interesting configuration space for classical dynamics. Therefore, it is worth to duplicate the previous construction, where now the role of the two groups is exchanged.

A convenient parametrization of $u \in SU(2)$ is $u = y^0 \mathbf{1} + 2y^i e_i$ with $y^\mu \in \mathbb{R}^4, \sum y^\mu y^\mu = 1$ and e_i the generators of $SU(2)$. From the left-invariant Maurer–Cartan one-form $u^{-1}du$, the three basis left-invariant one-forms are obtained

$$
\eta^i = 2 \left(y^0 dy^i - y^i dy^0 - \varepsilon^i{}_{jk} y^j dy^k \right) \qquad (4.1)
$$

with dual vector fields

$$
W_i = \frac{1}{2} \left(y^0 \partial_i - y^i \partial_0 - \varepsilon_{ij}{}^k y^j \partial_k \right). \qquad (4.2)
$$

Analogously, from $du\, u^{-1}$, the right-invariant forms are found to be

$$
\omega^i = 2 \left(y^0 dy^i - y^i dy^0 + \varepsilon^i{}_{jk} y^j dy^k \right), \qquad (4.3)
$$

with dual vector fields

$$
V_i = \frac{1}{2} \left(y^0 \partial_i - y^i \partial_0 + \varepsilon_{ij}{}^k y^j \partial_k \right). \qquad (4.4)
$$

4.1. *Poisson–Lie bracket on SU(2)*

The Poisson–Lie bracket (3.6) is defined on the whole group $SL(2,\mathbb{C})$. For

$$SL(2,\mathbb{C}) \ni \gamma = \begin{pmatrix} \alpha & \beta \\ \delta & \zeta \end{pmatrix} \tag{4.5}$$

with $\alpha\zeta - \beta\delta = 1$, $\alpha, \beta, \delta, \zeta$ complex variables, we obtain

$$\begin{aligned} &\{\alpha, \beta\} = -i\alpha\beta, \quad \{\alpha, \delta\} = -i\alpha\delta, \quad \{\alpha, \zeta\} = -2i\beta\delta \\ &\{\beta, \delta\} = 0, \qquad \{\beta, \zeta\} = -i\beta\zeta, \quad \{\delta, \zeta\} = -i\delta\zeta. \end{aligned} \tag{4.6}$$

Specializing it to $u \in SU(2)$, we find

$$\begin{aligned} &\{y_0, y_1\} = -(y_2^2 + y_3^2), \quad \{y_0, y_2\} = y_1 y_2, \quad \{y_0, y_3\} = y_1 y_3 \\ &\{y_1, y_2\} = -y_0 y_2, \qquad \{y_1, y_3\} = -y_0 y_3, \quad \{y_2, y_3\} = 0. \end{aligned} \tag{4.7}$$

The Poisson tensor which endows $SU(2)$ with a Poisson–Lie structure is therefore

$$Q = (y_1 \partial_0 - y_0 \partial_1) \wedge (y_2 \partial_2 + y_3 \partial_3) - (y_2^2 + y_3^2)\partial_0 \wedge \partial_1. \tag{4.8}$$

4.2. *Dressing actions of SB(2,$\mathbb{C}$) on SU(2)*

By repeating the calculation performed in Sec. 3.2 for the group $SB(2,\mathbb{C})$, it is possible to get the dressing actions (left and right) of the latter on $SU(2)$. In comparison with the standard momentum map picture, the left-invariant Maurer–Cartan one-forms η^i (resp. right-invariant) derived above play now the role of the exact one-forms $d\ell_i$, whereas the Konstant–Souriau–Kirillov bracket Λ is replaced by the Poisson–Lie tensor (4.8). The generators of the right dressing action (respectively left dressing action) of $SB(2,\mathbb{C})$ on $SU(2)$ are therefore retrieved according to $B^\ell = -Q(\eta^\ell)$, which gives

$$B^1 = -2(y_2 \partial_2 + y_3 \partial_3), \quad B^2 = 2(y_2 \partial_1 - y_3 \partial_0), \quad B^3 = 2(y_2 \partial_0 + y_3 \partial_1) \tag{4.9}$$

where use has been made of the constraints on the parameters of $SU(2)$, $y_\mu y^\mu = 1$ implying $y^\mu \partial_\mu = 0$. They satisfy the commutation relations of the Lie algebra $\mathfrak{sb}(2)$,

$$[B^1, B^2] = -2B^2, \quad [B^1, B^3] = -2B^3, \quad [B^2, B^3] = 0. \tag{4.10}$$

Analogously, from the right-invariant one-forms,

$$\omega^i = 2 \left(y^0 dy^i - y^i dy^0 + \varepsilon^i{}_{jk} y^j dy^k \right).$$ (4.11)

We get the vector fields $\widehat{B}^\ell = -Q(\omega^\ell)$ so that

$$\widehat{B}^1 = B^1, \quad \widehat{B}^2 = 2(y_2 \partial_1 + y_3 \partial_0), \quad \widehat{B}^3 = 2(y_3 \partial_1 - y_2 \partial_0)$$ (4.12)

which close the same Lie algebra as the previous one, therefore, they generate the left dressing action of $SB(2, \mathbb{C})$ on $SU(2)$.

5. Dressing Actions of $SL(2, \mathbb{C})$ on the Hyperboloid

In order to obtain the action of the Lorentz group on the mass hyperboloid, it is natural to exploit the isomorphism of its Lie algebra with $\mathfrak{sl}(2, \mathbb{C})$ and look for its dressing action on the dual. The group $SL(2, \mathbb{C})$ is itself a Poisson–Lie group with dual group another copy of $SL(2, \mathbb{C})$, which shall be indicated by $SL(2, \mathbb{C})^*$ to keep track of the doubling (and to stress that it is the exponentiation of the dual Lie algebra). Therefore, the Poisson–Lie bracket (3.6) may be used to derive the left and right dressing actions of $SL(2, \mathbb{C})$ on its dual. The hyperboloid (identified with the group manifold of $SB(2, \mathbb{C})$) may be regarded as the quotient $SL(2, \mathbb{C})^*/SU(2)$. Depending on the decomposition chosen, namely $SL(2, \mathbb{C}) \sim SB(2, \mathbb{C}) \cdot SU(2)$ or $SL(2, \mathbb{C}) \sim SU(2) \cdot SB(2, \mathbb{C})$, the hyperboloid is a right coset or a left coset.[a] Therefore, once a decomposition is chosen, the infinitesimal generators of the dressing action of $SL(2, \mathbb{C})$ on its dual have to be computed consistently. In this way, they will pass to the quotient and give the appropriate (left or right) dressing action of the whole Lorentz group on the hyperboloid.[b]

First of all, we choose to represent $\gamma \in SL(2, \mathbb{C})$ as the product $\gamma = \lambda u$, with $\lambda \in SB(2, \mathbb{C})$ and $u \in SU(2)$, parametrized as above. Then, in order to write the Poisson–Lie bi-vector field in the chosen parametrization,

---

[a]Dually, $SU(2)$ is a left or right quotient with respect to $SB(2, \mathbb{C})$. This remark and the next are relevant when, as in previous section, it is the group manifold of $SU(2)$ which is associated with the carrier space of dynamics, and, therefore, one is interested in the action of the Lorentz group on it.

[b]Analogously, the appropriate dressing action of the Lorentz group on $SU(2)$ is obtained.

we compute the LHS of (3.6)

$$\{\gamma_1, \gamma_2\} = \lambda_1\{u_1, \lambda_2\}u_2 + \lambda_1\lambda_2\{u_1, u_2\} + \{\lambda_1, \lambda_2\}u_1u_2 + \lambda_2\{\lambda_1, u_2\}u_1$$
$$= \lambda_1\{u_1, \lambda_2\}u_2 + \lambda_1\lambda_2(ru_1u_2 - u_1u_2r) + (r\lambda_1\lambda_2 - \lambda_1\lambda_2r)u_1u_2$$
$$+ \lambda_2\{\lambda_1, u_2\}u_1$$

where use has been made of the Poisson brackets for the $SU(2)$ and $SB(2, \mathbb{C})$ variables, $\{u_1, u_2\} = [r, u_1u_2]$ and $\{\lambda_1, \lambda_2\} = [r, \lambda_1\lambda_2]$. The RHS of (3.6) yields instead

$$r\gamma_1\gamma_2 - \gamma_1\gamma_2r = r\lambda_1u_1\lambda_2u_2 - \lambda_1u_1\lambda_2u_2r. \tag{5.1}$$

Comparing the two, we get $\lambda_1\{u_1, \lambda_2\}u_2 + \lambda_2\{\lambda_1, u_2\}u_1 = 0$, from which, by consistency,

$$\{u_1, \lambda_2\} = \lambda_2 r_{12}u_1 = e_i u \otimes \lambda f^i, \quad \{\lambda_1, u_2\} = -\lambda_1 r_{12}u_2 = -\lambda e_i \otimes f^i u.$$

Explicitly, they yield

$$\{y_0, a\} = -\frac{a}{2}y_1 \qquad \{y_1, a\} = \frac{a}{2}y_0 \qquad \{y_2, a\} = -\frac{a}{2}y_3 \qquad \{y_3, a\} = \frac{a}{2}y_2$$

$$\{y_0, b\} = \frac{b}{2}y_1 - ay_2 \quad \{y_1, b\} = -\frac{b}{2}y_0 + ay_3 \quad \{y_2, b\} = \frac{b}{2}y_3 + ay_0 \quad \{y_3, b\} = -\frac{b}{2}y_2 - ay_1$$

$$\{y_0, c\} = \frac{c}{2}y_1 - ay_3 \quad \{y_1, c\} = -\frac{c}{2}y_0 - ay_2 \quad \{y_2, c\} = \frac{c}{2}y_3 + ay_1 \quad \{y_3, c\} = -\frac{c}{2}y_2 + ay_0.$$
$$\tag{5.2}$$

Therefore, the Poisson–Lie tensor for $SL(2, \mathbb{C})$ may be written as $\mathbb{P} = P + Q + M$, with P and Q respectively given by Eqs. (3.11) and (4.8), while M is to be read from (5.2), that is,

$$M = \left[\frac{a}{2}y_1\partial_a + (ay_2 - \frac{b}{2}y_1)\partial_b + (ay_3 - \frac{c}{2}y_1)\partial_c\right] \wedge \partial_{y_0}$$

$$- \left[\frac{a}{2}y_0\partial_a + (ay_3 - \frac{b}{2}y_0)\partial_b - (ay_2 + \frac{c}{2}y_0)\partial_c\right] \wedge \partial_{y_1}$$

$$+ \left[\frac{a}{2}y_3\partial_a - (ay_0 + \frac{b}{2}y_3)\partial_b - (ay_1 + \frac{c}{2}y_3)\partial_c\right] \wedge \partial_{y_2}$$

$$- \left[\frac{a}{2}y_2\partial_a - (ay_1 + \frac{b}{2}y_2)\partial_b + (ay_0 - \frac{c}{2}y_2)\partial_c\right] \wedge \partial_{y_3}. \tag{5.3}$$

Note that, for simplicity, Eqs. (5.2) and (5.3) are not explicitly written in the light-front coordinates of the hyperboloid. We shall see, however, that the parametrization will homogenize after projection.

We now observe that a basis of left-(resp. right-)invariant one-forms for $SL(2,\mathbb{C})$ is represented by (η^i, α_i) (resp. (ω^i, θ_i)). Therefore, we may compute the dressing action of the Lorentz group on its dual by following the same procedure (the generalized momentum map) as in previous sections. Then, by considering the quotient with respect to each subgroup, we shall obtain the dressing action of the Lorentz group on each of its factors.

Note that, having chosen the parametrization $\gamma = \lambda u$, the group $SB(2,\mathbb{C})$ is a right coset with respect to the $SU(2)$ action, while $SU(2)$ is a left coset with respect to the $SB(2,\mathbb{C})$ action. Therefore, we have to consider the pair (ω^i, α_i), namely right-invariant forms of $SU(2)$ and left-invariant forms of $SB(2,\mathbb{C})$, in order to retrieve the dressing action correctly, upon projection. Let us start with $\mathbb{P}(\omega^i)$. We have to evaluate

$$\mathbb{P}(\omega^i) = P(\omega^i) + Q(\omega^i) + M(\omega^i). \tag{5.4}$$

$P(\omega^i) = 0$ trivially while $Q(\omega^i) = -\hat{B}^i$ according to Eq. (4.12), Eqs. (??)–(??), namely the generators of the left dressing action of $SB(2,\mathbb{C})$ on $SU(2)$. Moreover, $M(\omega^i) = -Y^i$, with Y^i the left-invariant vector fields of $SB(2,\mathbb{C})$ already computed in (3.3). Namely, $\mathbb{P}(\omega^i) = -\hat{B}^i - Y^i$, yielding in turn

$$\pi^R_* \mathbb{P}(\omega^i) = -Y_i, \quad \pi^L_* \mathbb{P}(\omega^i) = -\hat{B}_i \tag{5.5}$$

where we have indicated with π^R, π^L respectively the projection of $SL(2,\mathbb{C})$ to the quotients $SL(2,\mathbb{C})/SU(2)$, $SL(2,\mathbb{C})/SB(2,\mathbb{C})$. Analogously,

$$\mathbb{P}(\alpha_i) = P(\alpha_i) + Q(\alpha_i) + M(\alpha_i) \tag{5.6}$$

with now $Q(\alpha_i) = 0$ trivially because Q is the Poisson–Lie tensor on $SU(2)$ while $\alpha_i \in \Omega^1(SB(2,\mathbb{C}))$. As for $M(\alpha_i)$, we find $M(\alpha_i) = y^i \partial_0 - y^0 \partial_i - \varepsilon^i_{jk} y^j \partial_k$, that is, $M(\alpha_i) = -V_i$, the right-invariant vector fields of $SU(2)$ computed in Eq. (4.4). $P(\alpha_i)$ in turn yields the generators of the right dressing action of $SU(2)$ on $SB(2,\mathbb{C})$ that we have already computed with Eq. (3.12), $P(\alpha_i) = L_i$. Namely, $\mathbb{P}(\alpha_i) = L_i - V_i$, yielding in turn

$$\pi^R_* \mathbb{P}(\alpha_i) = L_i, \quad \pi^L_* \mathbb{P}(\alpha_i) = -V_i. \tag{5.7}$$

Summarizing, by contracting the Poisson–Lie tensor $\mathbb{P}$ of $SL(2,\mathbb{C})$ with an appropriate basis of one-forms of its dual, $(\alpha_i, \omega^i) \in \Omega^1(SL(2,\mathbb{C})^*)$, we have obtained a set of vector fields, with right and left projections given respectively by $(Y^i, L_i) \in \mathfrak{X}(SB(2,\mathbb{C}))$ and $(\hat{B}^i, V_i) \in \mathfrak{X}(SU(2))$.

It may be verified that Y^i, L_i verify the Lie bi-algebra relations

$$[L_i, Y^j] = Y^k \epsilon_{ki}{}^j + 2\epsilon_{1i\ell}\epsilon^{\ell jk} L_k \tag{5.8}$$

besides $[L_i, L_j] = \epsilon_{ij}{}^k L_k$, $[Y^i, Y^j] = f^{ij}{}_k Y^k$, where $f^{ij}{}_k = 2\epsilon^{1i}{}_\ell \epsilon^{\ell j}{}_k$, therefore representing the infinitesimal action of the Lorentz group on the hyperboloid.

Dually, $\hat{B}_i, V_i$ obey the same bi-algebra relations but are vector fields on the group manifold of $SU(2)$; therefore, they represent the infinitesimal action of the Lorentz group on the three sphere.

6. Conclusions

$SL(2, \mathbb{C})$, the universal covering of the Lorentz group, is a fantastic arena where to probe our understanding of Poisson–Lie geometry, Manin triples and dually related groups. We have proposed in this short contribution, with no pretention of rigor, a hopefully pedagogical approach.

Besides describing in detail the geometric structures involved, we have singled out two physically interesting applications of the formalism: one is already well known and widely studied and the other is new, to our knowledge. The former deals with the identification of the group manifold of $SU(2)$ with the target space of some dynamical system (such as the rigid rotor, or the sigma model with and without Wess–Zumino term, see, for example, Refs. 5–8). The latter explores the lesser known group manifold of $SB(2, \mathbb{C})$, which may be naturally associated with the mass hyperboloid of relativistic particles, in the light-front formalism. This second example might have interesting applications which are still under investigation.

Acknowledgements

This contribution is dedicated to A. P. Balachandran, on the occasion of his 85th birthday.

One of us (SGR) remembers arriving in Syracuse many years ago as a new graduate student. On my first day, Bal placed on my desk two stacks of papers to read. One was on the Skyrme model and the other on string theory. Neither was a popular topic back then. Indeed, when I went to a summer school at the ICTP Trieste, many experts told me that I would perish unless I worked on something "relevant" (which in those days meant supergravity). I am lucky to have had such a talented and generous mentor. I learned from Bal to make my own choices, even if mine may not have been as insightful as his.

The second author (PV) is indebted to Bal as a "second-generation" student. I have been a student of Marmo and Rajeev. I remember the day when, while preparing my master thesis (the italian "laurea"), I was invited in the office of my advisor, Beppe Marmo, to discuss some critical points of my work. There, I met Bal for the first time. He had just arrived in Napoli, I had never discussed physics in English before, and Beppe made me explain to Bal what I was doing. Bal of course asked a lot of questions, for most of which I didn't know the answer: it was a tough experience, but important. This was the first of many other occasions where Bal taught to me that, as a physicist, I should never forget the physical motivations of my work, and I thank him for that.

References

1. L. A. Takhtajan, *Lectures on Quantum Groups* in Introduction to Quantum Groups and Intergable Massive models of Quantum Field Theory, eds, M.-L. Ge and B.-H. Zhao, Nankai, Lectures on Mathematical Physics, World Scientific, 1990, pp. 69–197.
2. V. G. Drinfel'd, *Quantum Groups*, Proceedings of the International Congress of Mathematicians (Berkeley, Calif., 1986), American Mathematical Society, Providence (1987) pp. 798–820.
3. V. Chari and A. N. Pressley, *A Guide to Quantum Groups*, Cambridge University Press (1995).
4. P. Sawyer, Computing the Iwasawa decomposition of the classical Lie groups of noncompact type using the QR decomposition, *Linear Algebra Appl.* **493** (2016) 573.
5. S. G. Rajeev, Non Abelian Bosonization without Wess-Zumino terms. 1. New current algebra, *Phys. Lett.* **B 217** (1989), 123.
6. V. E. Marotta, F. Pezzella and P. Vitale, Doubling, T-duality and generalized geometry: A simple model, *JHEP* **08** (2018), 185.
7. V. E. Marotta, F. Pezzella and P. Vitale, T-dualities and doubled geometry of the principal chiral model, *JHEP* **11** (2019), 060.
8. F. Bascone, F. Pezzella and P. Vitale, Poisson-lie T-duality of WZW model via current algebra deformation, *JHEP* **09** (2020), 060.

https://doi.org/10.1142/9789811270437_0017

Chapter 17

TBM Mixing in Asymmetric Textures

Pierre Ramond

University of Florida, Gainesville, USA
pierre.ramond@gmail.com

In honor of A.P. Balachandran's 85th Birthday

1. Syracuse to Trieste

When accepted to graduate school at Syracuse University, I wanted to study General Relativity with Peter Bergmann. After an arduous first year in transition from engineering to physics, I took a wonderful course from Professor MacFarlane on Advanced Quantum Mechanics which turned me into a budding Particle Physicist. George Sudarshan, my first adviser, asked me this simple question: "Does the Spin–Statistic Theorem apply to infinite component wave equations?" and left shortly after for a sabbatical in India. This is how I got acquainted with Majorana's equation and Streater and Whightman's "PCT, Spin and Statistics, and All That". Sudarshan came back from India and told me that I. Todorov had built a counter example. So much for that!

Left adrift, I was rescued by Professor Aiyalam Balachandran who, I found out, lived in momentum space. He proposed a problem: Given a four-point S-matrix for massive particles, is it possible to continue into the Mandelstam triangle and emerge to a different channel? This was a concrete task, where I learned of the Appell–Kampé de Fériet's orthogonal polynomials in the triangle, but the physics problem was much too difficult because of the cuts, and I was left with an incomplete answer, much to my dismay. During this process, Bal continued his encouragements and when away, with helpful, lengthy and detailed letters. With Bal, my fellow graduate student Bill Meggs and Jean Nuyts, I participated in three published papers. This was almost not enough: after four years, I was on the job market but was not picked until much later when Bob Wilson decided at the last moment to initiate a theory group at the newly approved National Accelerator Laboratory (NAL), later known as Fermilab.

I will be forever grateful to Bal for securing me a summer appointment at Trieste's International Center for Theoretical Physics that Professor Salam had founded. There, I had the good fortune to meet my collaborator Jean Nuyts and Hirotaka Sugawara, who introduced me to the amazing mathematical structure behind the Veneziano model. Hirotaka and I worked hard to extract the triple Reggeon vertex using tensor methods, only to find Sciuto's elegant version using ladder operators — thankfully, we never published our version, but I was hooked on Veneziano's dual resonance model.

I am honored to contribute to this chapter in Bal's honor. Without him, I would have gone back to engineering, and perhaps been wealthier, but surely not as fulfilled and happy!

2. Asymmetric Texture

The presence of two large angles in the neutrino mixings is, I believe, the most important, yet most neglected puzzle, of particle physics. Whenever nature displays a breakdown in previously accepted patterns, there lies the path to further progress. Quarks and leptons, with similar gauge couplings, show totally different mixings. Quarks follow small mixing angles while neutrinos display only one small mixing angle and two large mixing angles whose values approximate "geometric" values. This led to an inspired approximation[1] of the lepton mixing matrix,

$$\mathcal{U}_{PMNS} = \mathcal{U}^{(-1)}\mathcal{U}_{TBM}, \quad \mathcal{U}_{\text{TBM}} = \begin{pmatrix} \sqrt{\dfrac{2}{3}} & \dfrac{1}{\sqrt{3}} & 0 \\ -\dfrac{1}{\sqrt{6}} & \dfrac{1}{\sqrt{3}} & \dfrac{1}{\sqrt{2}} \\ \dfrac{1}{\sqrt{6}} & -\dfrac{1}{\sqrt{3}} & \dfrac{1}{\sqrt{2}} \end{pmatrix},$$

where $\mathcal{U}^{(-1)}$ comes from the diagonalization of the charged lepton Yukawa matrix. In the SU_5 Grand Unified extension of the Standard Model, $\mathcal{U}^{(-1)}$ should generate a "Cabibbo haze" that fully accounts for the third (reactor) angle. In SU_5, which is summarized by the elegant Georgi–Jarlskog[2] construction,

$$\mathcal{U}^{(-1)} = \mathcal{U}_{\mathrm{CKM}}(c \rightarrow -3c), \tag{2.1}$$

where c is the (22) element of the down quarks Yukawa matrix, and the GUT scale relations

$$m_\tau = m_b, \ m_\mu = 3m_s, \ m_e = \frac{1}{3}m_d, \ \longrightarrow \ \det Y^{(-1/3)} = \det Y^{(-1)}.$$

Before the reactor angle had been measured, a plethora of models[3] satisfying these relations appeared in the literature, with predictions for the reactor angle around one-third of the Cabibbo angle. When measured[4] to be twice this value, the TBM approximation was (too) quickly forgotten.

As a superstring person, I always look for mathematical beauty in physics, following my hero Paul Dirac;[5] the TBM matrix is beautiful since it makes one think of angles between crystal faces, finite group theory, etc.

Closer examination reveals that these models assume that all Yukawa matrices are symmetric, an assumption based not on theory but on convenience. The Georgi–Jarlskog relation of Eq. (1) no longer applies if the charged lepton and down quarks Yukawa matrices are *asymmetric*.

The search for asymmetric Yukawa matrices that reproduce the same G-J relations while violating Eq. (1) was conducted[6] in the "Damn The Torpedoes" mode, an expression often used by Feynman (after Farragut). It meant do the calculation and do not worry — if difficulties arise, fix them later if possible. The Yukawa matrices turn out to be fine-tuned and models will be required to explain their fine-tunings.

Such matrices exist[6] but in a highly constrained form,

$$Y^{(2/3)} = \begin{pmatrix} \lambda^8 & 0 & 0 \\ 0 & \lambda^4 & 0 \\ 0 & 0 & 1 \end{pmatrix}, \quad Y^{(-1/3)} = \begin{pmatrix} bd\lambda^4 & a\lambda^3 & b\lambda^3 \\ a\lambda^3 & c\lambda^2 & g\lambda^2 \\ d\lambda & g\lambda^2 & 1 \end{pmatrix},$$

$$Y^{(-1)} = \begin{pmatrix} bd\lambda^4 & a\lambda^3 & d\lambda \\ a\lambda^3 & -3c\lambda^2 & g\lambda^2 \\ b\lambda^3 & g\lambda^2 & 1 \end{pmatrix},$$

where $a = c = 1/3, b = 0.306, g = 0.811, d = 2a/g$ are Wolfenstein-like parameters. We find that our "asymmetric texture", with TBM hypothesis, overshoots its PDG value:

$$\theta_{13} = \theta_{\mathrm{reactor}} = 0.184, \quad \text{compared to} = \theta_{\mathrm{reactor}}^{PDG} = 0.145.$$

It is fortunate that an overshoot can be corrected by adding to TBM a CP-violating phase

$$\mathcal{U}_{\text{TBM}} \longrightarrow U_{\text{TBM}}(\delta) = \text{Diag}(1, 1, e^{i\delta})\mathcal{U}_{\text{TBM}},$$

with

$$\sin|\theta_{\text{reactor}}| \to \sin|\theta_{\text{reactor}}(\delta)| = \frac{\lambda}{3\sqrt{2}}\left|1 + \frac{2e^{i\delta}}{g}\right| < \sin|\theta_{\text{reactor}}|.$$

The phase is thus determined by setting θ_{reactor} at its PDG value. Up to a sign, it comes out close:

$$\delta_{CP} = \pm 1.32\pi, \quad \text{compare with global fits} \quad \delta_{CP}^{PDG} = 1.36^{+0.20}_{-0.16}\pi,$$

and yields consistent values for the large mixing angles,

$$\theta_{\text{Atm}} = 44.9° \ (0.66° \ below \ PDG), \quad \theta_{\text{Solar}} = 34.16° \ (0.51° \ above \ PDG).$$

3. Family Symmetry

As mentioned above, the asymmetric texture was designed to satisfy the G-J SU_5 relations, and the CKM matrix for quark mixings and the Gatto relation. With a phased TBM, it predicts the magnitude of the phase. It also contains many fine-tuned patterns and values. To alleviate these problems, a discrete family symmetry is required. The general features of this model[7] are as follows:

– standard model chiral fields transform as SU_5 irreps : $\bar{\mathbf{5}}$, $\mathbf{10}$, $\mathbf{1}$,
– no tree level electroweak Yukawa couplings,
– vector-like Fermion messengers coupling to SM fields and "familons",
– familons: scalar SM singlets with only family charges,
– non-abelian family symmetry with CG coefficients to single out asymmetry.

The family symmetry of choice is the non-abelian "Frobenius" discrete group, the semidirect product $\mathcal{Z}_3$, $T_{13} = (Z_{13} \rtimes Z_3)$, with two inequivalent triplet representations

$$Z_{13} \rtimes Z_3 \ \text{Irreps}: \ \mathbf{3_1}, \mathbf{3_2}, \mathbf{1'}, \bar{\mathbf{3}}_1, \bar{\mathbf{3}}_2, \bar{\mathbf{1}}', \mathbf{1}.$$

It arises from continuous G_2 via the simple group $PSL(2, 13)$:

$$Z_{13} \rtimes Z_3 = T_{13} \subset PSL(2, 13) \subset G_2$$

– the chiral matter is assigned to different family triplets: $(\bar{5}, 3_1) + (10, 3_2)$,
– electroweak breaking Higgs fields: $H \sim (\bar{5}, 1)$, $H' \sim (\overline{45}, 1)$,
– vector-like messengers:

$\Delta \sim (5, 3_2)$ explains charged lepton (down-quarks) Yukawa with H,
$\Sigma \sim (10, 3_1)$ generates the G-J term with H',
$\Gamma \sim (\overline{10}, 3_2)$, $\Theta \sim (\overline{10}, \bar{3}_1)$ generate up-quarks Yukawas,

– familons: four $\sim (1, 3_2)$, two $\sim (1, 3_1)$,
– familon vacuum values lie along simple directions:

$$(1, 0, 0), (0, 1, 0), (0, 0, 1), (0, 1, 1,), (1, 0, 1).$$

The effect of the model is to explain the fine-tunings in terms of simple direction in the familons' vacuum manifold.

The next step is to generate the seesaw sector. To wit, as in the original seesaw, one upgrades from SU_5 to SO_{10} where the right-handed (sterile) neutrinos appear:

$$SO_{10} \supset SU_5 \times U(1): \quad 16 = \bar{5} + 10 + 1; \quad 10_v = \bar{5} + 5.$$

The chiral assignment upgrades naturally to

$$SU_5 \times T_{13}: \quad (10, 3_2) + (\bar{5}, 3_1) \quad \longrightarrow \quad SO_{10} \times T_{13}: \quad (16, 3_2) + (10_v, 3_1).$$

It reproduces the original chiral fields once the conjugate quintet in the spinor 16 couples to the quintet in the vector 10_v generate a vector-like mass through a familon interaction and three sterile neutrinos $\overline{N}$ which transform as $(1, 3_2)$ under $SU_5 \times T_{13}$.

Assume a dimension-four coupling of the sterile neutrinos to one familon: $\overline{N}\,\overline{N}\varphi_{\mathcal{M}}$. T_{13} invariance allows $\varphi_{\mathcal{M}} \sim (1, 3_2)$.

In the vacuum direction, $\langle \varphi_{\mathcal{M}} \rangle_0 = M(1, -1, 1)$, the Majorana mass of the three sterile neutrinos is diagonalized by the TBM matrix:

$$\mathcal{U}_{TBM}^t \frac{1}{\mathcal{M}} \mathcal{U}_{TBM} = \frac{1}{M} \mathrm{Diag}\left(1, -\frac{1}{2}, 1\right).$$

With T_{13}, a simple vacuum structure yields TBM diagonalization.

The seesaw mechanism[8] requires the diagonalization of

$$\mathcal{S} = \mathcal{D} \frac{1}{\mathcal{M}} \mathcal{D}^t,$$

where $\mathcal{D}$ is the Dirac mass stemming from the coupling of $\overline{F} \sim (\bar{5}, 3_1)$ with $\overline{N}$: $\overline{F}\,\overline{N}\varphi_{\mathcal{D}}$, with $\varphi_{\mathcal{D}} \sim \bar{3}_1$.

The simple familon vacuum $\langle \varphi_{\mathcal{D}} \rangle = a(1, -1, 1)$, $\mathcal{S}$ is diagonalized by TBM,

$$\mathcal{S} = \frac{a^2}{M} \mathcal{U}_{\text{TBM}} \text{Diag} \left(1, -\frac{1}{2}, 1 \right) \mathcal{U}_{\text{TBM}}^t,$$

from which we extract the light neutrino masses $m_{\nu_3} = 2m_{\nu_2}$ implies normal ordering, but the degeneracy $m_{\nu_1} = m_{\nu_3}$ implies no $(\nu_1 - \nu_3)$ oscillations, *in obvious contradiction with experiment!*

Could there be more sterile neutrinos? The sequence from $SU_5 \to SO_{10}$ continues to the exceptional group E_6 as we showed long ago.

Fourth Sterile Neutrino: $SO_{10} \to E_6$ and the group theory is as follows:

– $E_6 \supset SO_{10} \times U_1$: $\mathbf{27} = \mathbf{16} + \mathbf{10} + \mathbf{1}$,
– T_{13}'s mother group $PSL(2, 13)$ contains a complex septet: $\mathbf{7} = \mathbf{3_1} + \mathbf{3_2} + \mathbf{1'}$.

A natural marriage of E_6 and $PSL(2, 13)$ ensues with chiral assignments

$$SO_{10} \times T_{13} : \quad (\mathbf{16}, \mathbf{3_2}) + (\mathbf{10}, \mathbf{3_1}) \longrightarrow E_6 \times T_{13} : \quad (\mathbf{16}, \mathbf{3_2}) + (\mathbf{10}, \mathbf{3_1}) + (\mathbf{1}, \mathbf{1'}).$$

With a fourth sterile neutrino, the seesaw matrix contains a new addition

$$\mathcal{S} \to \mathcal{S}' = \mathcal{S} + \mathcal{W}\mathcal{W}^t,$$

where $\mathcal{W}\mathcal{W}^t$ has two zero eigenvalues. Of the three choices, only one agrees with oscillations, with required vacuum value $< \mathcal{W}^t >_0 \propto (2, e^{i\pi}, e^{i\delta})$.

Using oscillations, this model predicts normal ordering of the light neutrino masses:

$$m_{\nu_1} = 27.6 \text{ meV}, \quad m_{\nu_2} = 28.9 \text{ meV}, \quad m_{\nu_3} = 57.8 \text{ meV}.$$

We found recently[9] a version of the asymmetric texture, with the smaller family symmetry $T_7 = \mathcal{Z}_7 \rtimes \mathcal{Z}_3$. Its ancestry originates in continuous $G_2 \supset PSL(2, 7) \supset T_7$.

$PSL(2, 7)$ contains a real septet irrep $\mathbf{7} = \mathbf{3} + \bar{\mathbf{3}} + \mathbf{1}$, which suggests a similar mixture based on E_6 and $PSL(2, 7)$:

$E_6 \times T_7$: $(\mathbf{16}, \mathbf{3}) + (\mathbf{10}, \bar{\mathbf{3}}) + (\mathbf{1}, \mathbf{1})$. These curious chiral assignments whose modular symmetries point to G_2 may construct a path to coset manifolds in 11 dimensions.

Acknowledgements

I wish to thank Bal again for allowing me to spend my life chasing the fabrics of our Universe. I thank Professors Pérez and Stuart, as well as Drs Rahat and Xu for helpful discussions. This research was supported in part by the Department of Energy under Grant No. DE-SC0010296.

References

1. P. F. Harrison, D. H. Perkins and W. G. Scott, *Phys. Lett. B* **530** (2002) 167 [hep-ph/0202074]; P. F. Harrison and W. G. Scott, *Phys. Lett. B* **535** (2002) 163 [hep-ph/0203209].
2. H. Georgi and C. Jarlskog, *Phys. Lett. B* **86** (1979) 297–300.
3. For a recent review, see F. Feruglio and A. Romanino, *Rev. Mod. Phys.* **93** (1) (2021) 015007.
4. Y. Abe *et al.*, *Phys. Rev. Lett.* **108** (2012) 131801; F. P. An *et al.*, *Phys. Rev. Lett.* **108** (2012) 171803; J. K. Ahn *et al.*, *Phys. Rev. Lett.* **108** (2012) 191802.
5. P. A. M. Dirac, The relation between Mathematics and Physics, lecture delivered on Presentation of the JAMES SCOTT prize, February 6, 1939.
6. M. H. Rahat, P. R. and B. Xu, *Phys. Rev. D* **98** (2018) 055030.
7. M. J. Pérez, M. H. Rahat, P. R., A. J. Stuart and B. Xu, *Phys. Rev. D* **100** (2019) 075008; *Phys. Rev. D* **101** (2020) 075018.
8. P. Minkowski, *Phys. Lett. B* **67** (1977) 421; M. Gell-Mann, P. Ramond, and R. Slansky, in *Sanibel talk*, retroprinted as hep-ph/9809459, and in *Supergravity*, North-Holland, Amsterdam (1979), PRINT-80-0576, retroprinted as [arXiv:1306.4669 [hep-th]]; T. Yanagida, in *Proceedings of the Workshop on Unified Theory and Baryon Number of the Universe*, KEK, Japan (1979).
9. M. J. Pérez, M. H. Rahat, P. R., A. J. Stuart and B. Xu, in Preparation.

© 2023 World Scientific Publishing Company
https://doi.org/10.1142/9789811270437_0018

Chapter 18

Entanglement Entropy in Quantum Mechanics: An Algebraic Approach

A. F. Reyes-Lega

Departamento de Física, Universidad de los Andes,
A.A. 4976-12340, Bogotá, Colombia
anreyes@uniandes.edu.co

An algebraic approach to the study of entanglement entropy of quantum systems is reviewed. Starting with a state on a C^*-algebra, one can construct a density operator describing the state in the GNS representation space. Applications of this approach to the study of entanglement measures for systems of identical particles are outlined. The ambiguities in the definition of entropy within this approach are then related to the action of unitaries in the commutant of the representation and their relation to modular theory explained.

1. Introduction

Entanglement entropy has increasingly become an important concept in physics. From black hole physics[1, 2] to conformal field theory and statistical mechanics,[3] to quantum information theory,[4] its prominent role cannot be overemphasized. In the particular case of bipartite quantum systems described by a tensor product structure of the underlying Hilbert space, the information-theoretic content of entanglement entropy is well known.[5] In the case of entanglement measures for systems of identical particles, there has been an intense debate during the last years.[6] In fact, many proposals have been put forward which try to provide sensible definitions of entanglement measures which take into account the fact that a separation of such a system into subsystems does not necessarily entail a factorization of the Hilbert space into a tensor product. It is the aim of this chapter to review the basic aspects of an approach to this problem that I have pursued for some time, together with A.P. Balachandran and collaborators.[7–10]

2. Observables and States

Quite generally, it is possible to say that any sensible physical theory should at least provide a definition of what is to be understood as *observable* and what is supposed to be regarded as a *state* of a given system. Results of measurements then can be thought of as the result of a "pairing" between observables and states.

As an example, let us consider the Hamiltonian description of a classical system of $n < \infty$ degrees of freedom. Assuming a phase space of the form T^*Q, with Q the configuration space manifold, we can use local coordinates for position (q^i) and momentum (p_i). Time evolution is given by the solution $(q(t), p(t))$ of Hamilton's equations for a given Hamiltonian $H(q, p)$, which is a smooth function on phase space, $H \in C^\infty(T^*Q)$. In this context, we say that the state of the system at a given time t_0 is given by the point $(q(t_0), p(t_0)) \in T^*Q$. We are then led to think of points in T^*Q as possible states of the system. On the other hand, observable quantities like energy, momentum, *etc.* are given by smooth functions on phase space. The "pairing" we alluded to above is given, in this case, by the evaluation of an observable $f \in C^\infty(T^*Q)$ at the point (q_0, p_0) specified by the state. If the number of degrees of freedom becomes significantly large (as is the case in classical statistical mechanics), a statistical approach becomes necessary. In that case, the state of the system is given by a probability distribution $\rho(q, p)$ and the pairing is now given by the evaluation of the expectation value of $f(q, p)$,

$$\langle f \rangle = \int f(q, p) \rho(q, p) d\mu, \tag{2.1}$$

where μ is the integration (Liouville) measure. Note that this is a more general notion of state. We can use it to describe, say, the canonical ensemble, for which ρ is given (up to a normalization factor) by $\exp(-H(q, p)/k_B T)$. But we can also recover our initial definition of state, by assigning a Dirac delta distribution to the point (q_0, p_0):

$$\rho(q, p) = \delta(q - q_0, p - p_0). \tag{2.2}$$

In summary, we have the following:

- *Observables* are elements of the algebra $C^\infty(T^*Q)$.
- *States* are positive, normalized functionals of the form

$$\omega_\rho : C^\infty(T^*Q) \longrightarrow \mathbb{R}$$

$$f \longmapsto \omega_\rho(f) = \int f(q,p)\rho(q,p)d\mu. \tag{2.3}$$

One of the reasons why the algebraic approach to quantum physics is so use-ful in order to elucidate the probabilistic structure of quantum theory and its relation to its classical counterpart is that the pairing between observ-ables and states runs, essentially, in parallel to what we have just described in the above example. The difference lies in the types of algebras considered which, as a rule, are noncommutative ones in quantum theory. Positivity and normalization of states (defined as functionals with those properties) allow one to construct probability distributions. Following the analogy nat-urally provided by Gelfand duality, the similarities and (most importantly) the differences between classical and quantum physics are made very clear (the reader may consult Ref. 11 for details).

In quantum physics, C^*-algebras and von Neumann algebras are the two kinds of algebras which are most relevant. Since every von Neumann algebra is, in particular, a C^*-algebra and since in the finite-dimensional case (the one we are interested in this chapter) the distinction disappears, we will only consider C^*-algebras. Let us remark, though, that in (local) quantum field theory, it is essential to take into account those properties that are particular to von Neumann algebras.[12]

Definition 1. A C^*-algebra $(\mathcal{A}, \|\cdot\|, *)$ is an involutive Banach algebra satisfying the following compatibility condition between the involution "$*$" and the norm "$\|\cdot\|$":

$$\|aa^*\| = \|a\|^2, \quad \forall a \in \mathcal{A}. \tag{2.4}$$

Definition 2. A state ω on a (unital) C^*-algebra $(\mathcal{A}, \|\cdot\|, *)$ is a positive, normalized linear functional $\omega : \mathcal{A} \to \mathbb{C}$.

A fundamental result due to Gelfand, Naimark and Segal (the GNS con-struction[12, 13]) allows one to associate a representation of any C^*-algebra $\mathcal{A}$ (by bounded operators on a Hilbert space) to each state $\omega : \mathcal{A} \to \mathbb{C}$.

The main idea behind the construction is to take advantage of the vector space structure of $\mathcal{A}$ and to use the product in order to define an action of $\mathcal{A}$ (as an algebra) on itself (but this time regarded as a vector space). In order to endow this vector space with an inner product, one makes use of the fact that $\omega(a^*b) = \overline{\omega(b^*a)}$ in order to define a sesquilinear form $\langle a, b \rangle := \omega(a^*b)$. Since there might be elements $a \in \mathcal{A} \setminus \{0\}$ for which $\omega(a^*a) = 0$, it is necessary to take the quotient by the ideal $\mathcal{N}_\omega$ generated by all such elements. The Hilbert space for the representation is then defined as the completion of the quotient space:

$$\mathcal{H}_\omega := \overline{\mathcal{A}/\mathcal{N}_\omega}. \tag{2.5}$$

The (bounded) operator representing $a \in \mathcal{A}$ is the operator $\pi_\omega(a)$ defined through

$$\pi_\omega(a)|[b]\rangle := |[ab]\rangle. \tag{2.6}$$

It is through the GNS construction that we can make contact with the standard formulation of quantum mechanics in terms of Hilbert spaces and linear operators acting on them.

3. Entanglement Entropy

Let us suppose we are given a composite (bipartite) quantum system described by a Hilbert space of the form $\mathcal{H} = \mathcal{H}_A \otimes \mathcal{H}_B$. If $|\psi\rangle \in \mathcal{H}$ is a pure state, then it is well known that $|\psi\rangle$ is an entangled state if and only if the von Neumann entropy of either of its reduced density matrices is greater than zero. The reduced density matrices are obtained by partial trace:

$$\rho_A := \mathrm{Tr}_{\mathcal{H}_B}(|\psi\rangle\langle\psi|), \quad \rho_B := \mathrm{Tr}_{\mathcal{H}_A}(|\psi\rangle\langle\psi|). \tag{3.1}$$

Using the Schmidt decomposition, one then shows that $S(\rho_A) = S(\rho_B)$ and that, furthermore, this (von Neumann) entropy is zero if and only if the state $|\psi\rangle$ is separable. The problem with systems of identical particles stems from the fact that the underlying Hilbert space is not a tensor product: many-particle states must be either totally symmetric or totally anti-symmetric tensors. This leads to problems when trying to use partial trace or the usual Schmidt decomposition in this context. Some time ago,[7,8] we proposed to use the more general notion of *restriction* of a state to a subalgebra instead of partial trace. If $\omega : \mathcal{A} \to \mathbb{C}$ is a state on $\mathcal{A}$ and $\mathcal{A}_0 \subset \mathcal{A}$ is a subalgebra of $\mathcal{A}$, then the restriction of ω to $\mathcal{A}_0$ is a state on $\mathcal{A}_0$ that can be used to study entanglement properties relative to the subsystem decomposition corresponding to this assignment of a subalgebra. The following example (taken from Refs. 7, 11) is very useful in order to clarify the connection between restrictions and partial traces.

Example 1. Let $\mathcal{H} = \mathcal{H}_A \otimes \mathcal{H}_B \equiv \mathbb{C}^2 \otimes \mathbb{C}^2$ and take $\mathcal{A} = M_2(\mathbb{C}) \otimes M_2(\mathbb{C})$ as the observable algebra. Consider now the following state vector:

$$|\psi_\lambda\rangle = \sqrt{\lambda}|+, -\rangle + \sqrt{(1-\lambda)}|-, +\rangle, \tag{3.2}$$

where $0 \le \lambda \le 1$. We can also regard it as an algebraic state ω_λ, in the sense of Definition 2, if we put $\omega_\lambda(O) := \langle \psi_\lambda | O | \psi_\lambda \rangle$, for all $O \in \mathcal{A}$. Let now $\mathcal{A}_0 \subset \mathcal{A}$ denote the subalgebra generated by elements of the form $\alpha \otimes \mathbb{1}_2$, with $\alpha \in M_2(\mathbb{C})$. By the definition, the restriction of ω_λ to $\mathcal{A}_0$ is given by

$$\omega_{\lambda,0}(\alpha \otimes \mathbb{1}_2) = \langle \psi_\lambda | \alpha \otimes \mathbb{1}_2 | \psi_\lambda \rangle = \lambda \langle +|\alpha|+\rangle + (1-\lambda)\langle -|\alpha|-\rangle. \tag{3.3}$$

On the other hand, the use of partial trace (with $\rho_A = \mathrm{Tr}_B |\psi_\lambda\rangle\langle\psi_\lambda|$) leads to

$$\rho_A = \begin{pmatrix} \lambda & 0 \\ 0 & 1-\lambda \end{pmatrix}. \tag{3.4}$$

It is readily checked that $\omega_{\lambda,0}(\alpha \otimes \mathbb{1}_2) = \mathrm{Tr}_{\mathcal{H}_A}(\rho_A \alpha)$.

As already remarked, the use of partial trace in the case of identical particles will be problematic, whereas the restriction of a state to a subalgebra is always a well-defined operation and physically sensible. Being a purely algebraic operation, it remains to see how to compute the von Neumann entropy of the reduced (restricted) state. This can be done using the GNS construction. The idea is quite simple. If we restrict a state ω on $\mathcal{A}$ to a subalgebra $\mathcal{A}_0$, then the entropy of $\omega_0 := \omega|_{\mathcal{A}_0}$ can be computed as follows. Construct the GNS space $\mathcal{H}_{\omega_0}$ and find a density matrix ρ_{ω_0} such that

$$\omega_0(\alpha) = \mathrm{Tr}_{\mathcal{H}_{\omega_0}}(\rho_{\omega_0} \pi_{\omega_0}(\alpha)). \tag{3.5}$$

This can be done by a careful manipulation of the projectors associated with the decomposition of $\mathcal{H}_{\omega_0}$ into irreducible representations. Then, we can define

$$S(\omega_0) := -\mathrm{Tr}(\rho_{\omega_0} \ln \rho_{\omega_0}). \tag{3.6}$$

Explicit examples of this procedure, illustrating its usefulness in the definition of an entanglement measure for systems of identical particles, can be found in Refs. 7, 9, 11.

4. An Unexpected Symmetry

The assignment of an entropy to a state ω through the GNS construction is based on the possibility of finding a density operator ρ_ω acting on the representation space $\mathcal{H}_\omega$ in such a way that

$$\omega(a) = \mathrm{Tr}_{\mathcal{H}_\omega}(\rho_\omega \pi_\omega(a)) \quad \text{for all} \quad a \in \mathcal{A}. \tag{4.1}$$

The relationship between ω and ρ_ω is, however, not unique. The choice that was made in Ref. 7 was based on the idea of using the projectors $P^{(k)}$ arising from the decomposition of $\mathcal{H}_\omega$ into irreducible subspaces, $\mathcal{H}_\omega = \bigoplus_k \mathcal{H}_\omega^{(k)}$, in order to define

$$\rho_\omega := \sum_k P^{(k)} |[\mathbb{1}_\mathcal{A}]\rangle \langle [\mathbb{1}_\mathcal{A}]| P^{(k)}. \tag{4.2}$$

This choice might be regarded as a "natural" one, but it is not the only one. In fact, since the decomposition of $\mathcal{H}_\omega$ into irreducible subspaces is in general not unique (this can only happen if each irreducible component appears with multiplicity at most one), there are many choices of projectors, which in turn lead to different choices of density operators compatible with (4.2). The ensuing entropy ambiguities were studied in Refs. 14, 15. In particular, in, Ref. 14 it was suggested that the ambiguity could be naturally explained in the context of Tomita–Takesaki modular theory. Although the natural context of modular theory is that of von Neumann algebras, an analysis of the entropy ambiguity for finite dimensional algebras using modular theory really sheds light on the problem. Such an analysis, valid for any finite dimensional C^*-algebra, has been carried out in Ref. 10. In the following, I will present a simple example[10, 16] that illustrates the general idea.

Example 2. Let $\mathcal{A} = M_n(\mathbb{C})$ and consider the state ω defined through

$$\omega(a) := \sum_i^n \lambda_i a_{ii}, \tag{4.3}$$

where the coefficients λ_i are assumed to be strictly positive and $\sum_{i=1}^n \lambda_i = 1$. The fact that $\lambda_i > 0$ for all i implies that there is no null space ($\mathcal{N}_\omega = \{0\}$) so that vectors in $\mathcal{H}_\omega$ have only one representative. We may, therefore, write simply $|a\rangle$ to denote elements in $\mathcal{H}_\omega$. The inner product is given by $\langle a|b \rangle = \omega(a^*b)$. As $\omega(a) = \langle \mathbb{1}_n | \pi_\omega(a) | \mathbb{1}_n \rangle$, we see that the state vector $|\mathbb{1}_n\rangle$ gives us a *purification* of the state ω. The algebra $\mathcal{A}$ (system **A**) is now faithfully represented as a subalgebra $\pi_\omega(\mathcal{A})$ of $\mathcal{L}(\mathcal{H}_\omega)$ — an emergent system we will denote as system **C** — and which in turn contains the commutant $\pi_\omega(\mathcal{A})'$ of $\pi_\omega(\mathcal{A})$, denoted here as system **B**. According to modular theory, observables in **B** are obtained from observables in **A** by the action of the modular conjugation J.[16] Indeed, if $\alpha \in \pi_\omega(\mathcal{A})$, then $J\alpha J \in \pi_\omega(\mathcal{A})'$. In our simple example, J can be described explicitly in terms of its action on a basis of matrix units e_{ij} for $\mathcal{A}$:

$$J|e_{ij}\rangle = \sqrt{\lambda_j/\lambda_i}|e_{ji}\rangle. \tag{4.4}$$

We refer the reader to Refs. 10, 16 for details. Let now $|\hat{e}_{ij}\rangle$ denote the vector $|e_{ij}\rangle$, properly normalized with respect to the GNS inner product. It follows that the projectors entering Eq. (4.2) can in this case be written as follows:

$$P^{(k)} = \sum_{i=1}^{n} |\hat{e}_{ik}\rangle\langle\hat{e}_{ik}|. \tag{4.5}$$

As remarked above, any other orthogonal decomposition of $\mathcal{H}_\omega$ providing an equivalent representation of $\mathcal{A}$ will lead to a density operator satisfying (4.1). In particular, given any unitary element of $\mathcal{A}$ ($g \in \mathcal{A}$, $g^*g = \mathbb{1}_n$), we can induce an equivalent decomposition by defining $|\hat{e}_{ik}(g)\rangle := U(g)|\hat{e}_{ik}\rangle$, where

$$U(g) := J\pi_\omega(g)J \in \pi_\omega(\mathcal{A})' \tag{4.6}$$

is a unitary operator that belongs to the commutant and which, therefore, can be regarded as a *gauge transformation*.

Let us denote with $\rho_\omega(g)$ the density operator obtained by replacing the projectors $\{P^{(k)}\}_{k=1,...,n}$ in (4.2) by the following ones:

$$P_g^{(k)} := J\pi_\omega(gp^{(k)}g)J, \quad k = 1, \ldots, n, \tag{4.7}$$

where g is a unitary element of $\mathcal{A}$ and $p^{(k)} := e_{kk}$. It can be readily shown that $\rho_\omega(g)$ still satisfies (4.1) and can be used to describe the original state ω in system **A**. But its entropy will in general differ from that of $\rho_\omega \equiv \rho_\omega(\mathbb{1}_n)$. In fact, we have (*cf.* Theorem 2 in Ref. 10)

$$S(\rho_\omega(g)) \geq S(\rho_\omega). \tag{4.8}$$

The above example can be generalized to the case of arbitrary finite dimensional C^*-algebras.[10] The emergent gauge symmetry appearing through the action of the unitaries $U(g)$ has been related to the action of the gauge group of the fiber bundle describing "molecular shapes" in Ref. 14. A generalization of this approach to the case of homogeneous spaces of compact Lie groups has been studied by Tabban in Ref. 16.

Remark 1. Recently, an alternative approach to the study of the entropy ambiguities discussed in this section has been proposed by Facchi and collaborators.[17] It would be interesting to relate the two approaches.

5. Working with Bal

I met Bal for the first time during a conference on the connection between spin and statistics that took place in Trieste back in 2008. During the last decade, I have had the pleasure and the honor of collaborating with him. The lessons learnt from his way of doing research are countless, but there is one which has been particularly important for me and my group in Bogotá, namely the group discussions. Although I did not take part in the legendary "Room-316 meetings", I was lucky enough to enjoy many discussions on different occasions that I adopted as part of my way of doing research with my students, some of whom are now my younger collaborators. Seeing how fast they can learn — and become accomplished researchers — has shown me the value that a generous and lively exchange of ideas (something that Bal has always fostered) can have.

References

1. L. Bombelli, R. K. Koul, J. Lee and R. Sorkin, Quantum source of entropy for black holes, *Phys. Rev. D.* **34** (1986) 373–383. doi: 10.1103/PhysRevD. 34.373.
2. S. N. Solodukhin, Entanglement entropy of black holes, *Living Rev. Relativ.* **14**(1) (2011) 1–96.
3. G. Vidal, J. I. Latorre, E. Rico and A. Kitaev, Entanglement in quantum critical phenomena, *Phys. Rev. Lett.* **90** (June, 2003) 227902. doi: 10.1103/ PhysRevLett.90.227902. http://link.aps.org/doi/10.1103/PhysRevLett.90. 227902.
4. M. A. Nielsen and I. L. Chuang, *Quantum Computation and Quantum Information*, Cambridge University Press (2010).
5. D. Janzing, *Entropy of Entanglement*, In D. Greenberger, K. Hentschel and F. Weinert (eds.), *Compendium of Quantum Physics*, pp. 205–209. Springer Berlin Heidelberg, Berlin, Heidelberg (2009). doi: 10.1007/ 978-3-540-70626-7_66. https://doi.org/10.1007/978-3-540-70626-7_66.

6. F. Benatti, R. Floreanini, F. Franchini and U. Marzolino, Entanglement in indistinguishable particle systems, *Phys. Rep.* **878** (2020) 1–27. doi: 10.1016/ j.physrep.2020.07.003. https://www.sciencedirect.com/science/article/pii/ S0370157320302520. Entanglement in indistinguishable particle systems.
7. A. P. Balachandran, T. R. Govindarajan, A. R. de Queiroz and A. F. Reyes-Lega, Entanglement and particle identity: A unifying approach, *Phys. Rev. Lett.* **110** (February, 2013) 080503. doi: 10.1103/PhysRevLett.110.080503. http://link.aps.org/doi/10.1103/PhysRevLett.110.080503.
8. A. P. Balachandran, T. R. Govindarajan, A. R. de Queiroz and A. F. Reyes-Lega, Algebraic approach to entanglement and entropy, *Phys. Rev. A.* **88**(2) (2013) 022301. doi: 10.1103/PhysRevA.88.022301. https://doi.org/10.1103/ PhysRevA.88.022301.
9. A. Balachandran, T. Govindarajan, A. R. De Queiroz and A. Reyes-Lega, Algebraic theory of entanglement, *Il Nuovo Cimento C.* **36**(3) (2013) 27–33.
10. A. P. Balachandran, I. M. Burbano, A. F. Reyes-Lega and S. Tabban, Emergent gauge symmetries and quantum operations, *J. Phys. A* **53**(6) (January, 2020) 06LT01. doi: 10.1088/1751-8121/ab6143. https://doi.org/10.1088/17 51-8121/ab6143.
11. A. F. Reyes-Lega, Some aspects of operator algebras in quantum physics. In L. Cano, A. Cardona, H. Ocampo, and A. Reyes-Lega (eds.), *Geometric, Algebraic and Topological Methods for Quantum Field Theory*, pp. 1–74, World Scientific (2016). doi: 10.1142/9789814730884_0001. http://www.wor ldscientific.com/doi/suppl/10.1142/9861/suppl_file/9861_chap01.pdf.
12. R. Haag, *Local Quantum Physics*, 2nd edn. Texts and Monographs in Physics, Springer-Verlag (1996). doi: 10.1007/978-3-642-61458-3.
13. D. Werner, *Funktionalanalysis*, 3rd edn. Springer (2000).
14. A. P. Balachandran, A. Queiroz and S. Vaidya, Quantum entropic ambiguities: Ethylene, *Phys. Rev. D.* **88** (July, 2013) 025001. doi: 10. 1103/PhysRevD.88.025001. https://link.aps.org/doi/10.1103/PhysRevD.88. 025001.
15. A. P. Balachandran, A. R. de Queiroz and S. Vaidya, Entropy of quantum states: Ambiguities, *Eur. Phys. J. Plus.* **128** (2013) 112. doi: 10.1140/epjp/ i2013-13112-3.
16. S. Tabban, *Modular theory and algebraic quantum physics.* PhD thesis, Universidad de los Andes (2022). http://hdl.handle.net/1992/54545.
17. P. Facchi, G. Gramegna and A. Konderak, Entropy of quantum states, *Entropy* **23**(6) (2021). ISSN 1099-4300. doi: 10.3390/e23060645. https:// www.mdpi.com/1099-4300/23/6/645.

Chapter 19

From Virasoro Algebra to Cosmology

Vincent G. J. Rodgers

Department of Physics and Astronomy
The University of Iowa, Iowa City, IA 52242, USA
vincent-rodgers@uiowa.edu

Earlier work of Balachandran and friends provided a map from algebras to field theories. These methods provide insight into quantum gauge theories and anomalies. In this chapter, we take the reader from the coadjoint representation of the Virasoro algebra to four- (and higher-) dimensional gravitation and cosmology. The protagonist in this story is a component of the projective connection, the diffeomorphism field, which straddles between the one-dimensional world of initial data in string theories to cosmology in four dimensions. We review mathematical intuition that ties projective geometry to the Virasoro algebra, the Thomas–Whitehead (TW) gravitational action that gives the diffeomorphism field dynamics and the building blocks for gauge projective Dirac action.

In Honor of A.P. Balachandran on His 85th Birthday

1. Introduction

While a post-doc at Stony Brook in 1986, I would often venture back to Syracuse and listen in at the original Room 316 discussions directed by Professor Balachandran. As always, the Room 316 meetings were intellectually rich and inundated with ideas and perhaps far too many ideas than one could humanly thrash out. On one occasion, Professor Balachandran introduced the idea of geometric actions that arise on coadjoint orbits of Lie groups. He and friends[1-4] had been able to show that starting from an algebra and its dual, one could construct geometric actions that correspond to core components in certain quantum field theories.

One of his students, Balram Rai, and I took up Bal's challenge to build the geometric actions from that Kac–Moody and Virasoro algebras.[5] We recovered the Wess–Zumino–Witten model from the Kac–Moody algebra and the two-dimensional Polyakov action from the Virasoro algebra. This work was later extended to the super-Virasoro algebra.[6] I remember running to Balachandran, out of breath, to tell him how, not only did his method of coadjoint orbits give the anomalous contribution to effective actions of these two-dimensional theories but it also got the couplings correct. I did not expect this. But Prof. Balachandran's heart barely skipped a beat and he said, with the coolness of Hercule Poirot, "Don't be so surprised. That's what I was expecting."

$$S = \frac{\tilde{c}}{2\pi} \int dx d\tau \left[\frac{\partial_x^2 s}{(\partial_x s)^2} \partial_\tau \partial_x s - \frac{(\partial_x^2 s)^2 (\partial_\tau s)}{(\partial_x s)^3} \right] - \int dx d\tau \; D(x) \frac{(\partial s / \partial \tau)}{(\partial s / \partial x)}. \tag{1.1}$$

2D Polyakov Action 1. The 2D Polyakov action that arises using the method of coadjoint orbits. The geometric action relates the coadjoint element $\mathcal{D} = (D, \tilde{c})$ to the field D and the coupling constant $\tilde{c}$ in the action. D, called the *diffeomorphism field*, is the protagonist in this origin story.

To sketch how this works, let us consider the Virasoro algebra[7–9] as the centrally extended algebra of vector fields in one dimension. Let (ξ, a) and (η, b) be two such vectors with respective central elements, a and b. The commutator extends to

$$[(\xi, a), (\eta, b)] = (\xi \circ \eta, ((\xi, \eta))_0), \tag{1.2}$$

where $\xi \circ \eta$ is defined by

$$\xi \circ \eta \equiv \xi^a \partial_a \eta^b - \eta^a \partial_a \xi^b \tag{1.3}$$

and $((\xi, \eta))_0$, the so-called Gelfand–Fuchs two-cocycle,[10] is defined by

$$((\xi, \eta))_0 \equiv \frac{c}{2\pi} \int (\xi \eta''') \, d\theta - (\xi \leftrightarrow \eta) \tag{1.4}$$

$$= \frac{c}{2\pi} \int \xi^a \nabla_a ([g^{bc} \nabla_b \nabla_c] \eta^m) g_{mn} d\theta^n - (\xi \leftrightarrow \eta), \tag{1.5}$$

and where g_{ab} represents one-dimensional metric. Equations (1.4) and (1.5) form an invariant pairing between ξ and η'''. In the Gelfand–Fuchs two-cocycle, we see that the vector field η has been mapped into a *quadratic differential*, $Q_{an}(\eta)$, by the insertion of a Laplacian, $g^{bc}\nabla_b\nabla_a$, into the Lie derivative. Explicitly, using abstract index notation,

$$\eta\partial_\theta \to \eta'''d\theta^2 = \nabla_a([g^{bc}\nabla_b\nabla_c]\eta^m)g_{mn}d\theta^a d\theta^n = Q_{an}(\eta)d\theta^a d\theta^n. \quad (1.6)$$

The Gelfand–Fuchs two-cocycle is the simplest example of an invariant pairing between a vector and a quadratic differential, D, that is also centrally extended by the element $\tilde{c}$, viz., $\mathcal{D} = (D, \tilde{c})$.

$$< (\xi, a)|(D, \tilde{c}) > \equiv \int (\xi D) d\theta + a\tilde{c} = \int (\xi^i D_{ij}) d\theta^j + a\tilde{c}. \quad (1.7)$$

The invariance of this pairing requires that

$$ad^*_{(\eta,d)}(D, \tilde{c}) = (\eta D' + 2\eta' D - \tilde{c}\eta''', 0), \quad (1.8)$$

defining the coadjoint representation of the Virasoro algebra.[9, 11] Now, Kirillov observed[11] a symplectic two-form on the coadjoint orbits of the Virasoro algebra. For each coadjoint element, $\mathcal{D} = (D, \tilde{c})$ and a pair of coadjoint elements $\mathcal{D}_1$ and $\mathcal{D}_2$ that are infinitesimally close to $\mathcal{D}$ through $ad^*_{(\xi,b)}(D, \tilde{c})$ and $ad^*_{(\eta,a)}(D, \tilde{c})$, respectively, one has the two-form, $\Omega_{\mathcal{D}}(\mathcal{D}_1, \mathcal{D}_2)$, on the orbit of $\mathcal{D}$ given by

$$\Omega_{\mathcal{D}}(\mathcal{D}_1, \mathcal{D}_2) = \frac{\tilde{c}}{2\pi} \int (\xi\eta''' - \xi'''\eta)\,dx + \frac{1}{2\pi} \int (\xi\eta' - \xi'\eta)D\,dx. \quad (1.9)$$

Ω is a symplectic two-form as its anti-symmetric, closed and non-degenerate. This observation[5] is what led to Eq. (1.1). The method of coadjoint orbits has dissolved the coadjoint element $\mathcal{D}$ into a coupling constant and a component of a two-dimensional field D, called the *diffeomorphism field*, in two dimensions. In Eq. (1), it appears as a background field coupled to the Polyakov metric, $\frac{(\partial s/\partial \tau)}{(\partial s/\partial x)}$, and is often interpreted as an anomalous energy–momentum tensor. A similar situation occurs in the Kac–Moody case where the coadjoint element there, $\mathcal{A} = (A, \hat{b})$, allows us to interpret A as the remaining component of a two-dimension gauge-fixed Yang–Mills connection A_μ. It corresponds to a background Yang–Mills field that is coupled to a Wess–Zumino current. We seek a similar geometric connection interpretation of D that, like Yang–Mills, can be represented in any dimension. In the spirit of a connection, one also sees that the Gelfand–Fuchs case, Eq. (1.4), lives in the "pure gauge" sector since it corresponds to the $\mathcal{D} = (0, c)$ coadjoint element.

An important clue as to the geometric natures of D comes from, yet another, observation by Kirillov. He observed[11] that Eq. (1.8) provides a correspondence to the space of Sturm–Liouville operators. Thus, there is a one-to-one correspondence,

$$(D, \tilde{c}) \Leftrightarrow -2\tilde{c}\frac{d^2}{dx^2} + D(x), \tag{1.10}$$

between the coadjoint element (D, c) and a Sturm–Liouville operator with weight $\tilde{c}$ and Sturm–Liouville potential, $D(x)$. This important observation ties the Virasoro algebra into projective geometry. Let's explain.

To see this correspondence,[7,12] let ϕ_A and ϕ_B be the two independent solutions of the Sturm–Liouville equation that span the space of solution,

$$\left(-2\tilde{c}\frac{d^2}{dx^2} + D(x)\right)\phi_A = 0, \quad \left(-2\tilde{c}\frac{d^2}{dx^2} + D(x)\right)\phi_B = 0. \tag{1.11}$$

Now define the ratio, $f(x) = \frac{\phi_A(x)}{\phi_B(x)}$. Then, one can show that

$$D(x) = \frac{\tilde{c}}{2}\left(\frac{f'''(x)}{f'(x)} - \frac{3}{2}\left(\frac{f''(x)}{f'(x)}\right)^2\right) = S(f(x)). \tag{1.12}$$

This is precisely the projective invariant, the Schwarzian derivative of $f(x)$ with respect to x,[7,12] $S(f(x))$. Therefore, we may consider $f(x)$ to be an affine parameter, $\tau \equiv f(x)$, on the projective line $\mathbb{P}^1$ of a one-parameter family of Strum–Liouville operators, viz.

$$L_\tau\phi = -2\tilde{c}\frac{d^2}{dx^2}\phi + D_\tau\phi = 0.$$

This above equation is invariant under the action of the vector field (η, d), and in particular, the Sturm–Liouville operator transformation, $ad^*_{(\eta,d)} L_\tau$, is as in Eq. (1.8). This begins the identification of D as a *projective connection*[11] and to be interpreted[13] as a component of the Thomas–Whitehead connection, $\tilde{\nabla}_\mu$.[14–16] We discuss the salient features of this presently and circle back to this identification.

2. Thomas–Whitehead and Projective Curvature

2.1. *The TW projective connection*

The simplest way to appreciate projective curvature is through the parameterization of geodesics and connections that yield the same geodesics. If $\hat{\nabla}$ and ∇ admit the same geodesics, they belong to the same projective equivalence class. Thomas showed how one can write a gauge theory over this projective symmetry.[14,15]

Consider a d-dimensional manifold $\mathcal{M}$ with coordinates x^a where italic latin indices $a, b, c, m, n, \cdots = 0, 1, \ldots, d - 1$. Let $\hat{\nabla}_a$ be a connection on $\mathcal{M}$ where ζ^a is geodetic, i.e.,

$$\zeta^b \hat{\nabla}_b \zeta^a = \frac{d^2 x^a}{d\tau^2} + \hat{\Gamma}^a{}_{bc} \frac{dx^b}{d\tau} \frac{dx^c}{d\tau} = 0. \tag{2.1}$$

Now consider another connection related to the previous connection by a one-form v_a. This is called a *projective transformation*,

$$\Gamma^a{}_{bc} = \hat{\Gamma}^a{}_{bc} + \delta^a{}_b v_c + \delta^a{}_c v_b. \tag{2.2}$$

The geodesic equation for this connection is then

$$\zeta^b \nabla_b \zeta^a = \frac{d^2 x^a}{d\tau^2} + \Gamma^a{}_{bc} \frac{dx^b}{d\tau} \frac{dx^c}{d\tau} = f(\tau) \frac{dx^a}{d\tau}, \tag{2.3}$$

where $f(\tau) = 2v_b \frac{dx^b}{d\tau}$. We can eliminate the right-hand side by suitable reparameterization of τ to some $u(\tau)$. Because of this, both Eqs. (2.1) and (2.3) admit the same geodesic curves. The connections belong to the same projective equivalence class, $\hat{\Gamma}^a{}_{bc} \sim \Gamma^a{}_{bc}$.

Thomas[14,15] developed a "gauge" theory of projectively equivalent connections in the following way. First, define the *fundamental projective invariant* $\Pi^a{}_{bc}$

$$\Pi^a{}_{bc} \equiv \Gamma^a{}_{bc} - \frac{1}{(d+1)} \delta^a{}_{(b} \Gamma^m{}_{c)m}. \tag{2.4}$$

This is clearly invariant under a projective transformation, Eq. (2.2). The geodetic equation associated with $\Pi^a{}_{bc}$ is

$$\frac{d^2 x^a}{d\tau^2} + \Pi^a{}_{bc} \frac{dx^b}{d\tau} \frac{dx^c}{d\tau} = 0, \tag{2.5}$$

which is projectively invariant but not covariant. This is because $\Pi^a{}_{bc}$ does not transform as a connection under a general coordinate transformation from $x \to x'(x)$ due to the last summand in

$$\Pi'^a{}_{bc} = J^a{}_f \left(\Pi^f{}_{de} \bar{J}^d{}_b \bar{J}^e{}_c + \frac{\partial^2 x^f}{\partial x'^b \partial x'^c} \right) + \frac{1}{d+1} \frac{\partial \log |J|}{\partial x^d} \left(\bar{J}^d{}_b \delta^a{}_c + \bar{J}^d{}_c \delta^a{}_b \right). \tag{2.6}$$

Here, $J^a{}_b = \frac{\partial x'^a}{\partial x^b}$, the Jacobian of the transformation and $J = \det(J^a{}_b)$. Thomas constructs a line bundle (also called volume bundle) over $\mathcal{M}$ to form a d + 1-dimensional manifold, $\mathcal{N}$, referred to as the Thomas cone.[17] The coordinates on the Thomas cone are $(x^0, x^1, \ldots, x^{d-1}, \lambda)$, where λ is a fiber denoting the volume coordinate. It is because the volume coordinate, λ, takes values from $0 < \lambda < \infty$ that $\mathcal{N}$ is called a cone. These coordinates transform as

$$x'^{\alpha} = (x'^0(x^d), x'^1(x^d), \ldots, x'^{d-1}(x^d), \lambda' = \lambda |J|^{-\frac{1}{d+1}}), \qquad (2.7)$$

and one can see this is a fibration as the λ coordinate is not independent. For every coordinate transformation on $\mathcal{M}$, there is a unique coordinate transformation on $\mathcal{N}$. We use Greek indices over coordinates on $\mathcal{N}$ and italic Latin indices over coordinates on $\mathcal{M}$. We reserve the index λ and the upright letter d to refer to the volume coordinate $x^d = x^\lambda = \lambda$.

To proceed, let Υ denote the fundamental vector on $\mathcal{N}$ and a companion one-form ω_α such that $\Upsilon^\alpha \omega_\alpha = 1$. For any function on $\mathcal{N}$, $\Upsilon^\alpha \nabla_\alpha f = \lambda \partial_\lambda f$. The projective connection, $\tilde{\nabla}_\alpha$, must be compatible with Υ, meaning that the covariant derivative on Υ returns the identity. Thus, Υ satisfies the normalized geodesic equation, viz.

$$\tilde{\nabla}_\alpha \Upsilon^\beta = \delta_\alpha^\beta \;\Rightarrow\; \Upsilon^\alpha \tilde{\nabla}_\alpha \Upsilon^\beta = \Upsilon^\beta. \qquad (2.8)$$

Explicitly, we can take $\Upsilon^\alpha = (0, 0, \ldots, \lambda)$, so we can write a generalized connection $\tilde{\Gamma}^\alpha{}_{\beta\gamma}$ [16, 18] on $\mathcal{N}$ as

$$\tilde{\Gamma}^\alpha{}_{\beta\gamma} = \begin{cases} \tilde{\Gamma}^\lambda{}_{\lambda a} = \tilde{\Gamma}^\lambda{}_{a\lambda} = 0 \\[4pt] \tilde{\Gamma}^\alpha{}_{\lambda\lambda} = 0 \\[4pt] \tilde{\Gamma}^a{}_{\lambda b} = \tilde{\Gamma}^a{}_{b\lambda} = \omega_\lambda \delta^a_b \\[4pt] \tilde{\Gamma}^a{}_{bc} = \Pi^a{}_{bc} \\[4pt] \tilde{\Gamma}^\lambda{}_{ab} = \Upsilon^\lambda \mathcal{D}_{ab}. \end{cases} \qquad (2.9)$$

We have only required $\mathcal{D}_{ab}$ to transforms on $\mathcal{M}$ as

$$\mathcal{D}'_{ab} = \frac{\partial x^m}{\partial x'^a} \frac{\partial x^n}{\partial x'^b} (\mathcal{D}_{mn} - \partial_m j_n - j_m j_n + j_c \Pi^c{}_{mn}) \qquad (2.10)$$

so that the connection $\tilde{\Gamma}^\mu{}_{\alpha\beta}$ transforms as an affine connection on $\mathcal{N}$.[19] Here, $j_a = \partial_a \log |J|^{-\frac{1}{d+1}}$. The covariant derivative operator now transforms covariantly on the Thomas cone, i.e., $\nabla'_\alpha = \frac{\partial x^\beta}{\partial x'^\alpha} \nabla_\beta$. In the above, $\mathcal{D}_{ab}$ generalizes the work of Thomas and is independent of $\Pi^a{}_{bc}$. This is the origin of the diffeomorphism field $\mathcal{D}$.

First Relation to Virasoro Algebra: In one dimension, $\mathcal{D}_{ab}$ transforms in one-to-one correspondence with the coadjoint element $\mathcal{D}$.[20] Indeed, in one dimension, Eq. (2.10) collapses to $\delta_\xi \mathcal{D}_{11} = 2\xi' \mathcal{D}_{11} + \mathcal{D}'_{11}\xi - \frac{1}{2}\xi'''$, for a vector field ξ. Now, let $\mathcal{D}_{11} = qD$, where $q = \frac{1}{2\tilde{c}}$. Then, $\delta_\xi(qD) \Rightarrow \delta_\xi D = 2\xi' D + D'\xi - \frac{1}{2q}\xi'''$. This is as in Eq. (1.8).

Furthermore, a vector field χ on $\mathcal{M}$ may be promoted to a vector field $\tilde{\chi}$ on $\mathcal{N}$ by writing $\tilde{\chi}^\alpha = (\chi^a, -\lambda x^b \kappa_b)$, where κ_a only needs to transform as

$$\kappa'_a = \frac{\partial x^m}{\partial x'^a}\kappa_m - \frac{1}{d+1}\frac{\partial \log J}{\partial x'^a}. \tag{2.11}$$

Similarly, a one-form v on $\mathcal{M}$ can be related to a projective one-form $\tilde{v}$ via $\tilde{v}_\beta = (v_b + \kappa_b, \frac{1}{\lambda})$. Thus, any metric, g_{ab} on $\mathcal{M}$ accompanied with its, $g_a \equiv -\frac{1}{d+1}\partial_a \log \sqrt{|g|}$, where can be promoted to a metric $G_{\alpha\beta}$ on $\mathcal{N}$ as,

$$G_{\alpha\beta} = \delta^a{}_\alpha \delta^b{}_\beta g_{ab} - \lambda_0^2 g_\alpha g_\beta \tag{2.12}$$

$$G^{\alpha\beta} = g^{ab}(\delta^\alpha_a - g_a \Upsilon^\alpha)(\delta^\beta_b - g_b \Upsilon^\beta) - \lambda_0^{-2} \Upsilon^\alpha \Upsilon^\beta, \tag{2.13}$$

with $g_\alpha \equiv (g_a, \frac{1}{\lambda})$ and λ_0 a constant.

Second Relation to Virasoro Algebra: The projective two-cocycle, defined by (inserting the projective Laplacian as in Eq. (1.6)) the two-cocycle,

$$< \xi, \eta >_{(\zeta)} = \tilde{c} \int_{C(\zeta)} \xi^\alpha (\tilde{\nabla}_\alpha [G^{\rho\nu} \tilde{\nabla}_\rho \tilde{\nabla}_\nu] \eta^\beta \ G_{\beta\mu}) \zeta^\mu d\sigma - (\xi \leftrightarrow \eta),$$

for a path C parameterized by σ. The vector $\zeta^\mu = (\zeta^b, -\lambda\zeta^a g_a)$ defines the path C. This collapses to the Kirillov two-cocycle, Eq. (1.9), for paths restricted to $\mathcal{M}$, $\zeta_0^\mu = (\zeta^b, 0)$.

$$< \xi, \eta >_{(\zeta_0)} = \tilde{c} \int \xi_1 \left(2\mathcal{D}_{11} - g_{11}\frac{1}{\lambda_0^2}\right) \eta'_1 dx + \tilde{c} \int \xi_1 \eta'''_1 dx - (\xi \leftrightarrow \eta).$$

Comparing this to Eq. (1.9), we make the observation that the projective connection and the coadjoint element $(D, \tilde{c})$ are in correspondence through

$$2\tilde{c} \mathcal{D}_{11} - \frac{\tilde{c}}{\lambda_0^2} g_{11} = D,$$

where the projective connection is shifted by the one-dimensional metric tensor g_{11}.

2.2. *Geodesics revisited*

For insight, let us revisit the geodesics on $\mathcal{M}$. Consider a geodetics on $\mathcal{N}$,

$$\zeta^\alpha \tilde{\nabla}_\alpha \zeta^\beta = 0. \tag{2.14}$$

Separating the $\mathcal{M}$ coordinates from the fiber λ, we have the expressions

$$\frac{d^2 x^a}{du^2} + \Pi^a{}_{bc}\frac{dx^b}{du}\frac{dx^c}{du} = -2\frac{1}{\lambda}\left(\frac{d\lambda}{du}\right)\frac{dx^a}{du}, \tag{2.15}$$

$$\frac{d^2\lambda}{du^2} + \lambda\mathcal{D}_{bc}\frac{dx^b}{du}\frac{dx^c}{du} = 0. \tag{2.16}$$

Together, these equations are covariant and projectively invariant. Now, reparameterize u to a parameter τ that is affine with respect to the projective invariant $\Pi^a{}_{bc}$ in Eq. (2.15). Then, with $u \to \tau(u)$, we find that the Schwarzian derivative of τ with respect to u, $S(\tau(u))$, must satisfy

$$\mathcal{D}_{bc}\frac{dx^b}{du}\frac{dx^c}{du} = \frac{1}{2}\frac{\frac{d\tau}{du}(\frac{d^3\tau}{du^3}) - \frac{3}{2}(\frac{d^2\tau}{du^2})^2}{(\frac{d\tau}{du})^2} \equiv \frac{1}{2}S(\tau(u)) . \tag{2.17}$$

As an example, if $\mathcal{D}_{bc}\frac{dx^b}{du}\frac{dx^c}{du}$ vanishes, then $\tau = \frac{au+b}{cu+d}$, where a, b, c and d are real numbers. This is the Möbius transformation. Another example is when $\mathcal{D}_{bc}\frac{dx^b}{du}\frac{dx^c}{du} = \frac{m^2-1}{2u^2}$. Then, $\tau = (\frac{au^m+b}{cu^m+d})$ is the requirement. Möbius transformations are one-dimensional projective transformations. $\Pi^a{}_{bc}$ and $\mathcal{D}_{bc}$ are the components of a *projective connection*, $\tilde{\nabla}_\beta$. From here, we can compute curvature, spin connections, as well as the Dirac operator on the Thomas cone.

3. Building Blocks for TW Gravity

Equipped with the connection $\tilde{\nabla}_\alpha$, we can compute the projective curvature tensor, $[\tilde{\nabla}_\alpha, \tilde{\nabla}_\beta]V^\gamma = \mathcal{K}^\gamma{}_{\rho\alpha\beta}V^\rho$. The only non-vanishing components are

$$\mathcal{K}^a{}_{bcd} = \mathcal{R}^a{}_{bcd} + \delta^a{}_{[c}\mathcal{D}_{d]b} \text{ and } \mathcal{K}^\lambda{}_{cab} = \lambda\partial_{[a}\mathcal{D}_{b]c} + \lambda\Pi^d{}_{c[b}\mathcal{D}_{a]d},$$

where $\mathcal{R}^a{}_{bcd}$ is the projective equivariant curvature[14] "tensor" that is constructed using $\Pi^a{}_{bc}$ instead of $\Gamma^a{}_{bc}$ in the Riemann curvature tensor. The *Thomas–Whitehead action*[13] can be constructed as $S_{\text{TW}} = S_{\text{PEH}} + S_{\text{PGB}}$, where the projective Einstein–Hilbert action is

$$S_{\text{PEH}} = -\frac{1}{2\bar{\kappa}_0\lambda_0}\int d\lambda \, d^dx\sqrt{|G|}\mathcal{K}^a{}_{bcd}(\delta^c{}_a g^{bd}) \tag{3.1}$$

and the projective Gauss–Bonnet action is

$$S_{\text{PGB}} = -\frac{\bar{J}_0 c}{\lambda_0}\int d\lambda \, d^dx\sqrt{|G|} \left(\mathcal{K}^\alpha{}_{\beta\gamma\rho}\mathcal{K}_\alpha{}^{\beta\gamma\rho} - 4\mathcal{K}_{\alpha\beta}\mathcal{K}^{\alpha\beta} + \mathcal{K}^2\right). \tag{3.2}$$

The fields $\mathcal{D}_{ab}$, $\Pi^f{}_{de}$ and g_{ab} are independent fields. This action will give dynamics to $\mathcal{D}_{ab}$ in both two and four dimensions and collapses to the Einstein–Hilbert action plus a topological term when $\Pi^f{}_{de}$ is compatible with g_{ab} and when $\mathcal{D}_{ab} = 0$. This is a Palatini[21] formalism of the action.

Third Relation to Virasoro Algebra: The interaction term in the two-dimensional Polyakov action arises from the projective Einstein–Hilbert. Let g_{ab} denote the covariant

Polyakov metric. Then, the interaction term in the two-dimensional Polyakov action becomes

$$S_{\text{PEH}} = -\frac{1}{2\tilde{\kappa}_0\lambda_0} \int d^2x \, d\lambda \sqrt{|G|} \mathcal{K}_{\alpha\beta} G^{\alpha\beta} = -\frac{1}{2\tilde{\kappa}_0\lambda_0} \int d^2x \, d\lambda \sqrt{|G|} \mathcal{K}$$

$$= \frac{1}{\kappa_0} \int dx d\tau \, \mathcal{D}(x) \frac{(\partial s/\partial \tau)}{(\partial s/\partial x)} = \frac{1}{\kappa_0} \int dx_+ dx_- \, \mathcal{D}_{++} h_{--},$$

(3.3)

and κ_0 absorbs an integration of λ over specific limits. The last equality expresses the coupling in light-cone coordinates.[5]

We can continue to add building blocks and promote fermions to the Thomas cone. The gamma matrices on (even-dimensional) $\mathcal{M}$ easily extend to $\tilde{\gamma}^\alpha$, consistent with the anti-commutation relations that give the metric, one finds that $\tilde{\gamma}^m = \gamma^m$ and $\tilde{\gamma}^\lambda = -\frac{\lambda}{\lambda_0}\left(i\gamma^5 + \lambda_0 g_m \gamma^m\right)$, where γ^5 is as usual on $\mathcal{M}$. We can also build the spin connection using the frame fields of the metric, $\tilde{\omega}^\mu{}_{\alpha\nu} = \partial_\nu \tilde{e}^\mu{}_\alpha + \tilde{\Gamma}^\mu{}_{\rho\nu} \tilde{e}^\rho{}_\alpha$. It contains the diffeomorphism field in its λ component and will lead to a pseudoscalar coupling to the fermions,[19]

$$\tilde{\omega}_{\underline{5b}m} = -\lambda_0 e^p{}_{\underline{b}} \left(\mathcal{D}_{pm} - \partial_m g_p + \Pi^n{}_{pm} g_n\right).$$

With this, we can define the covariant derivative acting on the spinors as

$$\tilde{\nabla}_\mu = \partial_\mu + \tilde{\mathcal{O}}_\mu, \quad \text{where} \quad \tilde{\mathcal{O}}_\mu = \frac{1}{4}\tilde{\omega}_{\underline{\alpha\beta}\mu}\tilde{\gamma}^{\underline{\alpha}}\tilde{\gamma}^{\underline{\beta}}.$$

The fermions, themselves can be promoted to the Thomas cone[19] through a simple λ-dependent matrix-valued phase with v and w corresponding to density and chiral density weights,

$$\Psi(x^a, \lambda) = \left(\frac{\lambda}{\lambda_0}\right)^{\frac{(d+1)}{2}(vI_4 + w\gamma^5)} \phi(x^a).$$

4. Results and Discussion

The coadjoint orbit methods, developed by Balachandran and collaborators, provided the initial insight into the meaning of the Virasoro algebra in higher-dimensional gravities. The correspondence between the Virasoro algebra and projective geometry is exactly the correspondence of Kac–Moody algebras to Yang–Mills geometry. Because the field $\mathcal{D}_{bc}$ acts as a source of geometric origin in Einstein's equations, TW gravity gives insight into the origins of dark energy,[22] candidates for dark matter and dark matter portals through the TW Dirac action[19] and the origins of the inflaton.[23] Future work is focused on the phenomenology of the fermion coupling, longitudinal modes in gravitational radiation, a projective geometric source for super massive black holes and the interpretation of the λ integration in the context of renormalization group.

Acknowledgements

This work is dedicated to the immense intellectual energy that Prof. Balachandran shares selflessly with all of his collaborators. VR is proud to be among his students and collaborators.

References

1. A. P. Balachandran, G. Marmo, B. S. Skagerstam and A. Stern, *Gauge Theories and Fibre Bundles — Applications to Particle Dynamics.* Vol. 188 (1983). doi: 10.1007/3-540-12724-0_1.
2. F. Zaccaria, E. C. G. Sudarshan, J. S. Nilsson, N. Mukunda, G. Marmo and A. P. Balachandran, Universal unfolding of Hamiltonian systems: From symplectic structure to fiber bundles, *Phys. Rev. D.* **27** (1983) 2327. doi: 10.1103/PhysRevD.27.2327.
3. A. P. Balachandran, H. Gomm and R. D. Sorkin, Quantum symmetries from quantum phases: Fermions from Bosons, a $Z(2)$ Anomaly and Galilean invariance, *Nucl. Phys.* **B281** (1987) 573–612. doi: 10.1016/0550-3213(87) 90420-2.
4. A. P. Balachandran, Wess-ZuminoTerms and quantum symmetries, In *1st Asia Pacific Conference on High-energy Physics: Superstrings, Anomalies and Field Theory Singapore, Singapore, June 21–28, 1987*, pp. 375–407 (1987).
5. B. Rai and V. G. J. Rodgers, From coadjoint orbits to scale invariant WZNW type actions and 2-D quantum gravity action, *Nucl. Phys.* **B341** (1990) 119–133. doi: 10.1016/0550-3213(90)90264-E.
6. G. W. Delius, P. van Nieuwenhuizen and V. G. J. Rodgers, The method of coadjoint orbits: An algorithm for the construction of invariant actions, *Int. J. Mod. Phys.* **A5** (1990) 3943–3984. doi: 10.1142/S0217751X90001690.
7. V. Ovsienko and S. Tabachnikov, Projective differential geometry old and new: From the schwarzian derivative to the cohomology of diffeomorphism groups, *Cambridge Tracts in Mathematics;* 165 (2005).
8. G. Segal, Unitarity representations of some infinite dimensional groups, *Commun. Math. Phys.* **80** (1981) 301–342. doi: 10.1007/BF01208274.
9. E. Witten, Coadjoint orbits of the Virasoro group, *Commun. Math. Phys.* **114** (1988) 1. doi: 10.1007/BF01218287.
10. I. M. Gelfand and D. B. Fuchs, Cohomologies of lie algebra of tangential vector fields of a smooth manifold, *Funct. Anal. Appl.* **3**(3) (1969).
11. A. A. Kirillov, Infinite dimensional lie groups; their orbits, invariants and representations. The geometry of moments, *Lect. Notes Math.* **970** (1982) 101–123. doi: 10.1007/BFb0066026.

12. V. Ovsienko and S. Tabachnikov, What is the schwarzian derivative? *Not. Am. Math. Soc.* **56** (January, 2009).
13. S. Brensinger and V. G. J. Rodgers, Dynamical projective curvature in gravitation, *Int. J. Mod. Phys.* **A33**(36) (2019) 1850223. doi: 10.1142/S0217751X18502238.
14. T. Y. Thomas, Announcement of a projective theory of affinely connected manifolds, *Proc. Nat. Acad. Sci. USA* **11** (1925) 588–589.
15. T. Y. Thomas, On the projective and equi-projective geometries of paths, *Proc. Nat. Acad. Sci. USA.* **11** (1925) 199–203.
16. J. Whitehead, The representation of projective spaces, *Ann. Math.* **32** (1931) 327–360.
17. M. Eastwood and V. S. Matveev, Metric connections in projective differential geometry in "Symmetries and Overdetermined Systems of Partial Differential Equations", *Math. Appl.* **144** (2007) 339–351.
18. M. Crampin and D. Saunders, Projective connections, *J. Geom. Phys.* **57** (2007) 691–727.
19. S. Brensinger, K. Heitritter, V. G. J. Rodgers and K. Stiffler, General structure of Thomas−Whitehead gravity, *Phys. Rev. D.* **103**(4) (2021) 044060. doi: 10.1103/PhysRevD.103.044060.
20. S. J. Brensinger, *Projective Gauge Gravity*, PhD thesis, The University of Iowa (2020).
21. A. Palatini, Deduzione invariantiva delle equazioni gravitazionali dal principio di hamilton, *Rend. Circ. Mat. Palermo* (1919).
22. S. Brensinger, K. Heitritter, V. G. J. Rodgers, K. Stiffler and C. A. Whiting, Dark energy from dynamical projective connections, *Class. Quant. Grav.* **37**(5) (2020) 055003. doi: 10.1088/1361-6382/ab685d.
23. M. Abdullah, C. Bavor, B. Chafamo, X. Jiang, M. H. Kalim, K. Stiffler and C. A. Whiting, Inflation from dynamical projective connections, *Phys. Rev. D* **106**(8) (2022), 084049. doi: 10.1103/PhysRevD.106.084049.

Chapter 20

A New Approach to Classical Einstein–Yang–Mills Theory

Donald Salisbury

Austin College, 900 North Grand Ave, Sherman, Texas 75090, USA
dsalisbury@austincollege.edu

The conventional Rosenfeld–Bergmann–Dirac constrained Hamiltonian algorithm applied to Einstein–Yang–Mills theory is shown to be equivalent to a local gauge theoretic extension of Cartan's invariant integral approach to classical mechanics. In addition, the Hamiltonian generators of Legendre-projectable space-time diffeomorphism and gauge symmetries are derived directly as vanishing Noether charges. This leads directly to their interpretation as delivering the correct symmetry variations of both configuration and momentum field variables.

1. Introduction

I still vividly recall learning about Yang–Mills gauge theory in numerous visits with Bal in the 1970s on Clarendon Avenue in Syracuse. This coupled with his introductory class in group theory was instrumental in my professional preparation. But above all, I want to express my gratitude to him in helping me understand the origins and vicissitudes of particle theoretic dual resonance models. I have long identified him in this respect as an informal secondary thesis advisor. I like to characterize my resulting thesis publication in which I constructed a fully relativistic free quantized string in four space-time dimensions as one of the most widely uncited publications in the physics literature. And I've just confirmed — much to my satisfaction — that Bal has distinguished himself with a unique reference to this paper!

This work was my initial step in a career focus on the Hamiltonian analysis
of general covariance, and in this contribution, I present two new approaches
to this topic applied to fully covariant general relativity with a Yang–Mills
field source.

In part one, I illustrate an alternative derivation of Hamiltonian equations based on an extension of Élie Cartan's invariant integral principle to
gauge field theory. The method actually constitutes an alternative to the
Rosenfeld–Bergmann–Dirac constrained Hamiltonian algorithm. In part
two, I present a new derivation of the gauge generators of this model,
based on a refinement of the vanishing Noether charge that follows from
the underlying general coordinate covariance. I show how the space-time
transformations must be altered and expanded to include not only metric dependent space-time diffeomorphisms but also related Yang–Mills
gauge variations. These results follow from the demand that the generators be projectable under the Legendre transformation from the tangent
to the cotangent bundle, i.e., from configuration–velocity space to phase
space. This work actually constitutes a generalization of the vacuum general relativistic model that is presented in Ref. 8. It turns out that metric-
dependent space-time diffeomorphism symmetries cannot by themselves be
implemented as phase space transformations. Sundermeyer and I actually
showed this long ago.[9] More recently, I and my collaborators extended
these results to a larger phase space that incorporated the arbitrary gauge
variables,[3] but we invoked a different procedure than that presented in this
paper.

2. An Extension of Cartan's Invariant Integral Principle

Élie Cartan showed in 1922 that mechanical dynamical equations could be
derive from a new invariance principle.[1] He supposed that positions $q^i(t)$
and velocities $v^i(t)$ could be introduced as independent variables, and they
in addition to the time itself could be taken to be functions of an independent parameter s such that as the parameter value extended say from
zero to one, the variables returned to their original value. So, for example,
$q^i(t(s); s)\,|_{s=0} = q^i(t(s); s)\,|_{s=1}$. He then formed a closed integral over s
and demanded that the integral be independent of t. Several examples of
an extension of this principle to gauge field theories are given in Ref. 8.
Here, I extend the idea to the Einstein–Yang–Mills model.

I begin with the Lagrangian

$$\mathfrak{L}_{GRYM} = N\sqrt{g}\left({}^3R + K_{ab}K^{ab} - (K^a{}_a)^2\right) - \frac{1}{4}N\sqrt{g}F^i_{\mu\nu}F^j_{\alpha\beta}g^{\mu\alpha}g^{\nu\beta}C_{ij}.$$

(2.1)

The first term is the usual ADM Lagrangian and I employ a version of the Yang–Mills tensor in which the time derivative of the field A^i_a is replaced by an independent function $\mathcal{V}^i_a$ so that

$$F^i_{0a} = \mathcal{V}^i_a - A^i_{0,a} - C^i_{jk}A^j_0 A^k_a,$$

(2.2)

and

$$F^i_{ab} = A^i_{b,a} - A^i_{a,b} - C^i_{jk}A^j_a A^k_b.$$

(2.3)

The C^i_{jk} are the structure constants of the Yang–Mills gauge group and C_{ij} is a non-singular, symmetric group metric. (In a semisimple group, C_{ij} is usually taken to be $C^s_{it}C^t_{js}$; in an abelian group, one usually takes $C_{ij} = \delta_{ij}$.)

The three-metric time derivatives are replaced by independent fields v_{ab}, and inserting these expressions into the canonical momenta $p^{cd} = \sqrt{g}\left(K^{ab} - K^c{}_c e^{ab}\right)$, with indices raised by the inverse spatial metric e^{ab}, and the p^{cd} now taken to be functions of the v_{ab}. I also replace what will become the time derivatives of the lapse and shift N^μ by the fields V^μ.

I first carry out the equivalent of a Legendre transformation yielding

$$p^{ab} = \frac{\partial \mathfrak{L}_{GRYM}}{\partial v_{ab}} = \sqrt{g}\left(K^{ab} - K^c{}_c e^{ab}\right),$$

(2.4)

where

$$K_{ab} = \frac{1}{2N}\left(v_{ab} - N_{a|b} - N_{b|a}\right),$$

(2.5)

where indices are raised by e^{ab}, the inverse of g_{ab}. In addition, we have

$$\mathcal{P}^a_i = \frac{\partial \mathfrak{L}_{GRYM}}{\partial \mathcal{V}^i_a} = \sqrt{g}\left(\mathcal{V}^i_b - A^i_{0,b} - C^i_{jk}A^j_0 A^k_b\right)N^{-1}e^{ab}.$$

(2.6)

I also assume I have the primary constraints

$$P_\mu = 0, \quad \mathcal{P}^0_i = 0,$$

(2.7)

which are respectively the momenta conjugate to N^μ and A^i_0.

The canonical Hamiltonian, expressed in terms of the independent functions v_{ab} and $\mathcal{V}_a^i$, takes the form

$$
\begin{aligned}
\mathcal{H}_{GRYM} &= p^{ab}v_{ab} + \mathcal{P}_i^a\mathcal{V}_a^i - \mathfrak{L}_{GRYM} \\
&= \frac{N}{\sqrt{g}}\left(p_{ab}p^{ab} - (p^a{}_a)\right) - N\sqrt{g}\,^3R + 2N^a_{|b}p^b{}_a \\
&\quad + \frac{N}{2\sqrt{g}}C^{ij}g_{ab}\mathcal{P}_i^a\mathcal{P}_j^b + N^a\mathcal{P}_i^bF_{ab}^i \\
&\quad + \frac{N\sqrt{g}}{4}C_{ij}e^{ac}e^{bd}F_{ab}^iF_{cd}^j + \mathcal{D}_aA_0^i\mathcal{P}_i^a.
\end{aligned}
\tag{2.8}
$$

Now, finally, I impose the generalization of Cartan's invariant integral. I require that the closed integral over s as it ranges from zero to one be independent of t, i.e.,

$$
\begin{aligned}
I_{GRYM} &= \oint d^3x\big(p^{ab}dg_{ab} + P_\mu dN^\mu + \mathcal{P}_i^a dA_a^i - \mathcal{H}_{GRYM}dt \\
&\quad - P_\mu V^\mu dt - \mathcal{P}_i^0\mathcal{V}_0^i dt\big),
\end{aligned}
\tag{2.9}
$$

where $dg_{ab} = \frac{dg_{ab}}{ds}ds$, $dN = \frac{dN}{ds}ds$, $dA_i^i = \frac{dA_i^i}{ds}ds$ and $dt = \frac{dt}{ds}ds$ and the closed integral is required to be invariant under independent δ variations of the field variables. I have

$$
\begin{aligned}
0 = \delta I_{GRYM} &= \oint d^3x\,\Big[\delta p^{ab}dg_{ab} - \delta g_{ab}dp^{ab} + \delta P_\mu dn^\mu - \delta N^\mu dP_\mu + \delta\mathcal{P}_i^a dA_a^i \\
&\quad - \delta A_a^i d\mathcal{P}_i^a \\
&\quad - \left(\frac{\delta\mathcal{H}_{GRYM}}{\delta g_{ab}}\delta g_{ab} + \frac{\delta\mathcal{H}_{GRYM}}{\delta N^\mu}\delta N^\mu + \frac{\delta\mathcal{H}_{GRYM}}{\delta p^{ab}}\delta p^{ab}\right. \\
&\quad \left. + \frac{\delta\mathcal{H}_{GRYM}}{\delta A_\mu^i}\delta A_\mu^i + \frac{\delta\mathcal{H}_{GRYM}}{\delta\mathcal{P}_i^a}\delta\mathcal{P}_i^a\right)dt \\
&\quad + \left(\frac{\delta\mathcal{H}_{ADM}}{dg_{ab}}dg_{ab} + \frac{\delta\mathcal{H}_{ADM}}{\delta N^\mu}dN^\mu + \frac{\delta\mathcal{H}_{ADM}}{\delta p^{ab}}dp^{ab}\right. \\
&\quad \left. + \frac{\delta\mathcal{H}_{GRYM}}{\delta A_\mu^i}dA_\mu^i + \frac{\delta\mathcal{H}_{GRYM}}{\delta\mathcal{P}_i^a}d\mathcal{P}_i^a\right)\delta t \\
&\quad - \delta P_\mu V^\mu dt - \delta V^\mu P_\mu dt + dP_\mu V^\mu\delta t + P_\mu dV^\mu\delta t \\
&\quad - \delta\mathcal{P}_i^0\mathcal{V}_0^i dt - \delta\mathcal{V}_0^i\mathcal{P}_i^0 dt + d\mathcal{P}_i^0\mathcal{V}_0^i\delta t + \mathcal{P}_i^0 d\mathcal{V}_0^i\delta t\Big].
\end{aligned}
\tag{2.10}
$$

From the required vanishing of the coefficient of δp^{ab}, we conclude that $\frac{\partial g_{ab}}{\partial t} = \frac{\delta \mathcal{H}_{GRYM}}{\delta p^{ab}}$, from δg_{ab} that $\frac{\partial p^{ab}}{\partial t} = -\frac{\delta \mathcal{H}_{GRYM}}{\delta g_{ab}}$, from δP_μ that $\frac{\partial N^\mu}{\partial t} = V^\mu$, from $\delta \mathcal{P}_i^a$ that $\frac{\partial A_a^i}{\partial t} = \frac{\delta \mathcal{H}_{GRYM}}{\delta \mathcal{P}_i^a}$, from δA_a^i that $\frac{\partial \mathcal{P}_i^a}{\partial t} = -\frac{\delta \mathcal{H}_{GRYM}}{\delta A_a^i}$, from $\delta \mathcal{P}_i^0$ that $\frac{\partial A_0^i}{\partial t} = V_0^i$, from δN^μ that $\frac{\partial P_\mu}{\partial t} = -\frac{\delta \mathcal{H}_{ADM}}{\delta N^\mu}$ and from δA_0^i that $\frac{\partial \mathcal{P}_i^0}{\partial t} = -\frac{\delta \mathcal{H}_{GRYM}}{\delta A_0^i}$. Since, the primary constraint must be conserved, these latter two relations give us the secondary constraints

$$\mathcal{H}_0 := \frac{1}{\sqrt{g}} \left(p_{ab}p^{ab} - (p^c{}_c)^2 \right) \sqrt{g}^3 R \right) + \frac{1}{2\sqrt{g}} C^{ij} g_{ab} \mathcal{P}_i^a \mathcal{P}_j^b$$

$$+ \frac{\sqrt{g}}{4} C_{ij} e^{ac} e^{bd} F_{ab}^i F_{cd}^j = 0, \tag{2.11}$$

$$\mathcal{H}_a := -2p^b{}_{a|b} + \mathcal{P}_i^b F_{ab}^i = 0, \tag{2.12}$$

$$\mathcal{G}_i := -\mathcal{D}_a \mathcal{P}_i^a = 0. \tag{2.13}$$

With these results, it turns out that the coefficient of δt is zero as required. Thus, the invariant integral approach has delivered the known Hamiltonian analysis of general relativity with a Yang–Mills field source.

3. The Generator of Diffeomorphism-Induced Plus Gauge Phase Space Transformations

I next present a new derivation of the generators of Legendre-projectable symmetry transformations. The procedure follows closely the second Noether theorem-based approach pioneered by Leon Rosenfeld in 1930.[4,a] Rosenfeld however did not address the projectability problem. But it is in recognition of Rosenfeld's generally unappreciated development of constrained Hamiltonian dynamics that I and my co-authors have identified the following procedure as the Rosenfeld–Bergmann–Dirac method.

I consider transformed solutions of the Einstein–Yang–Mills equations obtained through an infinitesimal active coordinate transformation $x'^\mu = x^\mu - \epsilon^\mu(x)$. Rather than simply displace the new solutions to $x^\mu - \epsilon^\mu(x)$, I map the old solutions to this new location via the active manifold map. Note that from this active perspective, the coordinates are not altered, rather, the solutions as functions of these coordinates are shifted. The variation under these circumstances is simply the Lie derivative with respect to

---

[a]See Ref. 5 for a translation into English of this paper by myself and Kurt Sundermeyer and Ref. 10 for a careful analysis of the article and its relation to later work.

ϵ^μ — and continuing the tradition that began with Noether, I will represent these variations by $\bar{\delta}$.[b] I obtain

$$\bar{\delta}g_{ab} = 2g_{c(b}\epsilon^c_{,a)} + 2g_{c(a}N^c\epsilon^0_{,b)} + \dot{g}_{ab}\epsilon^0 + g_{ab,c}\epsilon^c, \tag{3.1}$$

$$\bar{\delta}N = N\dot{\epsilon}^0 - NN^a\epsilon^0_{,a} + \dot{N}\epsilon^0 + N_{,a}\epsilon^a, \tag{3.2}$$

$$\bar{\delta}N^a = N^a\dot{\epsilon}^0 - (N^2e^{ab} + N^aN^b)\epsilon^0_{,b} + \dot{\epsilon}^a - N^b\epsilon^a_{,b} + \dot{N}^a\epsilon^0 + N^a_{,b}\epsilon^b \tag{3.3}$$

and

$$\bar{\delta}A^i_\mu = A^i_\nu\epsilon^\nu_{,\mu} + A^i_{\mu,\nu}\epsilon^\nu. \tag{3.4}$$

Before inserting these variations in the action, I will first derive the corresponding vanishing charge that follows from Noether's second theorem. Noether's second theorem is applicable in this case since the action is invariant under the active diffeomorphism-induced field transformations. This is a consequence of the fact that Lagrangian transforms as a scalar density under these transformations — excepting for variations at spatial infinity which can be taken to vanish. Consequently, we have

$$\bar{\delta}\mathcal{L}_{GRYM} \equiv (\mathcal{L}_{GRYM}\epsilon^\mu)_{,\mu}. \tag{3.5}$$

Therefore, given that solutions are transformed into solutions, we have according to (3.5)

$$
\begin{aligned}
0 &= \int d^4x \left[\bar{\delta}\mathcal{L}_{GRYM} - (\mathcal{L}_{GRYM}\epsilon^\mu)_{,\mu}\right] \\
&= \int d^4x \left(\frac{\partial\mathcal{L}_{GRYM}}{\partial g_{ab,\mu}}\bar{\delta}g_{ab} + \frac{\partial\mathcal{L}_{GRYM}}{\partial N^\nu_{,\mu}}\bar{\delta}N^\nu + \frac{\partial\mathcal{L}_{GRYM}}{\partial A^i_{\nu,\mu}}\bar{\delta}A^i_\nu - \mathcal{L}_{GRYM}\epsilon^\mu\right)_{,\mu} \\
&= \int d^3x \left(\frac{\partial\mathcal{L}_{ADM}}{\partial g_{ab,0}}\bar{\delta}g_{ab} + \frac{\partial\mathcal{L}_{ADM}}{\partial N^\nu_{,0}}\bar{\delta}N^\nu + \frac{\partial\mathcal{L}_{GRYM}}{\partial A^i_{\nu,0}}\bar{\delta}A^i_\mu - \mathcal{L}_{GRYM}\epsilon^0\right)\Bigg|_{x^0_i}^{x^0_f} \\
&= \int d^3x \left(p^{ab}\bar{\delta}g_{ab} + \mathcal{P}^a_i\bar{\delta}A^i_a + P_\mu\bar{\delta}N^\nu + \mathcal{P}^0_i\bar{\delta}A^i_0 - \mathcal{L}_{GRYM}\epsilon^0\right)\Bigg|_{x^0_i}^{x^0_f}. \tag{3.6}
\end{aligned}
$$

It is noteworthy that Rosenfeld actually derived the equivalent conserved quantity for a general relativistic dynamical tetrad field in interaction with electrodynamic and spinorial fields.

---

[b]See Ref. 6 for Bergmann's use of Noether's notation. Note also that this is a special case of the variations that were represented by δ_0 in Ref. 7.

There is however a problem. It is not projectable to phase space because of the appearance of the time derivatives $\dot{N}^\mu$ and $\dot{A}^i_0$ as we note in substituting the variations (3.1) through (3.4). Representing what we have called the Rosenfeld–Noether vanishing charge density by $\mathfrak{C}^0_{GRYM}$, we have

$$
0 = \int d^3x\, \mathfrak{C}^0_{GRYM}
$$

$$
= \int d^3x \left[p^{ab} \left(2g_{c(b}\epsilon^c_{,a)} + 2g_{c(a}N^c\epsilon^0_{,b)} + \dot{g}_{ab}\epsilon^0 + g_{ab,c}\epsilon^c \right) \right.
$$

$$
+ \mathcal{P}^a_i \left(A^i_\nu \epsilon^\nu_{,a} + A^i_{a,b}\epsilon^b + \dot{A}^i_a \epsilon^0 \right) - \mathcal{L}_{GRYM}\epsilon^0 \tag{3.7}
$$

$$
+ P_0 \left(N\dot{\epsilon}^0 - NN^a\epsilon^0_{,a} + \dot{N}\epsilon^0 + N_{,a}\epsilon^a \right) \tag{3.8}
$$

$$
+ P_a \left(N^a\dot{\epsilon}^0 - (N^2e^{ab} + N^aN^b)\epsilon^0_{,b} + \dot{\epsilon}^a - N^b\epsilon^a_{,b} + \dot{N}^a\epsilon^0 + N^a_{,b}\epsilon^b \right) \tag{3.9}
$$

$$
\left. + \mathcal{P}^0_i \left(A^i_\nu \dot{\epsilon}^\nu + \dot{A}^i_0 \epsilon^0 + A^i_{0,a}\epsilon^a \right) \right]. \tag{3.10}
$$

I concentrate first on (3.8) and (3.9) where the unprojectable time derivatives of the lapse and shift fields appear. As was shown in Ref. 2, these can and must be eliminated through gravitational field-dependent infinitesimal transformations $\epsilon^\mu = n^\mu \xi^0 + \delta^\mu_a \xi^a$, where $n^\mu = \left(N^{-1}, -N^{-1}N^a \right)$ is the orthonormal to the constant time surfaces. It follows that

$$
\dot{\epsilon}^0 = -N^{-2}\dot{N}\xi^0 + N^{-1}\dot{\xi}^0, \tag{3.11}
$$

$$
\epsilon^0_{,a} = -N^{-2}N_{,a}\xi^0 + N^{-1}\xi^0_{,a}, \tag{3.12}
$$

$$
\dot{\epsilon}^a = N^{-2}\dot{N}N^a\xi^0 - N^{-1}\dot{N}^a\xi^0 - N^{-1}N^a\dot{\xi}^0 + \dot{\xi}^a \tag{3.13}
$$

and

$$
\epsilon^a_{,b} = N^{-2}N_{,b}N^a\xi^0 - N^{-1}N^a_{,b}\xi^0 - N^{-1}N^a\xi^0_{,b} + \xi^a_{,b}. \tag{3.14}
$$

Substitution into (3.8) and (3.9) yields

$$
P_0 \left(\dot{\xi}^0 - N^a\xi^0_{,a} + N_{,a}\xi^a \right) \tag{3.15}
$$

and

$$
P_a \left(N_{,b}e^{ab}\xi^0 - Ne^{ab}\xi^0_{,b} + \dot{\xi}^a - N^b\xi^a_{,b} + N^a_{,b}\xi^b \right). \tag{3.16}
$$

Next, substituting $\epsilon^\mu = n^\mu \xi^0 + \delta_a^\mu \xi^a$ into (3.7), I obtain first

$$\left(p^{ab}\dot{g}_{ab} + \mathcal{P}_i^a \dot{A}_a^i - \mathcal{L}_{GRYM}\right)\epsilon^0 = \mathcal{H}_0 \xi^0, \tag{3.17}$$

where $\mathcal{H}_0$ is the secondary constraint given by (2.11). The remaining terms in (3.7) are

$$-2p^{ab}g_{cb}N_{,a}^c N^{-1}\xi^0 - p^{ab}g_{ab,c}N^c N^{-1}\xi^0$$

$$+2p^{ab}g_{cb}\xi_{,a}^c + p^{ab}g_{ab,c}\xi^c + 2p^{ab}N_{a|b}N^{-1}\xi^0. \tag{3.18}$$

But we still have in addition to the unprojectable $\dot{A}_0^i$ term in (3.10) the return of time derivatives of the lapse and shift that result from the time derivative of ϵ^μ. Fortunately, there is a way of eliminating all of these unprojectable time derivatives by making use of the additional Yang–Mills local gauge symmetry of the form

$$\delta_\Lambda A_\mu^i = -\Lambda_{,\mu}^i - C_{jk}^i \Lambda^j A_\mu^k. \tag{3.19}$$

The action is invariant under this transformation, and there is therefore a corresponding vanishing Noether charge density which can be derived in a manner similar to (3.6), namely this equation simplifies in this case to

$$0 = \int d^3x \left(\mathcal{P}_i^a \bar\delta_\Lambda A_a^i + \mathcal{P}_i^0 \bar\delta_\Lambda A_0^i\right)\Big|_{x_i^0}^{x_f^0}. \tag{3.20}$$

Then, because the time dependence of Λ^i is arbitrary, we deduce the existence of the vanishing Noether charge density

$$\mathfrak{C}_\Lambda = -\mathcal{P}_i^\mu \left(\Lambda_{,\mu}^i + C_{jk}^i \Lambda^j A_\mu^k\right). \tag{3.21}$$

Fortunately, the form of the diffeomorphism variation given in (3.4) immediately suggests an appropriate additional gauge transformation that will eliminate the time derivatives of ϵ^ν. We add a gauge variation with $\Lambda = A_\nu^i \epsilon^\nu = A_\nu^i n^\nu \xi^0 + A_a^i \xi^a =: {}^\xi\Lambda$ which delivers a net variation

$$\delta A_\mu^i := \bar\delta A_\mu^i + \delta_{\xi\Lambda} A_\mu^i = F_{\nu\mu}^i \epsilon^\nu, \tag{3.22}$$

i.e.,

$$\delta A_0^i = -F_{0a}\left(-N^{-1}N^a \xi^0 + \xi^a\right)$$

$$= C^{ij}g_{ab}\mathcal{P}_j^b N^a \xi^0 + \left(NC^{ij}g_{ab}\mathcal{P}_j^b + N^b F_{ba}^i\right)\xi^a \tag{3.23}$$

$$\delta A_a^i = F_{0a}^i N^{-1}\xi^0 - F_{ab}\left(-N^{-1}N^b \xi^0 + \xi^b\right)$$

$$= \left(\dot{A}_a^i - D_a A_0^i\right)\xi^0 - F_{ab}^i \xi^b, \tag{3.24}$$

where in (3.23), I used the result from the field equations that $F_{0a}^i = NC^{ij}g_{ab}\mathcal{P}_j^b + N^b F_{ba}^i$ and in (3.24), the fact that it is also true that $F_{0a}^i = \dot{A}_a^i - \mathcal{D}_a A_0^i$.

Returning to the vanishing integrand of (3.6), I replace the $\bar{\delta}$ variations by the projectable δ variations, thereby delivering the vanishing Legendre projectable Rosenfeld–Noether charge density. I omit some calculation details here dealing with the metric field variations. These computations can be found in Ref. 8. I obtain the vanishing generator

$$
\begin{aligned}
\mathcal{C}(\xi) &= p^{ab}\delta g_{ab} + \mathcal{P}_i^a \delta A_a^i + P_\mu \delta N^\nu + \mathcal{P}_i^0 \delta A_0^i - \mathcal{L}_{GRYM} N^{-1}\xi^0 \\
&= \mathcal{H}_0 \xi^0 - 2p^{ab}g_{cb}N_{,a}^c N^{-1}\xi^0 - p^{ab}g_{ab,c}N^c N^{-1}\xi^0 \\
&\quad + 2p^{ab}g_{cb}\xi_{,a}^c + p^{ab}g_{ab,c}\xi^c + 2p^{ab}N_{a|b}N^{-1}\xi^0 \\
&\quad + P_0\left(\dot{\xi}^0 - N^a\xi_{,a}^0 + N_{,a}\xi^a\right) \\
&\quad + P_a\left(N_{,b}e^{ab}\xi^0 - Ne^{ab}\xi_{,b}^0 + \dot{\xi}^a - N^b\xi_{,b}^a + N_{,b}^a\xi^b\right) \\
&\quad - \mathcal{P}_i^a\left(\mathcal{D}_a A_0^i \xi^0 + F_{ab}^i \xi^b\right) \\
&\quad + \mathcal{P}_i^0\left(C^{ij}g_{ab}\mathcal{P}_j^b N^a\xi^0 + \left(NC^{ij}g_{ab}\mathcal{P}_j^b + N^b F_{ba}^i\right)\xi^a\right).
\end{aligned}
\tag{3.25}
$$

This is not entirely equivalent to the generator that was obtained in Ref. 3 through a different route and that group-theoretic calculation needs to be revisited. There is here, however, no doubt that this is the correct generator. It is based directly on the known transformation rules for the configuration field variables. And indeed, given that it relies directly on the phase space one-form $p_a dq^a$, it is clear that it also generates the correct variations of the momentum variables. This was actually proven already in 1930 by Rosenfeld but only for the Legendre-projectable case.

References

1. E. Cartan, Leçons sur les Invariants Intégraux. Librairie Scientifique A. Hermann et Fils (1992).
2. J. Pons, D. Salisbury and L. Shepley, Gauge transformations in the Lagrangian and Hamiltonian formalisms of generally covariant theories. *Phys. Rev. D* **55** (1997) 658–668.
3. J. Pons, D. Salisbury and L. Shepley, Gauge transformations in Einstein-Yang-Mills theories. *J. Math. Phys.* **41**(8) (2000) 5557–5571.
4. L. Rosenfeld, Zur Quantelung der Wellenfelder. *Annalen der Physik* **5** (1930) 113–152.

5. L. Rosenfeld, On the quantization of wave fields. *Eur. Phys. J. H* **42** (2017) 63–94.

6. D. Salisbury, Toward a quantum theory of gravity: Syracuse 1949–1962. In Blum, A., Lalli, R., and Renn, J., editors, *The Renaissance of General Relativity in Context*, pp. 221–255. Birkhäuser (2020).

7. D. Salisbury, A history of observables and Hamilton-Jacobi approaches to general relativity. *Eur. Phys. J. H* **47** (2022) 7-1–38.

8. D. Salisbury, J. Renn and K. Sundermeyer, Cartan rediscovered in general relativity, *Gen. Rel. Grav.* **54** (2022) 116.

9. D. Salisbury and K. Sundermeyer, Local symmetries of the Einstein-Yang-Mills theory as phase space transformations. *Phys. Rev. D* **27**(4) (1983) 757–763.

10. D. Salisbury and K. Sundermeyer, Léon Rosenfeld's general theory of constrained Hamiltonian dynamics, *Eur. Phys. J. H* (2017) 1–39.

Chapter 21

Renormalization Group and String Theory

Balachandran Sathiapalan[*][‡] and Homi Bhabha[†]

Institute of Mathematical Sciences, CIT Campus, Tharamani
Chennai 600113, India
†*National Institute, Training School Complex, Anushakti Nagar*
Mumbai 400085, India
‡*bala@imsc.res.in*

The renormalization group is one of the foundational concepts in quantum field theory. We discuss the role played by the renormalization group in string theory. One concrete example is described that involves an exact form of the renormalization group equation written down by Wilson and is commonly known as Exact Renormalization Group (ERG). This example is in the context of AdS/CFT correspondence in string theory. Starting from the ERG equation for the boundary conformal field theory, it is shown that one can obtain what has come to be called the holographic renormalization group equation. This procedure can lead to a derivation of the AdS/CFT correspondence. Another application of the ERG formalism is the derivation of gauge invariant and background invariant equations of motion for the fields of open and closed strings. Finally, a speculation that some version of the renormalization group is the underlying symmetry principle of string theory is also outlined.

1. Reminiscences

My first meeting with Bal was at an MRST conference held in Syracuse in the early 1990 s. I was working at Penn State University in Scranton and I drove there with my wife. At the conference, I was talking to Parameswaran Nair and S.G. Rajeev, both of whom I knew already, when Bal joined the group. I introduced myself and we started talking. Since we are all from Kerala, the conversation was in a mixture of Malayalam and English. I remember that he was very amused when

I gave a Malayalam translation for "electronic bulletin board"! I was really impressed by his down to earth personality and warmth.

He encouraged me to visit Syracuse. I used to drive there once a week and park my car in his driveway and go with him to the office. He had a bright student, L. Chandar, who was working on quantum Hall effect. (Chandar is now in the industry in Bangalore and we continue to be in touch.) The three of us collaborated and wrote a couple of nice papers on the topic. During those visits, we would all sit in one room (the legendary room 316) and all his students and post-docs would describe their ongoing work. I found the atmosphere very friendly and stimulating. Whenever I meet his students, they all fondly remember these lively sessions.

After I returned to Matscience in Chennai, we kept in touch through email and his occasional visits. Even when our research interests don't coincide, it is always a pleasure to describe my work to him and hear about his work.

His deep mathematical knowledge enabled him to think of issues that in a sense were "ahead of the time" for many of us. Referring to the isomorphism of Hilbert spaces, he once said "How do you tell a hydrogen atom from a hole in the ground?" That was a new line of thought for me. Again, Bal has been telling me about algebraic formulations of quantum field theory for many years now. Nowadays, I see many others (including string theorists) actively thinking about these issues.

When I meet Bal, we talk of many things besides Physics. Politics, history and literature are a staple for him. Also being from Kerala, we share a cultural background and he is very well read and knowledgeable in Malayalam literature (and also in fact literature in many other languages) and I have often followed his recommendations about novels that are worth reading.

I look forward to many more informative and entertaining discussions with Bal in Physics and outside Physics. On the occasion of his birthday, I wish him many more years of active intellectual life!

2. Introduction

The renormalization group has played a foundational role in quantum field theory.[1] In this chapter, we describe an application of the renormalization group in string theory. The application that we describe in some detail is in the context of AdS/CFT correspondence in string theory.[2] According to this, a CFT living on the boundary of AdS space has a dual description as a bulk gravity (actually string) theory in AdS. The extra radial dimension has an interpretation as the scale of the CFT. This is evident from the following form of the AdS metric:

$$ds^2 = \frac{dz^2 + dx^i dx_i}{z^2}, \qquad (2.1)$$

where z is the radial coordinate and x^i are the CFT coordinates. If the AdS/CFT conjecture is correct, then the bulk radial evolution should correspond to an RG evolution of the boundary. This has been termed "holographic RG".[3,a]

We start from the ERG equation of a D-dimensional boundary CFT. We show that the evolution operator for this equation can be written as a functional integral of a field theory in a D+1-dimensional bulk AdS space. The extra radial coordinate of the AdS space is the RG scale Λ of the boundary CFT. This thus makes contact with the idea of holographic RG. The procedure described here gives a derivation of holographic RG that does not depend on the AdS/CFT conjecture.[5–8] This is described in Section 3. When this procedure is applied to a perturbation of the boundary by the energy–momentum tensor, one finds bulk equations for the gravitational field. Thus, dynamical gravity comes out of ERG!

Another application of ERG not described here is that it gives gauge invariant equations for the modes of the open and closed string when applied to the string world sheet theory. This uses a "loop variable" formalism. The gauge transformation of string theory fields have the form of a space-time scale transformation in this formulation. This also motivates a speculation about the underlying symmetry principle of string theory. This is mentioned in Section 4.

3. Exact RG and Holographic RG

We begin by describing Polchinski's ERG equation and its evolution operator for a free scalar field theory.[4] Then, we show how it can be mapped to AdS space at which point it becomes the holographic RG that comes out of the holographic duality conjecture. The radial coordinate of the bulk AdS space plays the role of the RG scale as expected in holographic RG. We then apply the same technique to the free theory perturbed by the

---

[a]For want of space, we have given very few references in this chapter. We refer the reader to Refs. 5, 8 and also Refs. 16, 17 for more references.

energy–momentum tensor. On mapping the evolution operator to AdS as before, we obtain the action for the gravitational field. Thus, we have the remarkable fact that dynamical gravity comes out of exact RG!

3.1. Polchinski's ERG, its Evolution Operator and map to AdS

The procedure mentioned above consists of three steps.

Step 1: ERG Equation

The starting point is Polchinski's ERG equation for the Wilson action (or equivalently for the generating functional) of a *fundamental* scalar field. We start with a scalar field (bare) theory defined at the scale Λ_0.

$$S_B = \int \frac{d^D p}{(2\pi)^D} [\frac{1}{2}\phi(p)\frac{p^2}{K(\Lambda_0)}\phi(-p)] + S_{B,I}[\phi], \qquad (3.1)$$

where $K(\Lambda)$ is a cutoff function that allows propagation of modes below Λ. For example,

$$K(\Lambda) = e^{-\frac{p^2}{\Lambda^2}}.$$

Introduce low- and high-energy modes with propagators Δ_l and Δ_h:

$$\phi(p) = \phi_l(p) + \phi_h(p); \quad \frac{K(\Lambda_0)}{p^2} \equiv \Delta = \Delta_l + \Delta_h. \qquad (3.2)$$

We can take

$$\Delta_l = \frac{K(\Lambda)}{p^2}, \quad \Delta_h = \frac{K(\Lambda_0) - K(\Lambda)}{p^2}. \qquad (3.3)$$

The original functional integral with the standard kinetic term can then be shown to be equivalent to a theory with two kinetic terms:

$$\frac{1}{2}\int_p \phi_l(p)\Delta_l^{-1}\phi_l(-p) + \frac{1}{2}\int_p \phi_h(p)\Delta_h^{-1}\phi_h(-p). \qquad (3.4)$$

For simplicity, $G = \Delta_l$ in the following.

The Wilson action obtained by integrating out ϕ_h is written as

$$S_\Lambda[\phi] = \frac{1}{2}\int \frac{d^D p}{(2\pi)^D}\phi(p)G^{-1}\phi(-p) + S_{\Lambda,I}[\phi], \qquad (3.5)$$

where

$$\int \mathcal{D}\phi_h \ e^{-\frac{1}{2} \int_p \phi_h(p) \Delta_h^{-1} \phi_h(-p) + S_{B,I}[\phi_l + \phi_h]} \equiv e^{-S_{\Lambda,I}[\phi_l]}. \tag{3.6}$$

It obeys Polchinski's ERG equation:

$$\frac{\partial}{\partial t} e^{-S_{\Lambda,I}[\phi]} = -\frac{1}{2} \int_p \dot{G}(p) \frac{\delta^2}{\delta\phi(p)\delta\phi(-p)} e^{-S_{\Lambda,I}[\phi]}. \tag{3.7}$$

Note that it looks like a Schrödinger (or diffusion) equation with a time-dependent mass.

Step 2: Holographic Form

Using the analogy with Schrödinger equation, one can write an evolution operator in the form of a functional integral.[5]

Defining $\psi[\phi, t] = e^{-S_{\Lambda,I}}$ with $\Lambda_0 = \Lambda e^t$, we have

$$\psi[\phi_f, t_f] = \int \mathcal{D}\phi_i(p) \int_{\phi(p,t_i)=\phi_i(p)} \mathcal{D}\phi(p,t) \ e^{-\frac{1}{2}\int dt \ \int_p \dot{G}^{-1}\dot{\phi}(p,t)\dot{\phi}(-p,t)} \psi[\phi_i(p), t_i].$$

$$\tag{3.8}$$

Thus, starting with a D-dimensional field theory, we obtain a $D + 1$-dimensional field theory as the ERG evolution operator. This is a holographic form of the (free) boundary theory. We now map this to AdS space.

Step 3: Mapping to AdS_{D+1} space-time

The AdS_{D+1} metric is taken to be

$$ds^2 = \frac{dz^2 + dx_i dx^i}{z^2}. \tag{3.9}$$

In Ref. 5, it was shown that the bulk scalar field action above can be mapped to a standard AdS_{D+1} free scalar field theory by a field redefinition: Writing $z = \Lambda^{-1}$ instead of t, define an AdS scalar field $y(p, z)$

$$\phi(p, z) = f(p, z)y(p, z), \tag{3.10}$$

where f is defined by

$$f^2 = -z^{-D}\dot{G} \tag{3.11}$$

and is chosen to satisfy

$$\left[\frac{d^2}{dz^2} + \frac{1}{z}\frac{d}{dz} - \left(p^2 + \frac{m^2}{z^2} \right) \right] \left(\frac{1}{f(p,z)} \right) = 0. \tag{3.12}$$

Then, the scalar field action becomes

$$S[y(p)] = \int dz \int_p z^{-D+1} \left[\frac{\partial y(p,z)}{\partial z} \frac{\partial y(-p,z)}{\partial z} + \left(p^2 + \frac{m^2}{z^2} \right) y(p,z)y(-p,z) \right].$$

(3.13)

f, G are given in terms of Bessel functions and are given in Ref. 5.

Thus, we have a D+1-dimensional AdS bulk space-time and a scalar field action exactly as dictated by the AdS/CFT correspondence.

We would like to apply this to obtain what is perhaps the most interesting case: gravity in the bulk.

3.2. *ERG for energy–momentum tensor*

We proceed to study a perturbation of the boundary fixed point theory by a conserved spin-2 tensor, i.e., the energy–momentum tensor $T_{\mu\nu}$. Being a composite operator, we introduce an auxiliary field $\sigma_{\mu\nu}$ to stand for the composite $T_{\mu\nu}$ as well as a source $h^{\mu\nu}$ for it and write down an ERG equation for the action $S_\Lambda[\sigma_{\mu\nu}]$ for $\sigma_{\mu\nu}$. The evolution operator for this ERG equation is a $D+1$-dimensional field theory as before. On mapping to AdS space, one expects to get an action for a massless spin-2 field, viz. the graviton, in the bulk. We work out only the quadratic part of the bulk action (kinetic term), as a proof of the principle that a dynamical graviton emerges out of exact RG without invoking an AdS/CFT conjecture.[8]

In the boundary CFT, we keep only the physical degrees of freedom. This has the consequence that one obtains a gauge-fixed version of the quadratic graviton action in AdS space. The connection with the gauge invariant action has been worked out in the literature.[9-13]

For the CFT at the boundary, we consider the simplest case of a theory of N free scalars. The form of the *quadratic* bulk graviton action is clearly independent of this choice.

We start with

$$Z[h_{\mu\nu}] = \int \mathcal{D}\phi^I e^{-\frac{1}{2} \int_x \partial_\mu \phi^I \partial^\mu \phi^I + \int_x h_{\mu\nu} \Theta^{\mu\nu}},$$

(3.14)

where $\Theta^{\mu\nu}$ is the *improved* energy–momentum tensor given by

$$\Theta_{\mu\nu} = \partial_\mu \phi^I \partial_\nu \phi^I - \frac{1}{2} \delta_{\mu\nu} \partial_\alpha \phi^I \partial^\alpha \phi^I - \frac{D-2}{4(D-1)} s_{\mu\nu}[\phi^2],$$

(3.15)

where

$$s_{\mu\nu}[\phi^2] \equiv (\partial_\mu \partial_\nu - \delta_{\mu\nu}\Box)\phi^2 \qquad (3.16)$$

is a transverse piece that has been added to make the EM tensor traceless. $\Theta_{\mu\nu}$ is conserved (due to diffeomorphism invariance),

$$\partial^\mu \Theta_{\mu\nu} = 0.$$

Therefore, there is an invariance under

$$\delta h_{\mu\nu} = \partial_\mu \xi_\nu + \partial_\nu \xi_\mu,$$

which allows us to set a gauge condition

$$\partial^\mu h_{\mu\nu} = 0.$$

It is also traceless because it is a CFT:

$$\Theta^\mu_\mu = 0.$$

So, we can choose $h^\mu_\mu = 0$.

3.2.1. *Auxiliary field*

Now, we can define an auxiliary field $\sigma^{\mu\nu}$ by

$$\sigma^{\mu\nu} = \Theta^{\mu\nu} \qquad (3.17)$$

in the bare theory. Thus, we can introduce it in the functional integral by

$$\int \mathcal{D}\sigma^{\mu\nu} \delta(\sigma^{\mu\nu} - \frac{\delta}{\delta h_{\mu\nu}}) Z[h_{\mu\nu}] = Z[h_{\mu\nu}].$$

The delta function can be implemented using Lagrange multipliers. An IR scale Λ is introduced as was done for the scalar, and finally, one obtains a Λ-dependent action for $\sigma_{\mu\nu}$:

$$Z_\Lambda[h_{\mu\nu}] = e^{W_\Lambda[h_{\mu\nu}]} = \int \mathcal{D}\sigma^{\mu\nu} e^{-S_\Lambda[\sigma^{\mu\nu}] + \int_x h_{\mu\nu}\sigma^{\mu\nu}}. \qquad (3.18)$$

Starting from the Polchinski's ERG equation in terms of ϕ_l, one obtains the following ERG equation for $S_\Lambda[\sigma_{\mu\nu}]$ at leading order[8]:

$$\frac{\partial}{\partial t} e^{-S_\Lambda[\sigma_{\mu\nu}]} = -\frac{1}{2} \int_x \int_y 4\dot{\Delta}_h(x-y)\partial^\mu_x \partial^\nu_x \partial^\rho_y \partial^\sigma_y \Delta_h(x-y)$$
$$\times \frac{\delta^2}{\delta\sigma^{\mu\nu}(x)\delta\sigma^{\rho\sigma}(y)} e^{-S_\Lambda[\sigma^{\mu\nu}]} + higher\ order.$$

Using the transversality properties, one obtains in momentum space an action of the form

$$\frac{\partial}{\partial t} e^{-S_\Lambda[\sigma^{\mu\nu}]} = \frac{1}{2} \int_p I(p^2) \frac{\delta^2}{\delta\sigma^{\mu\nu}(p)\delta\sigma_{\mu\nu}(-p)} e^{-S_\Lambda[\sigma^{\mu\nu}]} + higher \ order.$$

(3.19)

The evolution operator for this ERG equation is known, and we get

$$e^{-S_\Lambda[\sigma_f^{\mu\nu}]} = \int \mathcal{D}\sigma_i^{\mu\nu}(p) \int_{\sigma^{\mu\nu}(p,t_i)=\sigma_i^{\mu\nu}(p)}^{\sigma^{\mu\nu}(p,t_f)=\sigma_f^{\mu\nu}(p)} \mathcal{D}\sigma^{\mu\nu}(p,t)$$

$$e^{-\frac{1}{2}\int_p \int_{t_i}^{t_f} dt \frac{\dot{\sigma}^{\mu\nu}(p,t)\dot{\sigma}_{\mu\nu}(p,t)}{I(p^2)}} e^{-S_{\Lambda_0}[\sigma_i^{\mu\nu}]}.$$

(3.20)

Here, we take $e^{-t_i} = \Lambda_0$, $e^{-t_f} = \Lambda$ in some units.

The action of radial evolution is then

$$S = \frac{1}{2} \int_p \int_{t_i}^{t_f} dt \frac{\dot{\sigma}^{\mu\nu}(p,t)\dot{\sigma}_{\mu\nu}(p,t)}{I(p^2)}.$$

(3.21)

3.2.2. Mapping to AdS

The radial evolution action (3.21) is that of $\frac{D(D-1)}{2} - 1$ scalars and we can use the same map as in Refs. 5, 7 and in the previous section to map this to AdS:

$$\sigma_{\mu\nu}(p,t) = y_{\mu\nu}(p,t)f(p,t),$$

(3.22)

with $f^2 = -\dot{I}z^{-D}$, and we need to choose $m^2 = 0$ to obtain the correct low-energy behavior of the propagator. f is then determined by (3.12). This maps the system to an action

$$S_{AdS} = \int dz \frac{d^D p}{(2\pi)^D} z^{-1+D} \left[\frac{\partial y_{\mu\nu}(p)}{\partial z} \frac{\partial y_{\rho\sigma}(-p)}{\partial z} + p^2 y_{\mu\nu}(p)y_{\rho\sigma}(-p) \right] \delta^{\mu\rho}\delta^{\nu\sigma}.$$

(3.23)

By defining $h_{\mu\nu}^B z^2 = y_{\mu\nu}$, one obtains an action for $h_{\mu\nu}^B$, the bulk metric perturbation:

$$\int dz \frac{d^D p}{(2\pi)^D} z^{1-D} \left[z^4 \left(\frac{\partial h_{\mu\nu}^B(p)}{\partial z} \frac{\partial h_{\rho\sigma}^B(-p)}{\partial z} + p^2 h_{\mu\nu}^B(p)h_{\rho\sigma}^B(-p) \right) \right.$$

$$\left. + 4z^3 h_{\mu\nu}^B \frac{\partial h_{\rho\sigma}^B}{\partial z} + 4z^2 h_{\mu\nu}^B h_{\rho\sigma}^B \right] \delta^{\mu\rho}\delta^{\nu\sigma},$$

(3.24)

and an equation of motion as also found in Ref. 13

$$(\partial_z^2 - p^2)h_{\mu\nu}^B(p) + \frac{5-D}{z}\partial_z h_{\mu\nu}^B(p) - \frac{2D-4}{z^2}h_{\mu\nu}^B(p) = 0. \qquad (3.25)$$

Thus, let us summarize: Starting from ERG in the boundary theory for perturbation involving the energy–momentum tensor, we obtain a quadratic action for gravity in the bulk AdS space. Taking account of higher-order terms in the ERG is expected to produce the interaction terms. Thus, dynamical gravity in AdS_{D+1} comes out of ERG of a D-dimensional CFT! Note that we have not used the AdS/CFT conjecture anywhere to obtain this. One can in principle do this for other higher spin composites in the CFT and obtain actions for higher spin fields in the bulk. In general, one expects a theory with an infinite number of higher spin fields in the bulk.[14, 15]

4. Conclusions and a Speculation

We have described in this chapter how ERG gives dynamical gravity as suggested by the AdS/CFT correspondence. The hope is that one can derive the AdS/CFT correspondence using such an approach.

Our interest in ERG is primarily the result of another application in string theory. One can start with world sheet CFT description of string theory and write the action in terms of "loop variables".[16, 17] It is well known that the equations of motion of the fields of the string are equivalent to the conditions of conformal invariance. Thus, they are obtained by using an ERG on the *world sheet*, and in the loop variable formalism, they are gauge invariant and also background independent. Interestingly, the gauge transformations associated with the various modes of the string can be packaged into a simple form that looks like a *space-time* scale transformation but local along the string.

This (loop variable) formalism was motivated by a speculation on the principle underlying string theory.[16] One can think of string theory as a way of regularizing field theory that maintains Lorentz invariance. This is like a lattice, but instead of treating it as a scaffolding to be discarded at the end, we would like to take the lattice seriously. One can imagine a dynamical lattice like structure for space-time with the links being dynamical strings. Thus, one gives up the space-time continuum hypothesis. On the other hand, it is natural to require that nothing should depend on the details of the lattice. Elevating this to a gauge principle leads to a *generalized local space-time renormalization group* symmetry (with finite cutoff) and would ensure that physics is invariant under deformations of the lattice. The form of the gauge transformations in the loop variable formalism provides some preliminary evidence for such a symmetry.

Acknowledgements

Bal has been concerned with foundational aspects of quantum field theory for a long time. Perhaps, this chapter will encourage him to consider the consequences, for quantum field theory, of giving up the space-time continuum.

Wish you many happy returns of the day, Bal!

References

1. K. G. Wilson and J. B. Kogut, The renormalization group and the epsilon expansion, *Phys. Rept.* **12** (1974) 75. doi:10.1016/0370-1573(74)90023-4.
2. J. M. Maldacena, The large N limit of superconformal field theories and supergravity, *Int. J. Theor. Phys.* **38** (1999) 1113 [*Adv. Theor. Math. Phys.* **2** (1998) 231] doi: 10.1023/A:1026654312961 arXiv:hep-th/9711200.
3. E. T. Akhmedov, A remark on the AdS/CFT correspondence and the renormalization group flow, *Phys. Lett.* **B442** (1998) 152–158, arXiv:hep-th/9806217 [hep-th].
4. J. Polchinski, Renormalization and effective Lagrangians, *Nucl. Phys.* **B231** (1984) 269. doi: 10.1016/0550-3213(84)90287-6.
5. B. Sathiapalan and H. Sonoda, A holographic form for Wilson's RG, *Nucl. Phys. B* **924** (2017) 603. doi: 10.1016/j.nuclphysb.2017.09.018 [arXiv: 1706.03371 [hep-th]].
6. B. Sathiapalan and H. Sonoda, Holographic Wilson's RG, *Nucl. Phys. B* **948** (2019) 114767. doi: 10.1016/j.nuclphysb.2019.114767 [arXiv:1902.02486 [hep-th]].
7. B. Sathiapalan, Holographic RG and exact RG in O(N) model, *Nucl. Phys. B* **959** (2020) 115142. doi: 10.1016/j.nuclphysb.2020.115142 [arXiv:2005.10412 [hep-th]].
8. P. Dharanipragada, S. Dutta and B. Sathiapalan, [arXiv:2201.06240 [hep-th]].
9. G. E. Arutyunov and S. A. Frolov, On the origin of supergravity boundary terms in the AdS/CFT correspondence, *Nucl. Phys. B* **544** (1999) 576–589. doi: 10.1016/S0550-3213(98)00816-5 [arXiv:hep-th/9806216 [hep-th]].
10. I. Y. Aref'eva and I. V. Volovich, On the breaking of conformal symmetry in the AdS/CFT correspondence, *Phys. Lett. B* **433** (1998) 49–55. doi: 10.1016/S0370-2693(98)00699-6 [arXiv:hep-th/9804182 [hep-th]].
11. H. Liu and A. A. Tseytlin, D = 4 superYang-Mills, D = 5 gauged supergravity, and D = 4 conformal supergravity, *Nucl. Phys. B* **533** (1998) 88–108. doi: 10.1016/S0550-3213(98)00443-X [arXiv:hep-th/9804083 [hep-th]].

12. W. Mueck and K. S. Viswanathan, The Graviton in the AdS-CFT correspondence: Solution via the Dirichlet boundary value problem, [arXiv:hep-th/9810151 [hep-th]].

13. D. Kabat, G. Lifschytz, S. Roy and D. Sarkar, Holographic representation of bulk fields with spin in AdS/CFT, *Phys. Rev. D* **86** (2012) 026004. doi: 10.1103/PhysRevD.86.026004 [arXiv:1204.0126 [hep-th]].

14. I. R. Klebanov and A. M. Polyakov, AdS dual of the critical O(N) vector model, *Phys. Lett. B* **550** (2002) 213. doi: 10.1016/S0370-2693(02)02980-5 [hep-th/0210114].

15. M. A. Vasiliev, Nonlinear equations for symmetric massless higher spin fields in (A)dS(d), *Phys. Lett. B* **567** (2003) 139. doi: 10.1016/S0370-2693(03)00872-4 [hep-th/0304049].

16. B. Sathiapalan, Loop variables, the renormalization group and gauge invariant equations of motion in string field theory, *Nucl. Phys. B* **326** (1989) 376–392. doi: 10.1016/0550-3213(89)90137-5.

17. B. Sathiapalan, Exact renormalization group and loop variables: A background independent approach to string theory, *Int. J. Mod. Phys. A* **30**(32) (2015) 1530055. doi: 10.1142/S0217751X15300550 [arXiv:1508.03692 [hep-th]].

Chapter 22

An Inside View of the Tensor Product

Rafael D. Sorkin

Perimeter Institute, 31 Caroline Street North, Waterloo ON,
N2L 2Y5, Canada
Raman Research Institute, C.V. Raman Avenue, Sadashivanagar,
Bangalore 560 080, India
Department of Physics, Syracuse University, Syracuse,
NY 13244-1130, USA
School of Theoretical Physics, Dublin Institute for Advanced Studies,
10 Burlington Road, Dublin 4, Ireland
rsorkin@perimeterinstitute.ca

Given a vector space V which is the tensor product of vector spaces A and B, we reconstruct A and B from the family of simple tensors $a \otimes b$ within V. In an application to quantum mechanics, one would be reconstructing the component subsystems of a composite system from its unentangled pure states. Our constructions can be viewed as instances of the category-theoretic concepts of functor and natural isomorphism, and we use this to bring out the intuition behind these concepts and also to critique them. Also presented are some suggestions for further work, including a hoped-for application to entanglement entropy in quantum field theory.

This chapter is dedicated to my friend, A.P. Balachandran,
on the occasion of his 85th birthday. With his knack for dis-
cerning concrete implications of abstract mathematical relation-
ships, maybe he'll think of an unexpected use in physics for the
conception of tensor product proposed herein!

1. Introduction

It would hardly be possible to review all the ways in which tensors enter into physics. General Relativity and Quantum Field Theory would not exist without certain individual tensors or tensor fields, like the Lorentzian metric, the Riemann curvature or the stress–energy tensor, but it is perhaps in abstract quantum theory where the concept of tensor product itself and of the corresponding *product space* is most prominent. The reason, of course, is that insofar as one deals with state-spaces of quantal "systems", the tensor product furnishes the construction that combines the state spaces of two or more subsystems into that of the larger compound system or "whole".

Given this role of tensor product, it could be unsettling that aside from its dimension, the resulting state space (call it V) appears to remember nothing about the constituent spaces whose product it is. For example, because $12 = 4 \times 3 = 2 \times 6$, a given state space of dimension 12 that arose by combining spin 3/2 with spin 1 might equally well be describing a composite of spins 1/2 and 5/2. Thus arises the following mathematical question.

Suppose that a certain vector space V is the tensor product of spaces V_1 and V_2. What extra information do you need in order to recover V_1 and V_2 from V? Or to put the question another way: What does it mean for V to carry the structure of a tensor product?

To bring this question into sharper focus, imagine that instead of being a tensor product, V were a direct sum, as it would be, for example, if V_1 described an ionized hydrogen atom, while V_2 described the same atom in unionized form. Then, the appropriate state space would be $V = V_1 \oplus V_2$, and the same equation would describe other mutually exclusive alternatives, like an alpha particle being inside vs. outside a nucleus or a molecule being orthohydrogen vs. parahydrogen. To our question about tensor products, the counterpart in such cases would be as follows: What does it mean for V to carry the structure of a direct sum? Here, however the answer is simple. One only needs to indicate V_1 and V_2 as *subspaces* of V. These two "parts" of V are contained bodily within the whole, and every $v \in V$ is uniquely a sum, $v = v_1 + v_2$. A "direct sum structure" for V, in other words, is nothing but a pair of complementary subspaces of V.

Why isn't the case of a tensor product space V equally straightforward? To appreciate what's different, let $V = V_1 \otimes V_2$ be a tensor product space. The first thing one can notice is that V_1 and V_2 are no longer contained within V in any obvious way. Moreover, the analog of $v = v_1 + v_2$ in a direct sum would be $v = v_1 \otimes v_2$, but how could one express this within

V, given that vector spaces are by definition endowed with a notion of sum but not of product! And lacking the operation $\otimes$ within V, how could you recover V from V_1 and V_2, even if two such subspaces of V could be identified? The development that follows will answer these questions and show that the space $\mathfrak{S}$ of *simple vectors* in V — elements of V of the form $v = v_1 \otimes v_2$ — can play the structural role that the complementary subspaces played in the case of direct sum.

Although our questions are physically inspired, they are purely mathematical, and the answers we will come upon must surely be well known in some circles, even if they haven't shown up in the literature I'm familiar with. Nor do I know whether the constructions we will explore have any deeper physical ramifications, but I do believe that they offer a more intrinsic way to conceive of the tensor product (an "inside view" as one might say) and a more intrinsic conception, more often than not, deepens one's intuition.

As anyone who has taught a course in relativity, differential geometry or "mathematical methods" can attest, there's something about the concept of tensor that intuition finds hard to grasp. Perhaps it's no coincidence then that quantum entanglement, the mathematics of which is that of tensor product spaces, also seems counter-intuitive to so many people. In this situation, the *square construction* on which our development will rest offers a complementary way to think about tensor products, a way that starts not with the individual factor spaces, V_1 and V_2, but with the product space V itself. Inasmuch as this more analytical approach shifts the main emphasis onto the simple vectors within V, and inasmuch as these simple vectors in a quantum context are precisely the unentangled state vectors, our development might make better contact with physical intuition than the more formal definitions one usually encounters. At the very least, it offers an alternative to more familiar ways of approaching the topic. After all, the more ways one has to think about a subject, the better the prospect that at least one of them will be able to provide the key to the deeper understanding that one is seeking.

The main ingredients of our constructions are presented in Sections 2–5. Sections 2 and 3 are preparatory, while the constructions themselves appear in Sections 4 and 5.

In Section 6, we ask whether the results of Sections 4 and 5 can be regarded as fully capturing the structure of V as a tensor product, and we show in some detail how to make this type of question precise in terms of the category-theoretic concepts of functor and natural transformation, to which we provide a brief introduction. In this connection, we also point out a certain shortcoming of the functor concept itself.

In Section 7, we suggest some extensions of our constructions to tensor products of three or more spaces or to products of a single space with itself, in which case symmetry conditions come into play (bosonic, fermionic or non-abelian). We also speculate that a suitable generalization of the notion of simple vector to infinite dimensions could help clear away the mathematical obstructions (involving type-III von Neumann algebras, where tensor products as normally defined are not available) that prevent one from understanding entanglement entropy (with its need for a cutoff) in terms of reduced density matrices.

In what follows, we will assume, unless otherwise specified, that all vector spaces are real and finite dimensional. Nothing would change if we replaced the field $\mathbb{R}$ with the field $\mathbb{C}$, but it's convenient for exposition to pick one or the other and stick with it. We will also assume without special mention that the spaces V_1 and V_2 are distinct from each other.

2. Posing the Problem

Our question asks for a more intrinsic definition of tensor product or as we worded it above: What could play the role of a tensor product structure for V? We will contemplate three possible answers to this question, and in doing so, we will always assume that the two spaces of which V is a product are distinct from each other. Often, this will not matter, but sometimes it would make a difference, notably in definition (1) of the paragraph after next and then farther below in the "Second answer" to our main question.

Before suggesting answers to our main question, however, it seems advisable to dwell for a moment on the more common definitions of tensor product. Different authors favor different ones, and the answers to our questions will tend to take on different forms, depending on which definition one has in view. What, then, are some of the popular definitions[1] of the space $V_1 \otimes V_2$ and of the tensors therein?

(1) An element of $V_1 \otimes V_2$ is a numerical matrix whose entries depend on a choice of bases for V_1 and V_2 and transform in a certain way when these bases are changed. (This might be the oldest definition.)

(2) An element of $V_1 \otimes V_2$ is a linear mapping between two vector spaces, for example, a linear mapping from V_2^* to V_1, where V_2^* is the dual space of V_2.

(3) An element of $V_1 \otimes V_2$ is an equivalence class of formal sums of symbols, $\alpha \otimes \beta$, where $\alpha \in V_1$ and $\beta \in V_2$.

(4) "The" space $V_1 \otimes V_2$ is any solution of a certain "universal mapping problem" involving bilinear functions from $V_1 \times V_2$ to an arbitrary vector space Y. (Thereby, a bilinear function from $V_1 \times V_2$ to Y induces a unique linear mapping from $V_1 \otimes V_2$ to Y.)

Remark: Note that definition (1) refers to independently chosen bases for V_1 and V_2. Were we to assume that V_1 and V_2 were literally the same space, only a single basis would come into play. Tensors in this vein are common in GR and differential geometry, with $V_1 = V_2$ being the tangent space to a point of space-time. In quantum mechanics on the other hand, distinct factor spaces are typical, albeit not in the case of indistinguishable particles.

In finite dimensions, all these definitions are provably equivalent. In infinite dimensions, one must distinguish between the so-called algebraic tensor product and various topological tensor products, not all of which are the same in general. For our purposes, entering into those subtleties would be too much of a distraction.[2] Instead, we will work throughout in finite dimensions. In the jargon of category theory, each of these definitions yields a "functor" from pairs of vector spaces to vector spaces, and the statement that (in finite dimensions) they are all equivalent asserts that between any two of the functors, there is an invertible "natural transformation". Now back to our main question and some possible answers to it.

First possible answer. The most direct and obvious answer, but at the same time, the least informative, is that a tensor product structure for V is an isomorphism between V and a space of the form $V_1 \otimes V_2$. This is a good start, but it has the drawback that the auxiliary spaces V_1 and V_2 are not derived from V, with the consequence that different choices of them would strictly speaking define different tensor product structures for V. We could address this difficulty by forming equivalence classes under isomorphisms of V_1 and V_2, but let's instead continue on to the second and third proposals.

Second answer. The second possible answer to our main question, already more concrete and "intrinsic", hearks back to definition (1) in the above list. In order to represent an element of $V = V_1 \otimes V_2$ as a numerical matrix, one needs a basis of V whose members are themselves organized into a matrix. Specifically, if a list of vectors $e_j \in V_1$ furnishes a basis for V_1 and a second list of vectors $f_k \in V_2$ furnishes a basis for V_2, then the products $e_j \otimes f_k$ furnish a basis for $V_1 \otimes V_2$ whose members array themselves in a rectangular matrix with rows labeled by j and columns by k. In such a basis, the matrix representing a simple vector $v_1 \otimes v_2$ will be a matrix product of the form, column vector × row vector. (Such a "special basis" is precisely an isomorphism between V and $\mathbb{R}^m \otimes \mathbb{R}^n$, where m and n are the respective dimensions of V_1 and V_2.)

We could thus answer that a tensor product structure for V is a basis for V organized into a rectangular matrix. Unfortunately, this won't quite do because many other bases will define the same product structure. First of all, one might swap rows with columns, which amounts to writing $V_2 \otimes V_1$ instead of $V_1 \otimes V_2$. This does nothing. However, one can also replace each of the two bases by some other basis for the same space, which doesn't change the spaces V_1 or V_2 themselves but only their representations. We are thus led to identify a tensor product structure for V as an equivalence class of bases, parameterized by $G = GL(n_1) \times GL(n_2)$, where $n_i = \dim(V_i)$. But even this is not quite correct since it is always possible to rescale the basis for V_1 by some factor, while rescaling the basis for V_2 in the opposite way. Since this doesn't affect the resulting basis for $V_1 \otimes V_2$, we conclude that G is really the quotient group $GL(n_1) \times GL(n_2) / GL(1)$. Here, of course, $GL(n)$ is the group of invertible $n \times n$ matrices.

The fact that true group is $GL(n_1) \times GL(n_2) / GL(1)$ and not simply $GL(n_1) \times GL(n_2)$ seems a detail, but it is actually telling us something that will show up again in our deliberations in the following. From a tensor product structure for V, we cannot fully reconstruct the factor spaces V_1 and V_2; we can obtain them only up to a joint scaling ambiguity.

Remark: In a classical (non-quantum) context, a composite system would be described by a cartesian product, $A \times B$. In that case, the counterpart of a rectangular basis would just be (for discrete spaces A and B) a rectangular list of elements of $A \times B$, and the story would more or less end there.[a] Tensor products are more subtle than cartesian products, however, and there's a third possible answer to our question which is still more concrete and intrinsic than an equivalence class of bases.

Third answer. The third possible answer to our question, and the one which the rest of this chapter will explore, is that the tensor product structure for $V = V_1 \otimes V_2$ can be taken to be the subspace $\mathfrak{S}$ of *simple vectors*:

$$\mathfrak{S} = \{\alpha \otimes \beta \mid \alpha \in V_1, \beta \in V_2\}. \tag{1}$$

---

[a]The counterpart of G would be the product of the permutation groups of A and B.

As we will see, there exist explicit constructions that take you from $\mathfrak{S} \subseteq V$ to (copies of) V_1 and V_2 and thence back to V.

Remark: In quantum language, $\mathfrak{S}$ would be the set of unentangled state-vectors. Obviously, there's something special about them, but mathematically, they are only the first in a hierarchy of successively more generic tensors, those of ranks 2, 3, etc., where the "rank" of v is the minimum number of simple vectors of which it is a sum (quantum mechanically, the number of terms in a Schmidt decomposition of v).[b]

3. The Space $\mathfrak{S}$ of Simple Vectors in V

Since it is related quadratically to V_1 and V_2, the set $\mathfrak{S}$ of simple vectors obviously will not be a linear subspace of V in general, but it will be foliated by two families of linear subspaces, which we will denote by $\mathcal{M}_1$ and $\mathcal{M}_2$.

Before demonstrating this, let us deal with two trivial cases that don't fit easily into the general pattern. In the most trivial case, both V_1 and V_2 are one-dimensional: $\dim(V_1) = \dim(V_2) = 1$. Both are then copies of $\mathbb{R}$, as also is $V = V_1 \otimes V_2$. In this case, every $v \in V$ is plainly a simple vector, and so $\mathfrak{S}$ is all of V. Conversely, given that $\dim(V) = 1$, and since we know in general that $\dim V_1 \otimes V_2 = \dim V_1 \times \dim V_2$, we know immediately that both V_1 and V_2 are isomorphic to V itself. In a reconstruction of V_1 and V_2, we can thus do no better than to take both to be copies of V, and this suffices. The only small subtlety shows up when, having passed from V to V_1 and V_2, we seek to reconstruct V as $V_1 \otimes V_2 = V \otimes V$ but have to face the fact that although V is isomorphic to $V \otimes V$, the isomorphism is not canonical.[c]

In the second trivial case, $\dim V_1 > 1$ while $\dim V_2 = 1$ (or vice versa). Here again, $\mathfrak{S}$ is trivially all of V. (By definition, any $v \in V$ is a sum of terms of the form $a \otimes b$ for $a \in V_1$ and $b \in V_2$, but since all non-zero b are proportional to each other, all the b can be taken equal, whence $v = a_1 \otimes b + a_2 \otimes b + \cdots = (a_1 + a_2 \cdots) \otimes b \in \mathfrak{S}$.) It follows that $V = V_1 \otimes V_2 \simeq V_1 \otimes \mathbb{R} = V_1$ (where '$\simeq$' signifies isomorphic-to). Conversely, whenever $\mathfrak{S} = V$, we can construct spaces V_1 and V_2 by taking V_1 to be V and V_2 to be any one-dimensional vector space, for example, the subspace of

---

[b]Algebraic geometry has given the name "Segre variety" not quite to $\mathfrak{S}$ itself but to the set of rays in $\mathfrak{S}$.

[c]In some sense, this is just "dimensional analysis". If the elements of V were "lengths", then those of $V \otimes V$ would be squared lengths.

V given by $\mathbb{R} \, b_0$, where b_0 is any fixed vector[d] in V. The now-familiar scaling ambiguity corresponds then to the undetermined normalization of b_0.

Note in these two examples that V_1 was identified with a maximal linear subspace of $\mathfrak{S}$. Although completely trivial in the two examples, this observation will be the basis of our reconstruction of V_1 and V_2 in the generic case. In seeking to understand the linear subspaces of $\mathfrak{S}$, we will need a few "obvious" facts about tensor products which we will now review in the spirit of definition (3) mentioned in the previous section.

Recall then that any tensor $T \in V_1 \otimes V_2$ is a sum of simple tensors, i.e., a sum of products of vectors from V_1 with vectors from V_2:

$$T = \sum_j a_j \otimes b_j. \tag{2}$$

To fully characterize the space $V_1 \otimes V_2$, however, one needs to specify which such sums are equal to which others or equivalently which expressions T equal the zero tensor. Intuitively, the answer is that $T = 0$ iff it is forced to vanish by the combining rules for the symbols $a \otimes b$ together with the linear dependences among the vectors of V_1 and V_2. This criterion is implicit in the aforementioned definition (4), but it is more useful to express it algorithmically.

Rule: Provided that the vectors a_j in (2) are linearly independent, $T = 0$ if and only if all of the b_j vanish.

(Obviously, the same rule will hold true if we exchange the roles of a_j and b_j.) As stated, the rule wants the a_j to be linearly independent. If they are not, then some of them can be expressed as linear combinations of the others, and one should do this before applying the rule. Thus, an *algorithm* for deciding whether $T = 0$ consists in first writing any redundant a_j in terms of the others, second expanding out the resulting expression to put T into the form (2) and third applying the rule as stated.

As a trivial consequence of this rule, we learn that $a \otimes b$ is non-zero if both a and b are. In stating the following further consequences, we will interpret $a \propto b$ to mean that either $a = \lambda b$ or $b = \lambda a$, $\lambda \in \mathbb{R}$.

Lemma 1. *If α and β are non-zero, then $\alpha \otimes \beta = \alpha' \otimes \beta' \Rightarrow \alpha' \propto \alpha$ and $\beta' \propto \beta$.*

---

[d]The reason for this particular choice will become clear soon. Note also that we could of course have exchanged the roles of V_1 and V_2.

Proof: Were α' is not proportional to α, they would be linearly independent. Our "Rule" would then imply that $\alpha \otimes \beta - \alpha' \otimes \beta'$ could not vanish. Therefore, $\alpha' \propto \alpha$, and by symmetry, $\beta' \propto \beta$.

Lemma 2. *Let $\alpha \otimes \beta \in \mathfrak{S}$ and $\alpha' \otimes \beta' \in \mathfrak{S}$ be non-zero simple vectors. If their sum is also simple, then either $\alpha' \propto \alpha$ or $\beta' \propto \beta$.*

Proof (by contradiction): Assume that $\alpha' \not\propto \alpha$ and $\beta' \not\propto \beta$. The four terms $\alpha \otimes \beta$, $\alpha' \otimes \beta'$, $\alpha \otimes \beta'$, $\alpha' \otimes \beta$ are then (by a simple application of the Rule) linearly independent. By hypothesis, $\alpha \otimes \beta + \alpha' \otimes \beta' = \gamma \otimes \delta$ for some γ and δ. Appealing once again to the Rule, and remembering that α' is independent of α, we conclude that γ must be a linear combination of α and α'; similarly, δ must be a linear combination of β and β'. But then $\gamma \otimes \delta$, when expanded out, could not contain the required terms, $\alpha \otimes \beta$ and $\alpha' \otimes \beta'$, without also containing terms in $\alpha \otimes \beta'$ and $\alpha' \otimes \beta$.

Returning now to the analysis of $\mathfrak{S}$, and recalling that we have already disposed of the possibility that either $\dim V_1 = 1$ or $\dim V_2 = 1$, we can assume for now that $\dim V_1 \geq 2$ and $\dim V_2 \geq 2$, this being where the typical structure of $\mathfrak{S}$ reveals itself, namely that of the two foliations $\mathcal{M}_1$ and $\mathcal{M}_2$ already alluded to but not yet defined. For the time being, we will define $\mathcal{M}_1$ and $\mathcal{M}_2$ as follows. Soon, we will define them intrinsically (meaning directly from V and $\mathfrak{S}$ alone), whereupon (3) and (4) will shed their status as definitions and become theorems. The members of $\mathcal{M}_1$ will be the subsets of $\mathfrak{S}$ of the form $V_1 \otimes \beta$ for $\beta \in V_2$ and likewise for $\mathcal{M}_2$:

$$\mathcal{M}_1 = \{V_1 \otimes \beta \,|\, \beta \in V_2\}, \tag{3}$$

$$\mathcal{M}_2 = \{\alpha \otimes V_2 \,|\, \alpha \in V_1\}. \tag{4}$$

(Here, of course, our notation means that, e.g., $V_1 \otimes \beta = \{\alpha \otimes \beta \,|\, \alpha \in V_1\}$.)

We want to prove first, that every M in either $\mathcal{M}_1$ or $\mathcal{M}_2$ is a *maximal linear subspace of* $\mathfrak{S}$; second, that $\mathcal{M}_1$ and $\mathcal{M}_2$ exhaust the maximal linear subspaces of $\mathfrak{S}$; third, that $M, N \in \mathcal{M}_1$ and $M \neq N \Rightarrow M \cap N = \{0\}$ (and likewise for $\mathcal{M}_2$); and fourth, that $M \in \mathcal{M}_1, N \in \mathcal{M}_2 \Rightarrow \dim(M \cap N) = 1$.

Why are the members of $\mathcal{M}_1$ (for example) maximal linear subspaces of $\mathfrak{S}$? That $M = V_1 \otimes \beta$ is a linear subspace is obvious, but why is it maximal? Well, any simple vector not in M must take the form $\alpha' \otimes \beta'$ with β' independent of β. Choose also an $\alpha \in V_1$ that is independent of α' (which is always possible since $\dim V_1 > 1$) and note that $\alpha \otimes \beta \in M$.

If we could adjoin $\alpha' \otimes \beta'$ to M, then $\alpha' \otimes \beta' + \alpha \otimes \beta$ would also have to be in M, and therefore simple, contrary to Lemma 2.[e]

And why does every maximal linear subspace of $\mathfrak{S}$ have to belong to either $\mathcal{M}_1$ or $\mathcal{M}_2$? Well, let M be such a subspace, and let $\alpha \otimes \beta \in M$. Certainly, $\alpha \otimes \beta$ alone is not maximal (it belongs to $V_1 \otimes \beta$, for example), so let $\alpha' \otimes \beta'$ be an independent member of M. By the same lemma, either α and α' are proportional or β and β' are proportional, say the latter. Then, as we just saw, every other member of M must also take the form $\gamma \otimes \beta$ for some $\gamma \in V_1$, in other words, $M \subseteq V_1 \otimes \beta \in \mathcal{M}_1$, whence $M = V_1 \otimes \beta$ since M is maximal.

Third, if $M, N \in \mathcal{M}_1$ are unequal, then $M = V_1 \otimes \beta$ and $N = V_1 \otimes \beta'$ with β independent of β'. Hence, any $v \in M \cap N$ must satisfy $v = \alpha \otimes \beta = \alpha' \otimes \beta'$ for some α and α'. But by Lemma 1, this is impossible unless $v = 0$.

Fourth, if $M \in \mathcal{M}_1, N \in \mathcal{M}_2$, then $M = V_1 \otimes \beta$, $N = \alpha \otimes V_2$ for some $\alpha \in V_1, \beta \in V_2$. If $v \in M \cap N$, then by definition, $v = \alpha' \otimes \beta = \alpha \otimes \beta'$ for some $\alpha' \in V_1, \beta' \in V_2$. The lemma just cited then informs us that $\alpha' \propto \alpha$ and $\beta' \propto \beta$, whence $v = \alpha' \otimes \beta \propto \alpha \otimes \beta$. In other words, $M \cap N$ is the 1-dimensional subspace, $\mathbb{R}\alpha \otimes \beta$.

The essential feature we have discovered is that *any two members of different foliations meet in a ray (a one-dimensional subspace of V) and any two distinct members of the same foliation are disjoint*. This lets us determine the foliations $\mathcal{M}_1$ and $\mathcal{M}_2$ simply from a knowledge of $\mathfrak{S} \subseteq V$, without any further recourse to how V arose as a tensor product: if M and N are elements of the set $\mathcal{M}$ of all maximal linear subspaces of $\mathfrak{S}$, then they belong to the same foliation if and only if they are disjoint, and this criterion is guaranteed to produce exactly two disjoint subsets of $\mathcal{M}$, which we can label as $\mathcal{M}_1$ and $\mathcal{M}_2$. Henceforth, we will adopt this *intrinsic definition* of $\mathcal{M}_1$ and $\mathcal{M}_2$, which we can record in the following two maps that associate with each simple vector in V the two maximal linear subspaces of $\mathfrak{S}$ to which it belongs.

Definition: Let $v \in \mathfrak{S}$. Then, $\pi_1(v)$ [resp. $\pi_2(v)$] is the unique maximal linear subspace of type $\mathcal{M}_1$ [resp. $\mathcal{M}_2$] that contains v.

Equations (3)–(4) are hereby no longer definitions but theorems which apply whenever we can exhibit vector spaces V_1 and V_2 such that $V = V_1 \otimes V_2$.

---

[e]What we are effectively proving could be reduced to a lemma to the effect that every linear subspace of $\mathfrak{S}$ has the form $W \otimes \beta$ or $\alpha \otimes W$, for some vector subspace W of V_1 or V_2.

With these observations, we have taken a first step in recovering the tensor product structure of V from $\mathfrak{S}$. In fact, one sees from (3) and (4) that each M_1 in $\mathcal{M}_1$ is a copy of V_1 and each M_2 in $\mathcal{M}_2$ is a copy of V_2. In the following section, we will build on our knowledge of $\mathcal{M}_1$ and $\mathcal{M}_2$ to recover fully the ray spaces associated with V_1 and V_2 and then to recover V_1 and V_2 themselves up to scale.

4. How to Recover V_1 and V_2 Up to Scale

Our ultimate aim is to find a construction that, relying on nothing more than the set $\mathfrak{S}$ of simple vectors in V, will resolve the latter into its two factors (as uniquely as possible) and then to discover how to rebuild V as the tensor product of these factors. This will take place in Section 5, and not everything from this section will be needed there. If you are reading these lines, you might thus want to skip over this section in order to appreciate the great simplicity of the final constructions. On the other hand, this section, as well as providing much of the background for Section 5, will also show how, in becoming aware of the two spaces $\mathcal{M}_1$ and $\mathcal{M}_2$, we have *already* recovered from $\mathfrak{S}$ the *rays* of V_1 and V_2, which in a quantum context means we have already recovered, if not the respective subsystems themselves, then at least their "pure states".

To appreciate this fact, recall that when $V = V_1 \otimes V_2$, any member M of $\mathcal{M}_2$ can be expressed in the form (4). But the subspace $M = \alpha \otimes V_2$ determines and is determined by the ray, $\mathbb{R}\,\alpha \subseteq V_1$. The points of $\mathcal{M}_2$ are thus in bijective correspondence with the rays of V_1 and likewise for $\mathcal{M}_1$ and V_2. Introducing the notation $\mathcal{P}V$ for the projective space formed from the rays of any vector space V, we can therefore assert that

$$\mathcal{P}V_1 = \mathcal{M}_2 \quad \text{and} \quad \mathcal{P}V_2 = \mathcal{M}_1. \tag{5}$$

Of course, there's more to it than this because so far, we have only introduced $\mathcal{M}_1$ and $\mathcal{M}_2$ as sets without further structure. In order to fully corroborate the claim that $\mathcal{P}V_1 = \mathcal{M}_2$, we need to present $\mathcal{M}_2$ as the set of rays of some intrinsically defined vector space, this being one way to equip it with a projective structure. In the course of doing so, we will also see how to get our hands on V_1 itself up to scale.

Let's first see the procedure *per se* and then return to see more fully why it works. To get started, select arbitrarily any $M \in \mathcal{M}_1$ and let P be the restriction of π_2 to M. It is not hard to see that $P : M \to \mathcal{M}_2$ sets up a one-to-one correspondence between the rays in M and the points of $\mathcal{M}_2$.

By definition, if $v \in M$, then $P(v) = \pi_2(v)$ is the unique maximal linear subspace in $\mathcal{M}_2$ that contains v; being linear, it also contains the entire ray, $\mathbb{R}\,v$. Furthermore, P is trivially surjective because for any $N \in \mathcal{M}_2$, $M \cap N$ is (as observed earlier) a ray ℓ in M that gets mapped by P to N itself. This also proves that P is injective (on the *rays* of M) because any other ray in M that was mapped to N by P would by definition have to lie in $M \cap N$ and therefore coincide with ℓ.

The mapping, $P : M \to \mathcal{M}_2$, is what we were looking for, but it remains to demonstrate that any other $M' \in \mathcal{M}_1$ would have induced the same projective structure on $\mathcal{M}_2$. For this, it suffices to find a linear isomorphism between M and M' that commutes with the corresponding projections. In other words, with P' taken to be the restriction of π_2 to M', we should seek an isomorphism $f : M \to M'$ such that $P = P' \circ f$. Such an f would induce an isomorphism $\hat{f} : \mathcal{P}M \to \mathcal{P}M'$, and it is actually easier to characterize this isomorphism intrinsically than to exhibit f itself. Let us therefore define $\hat{f}$ first and only then consider how to lift it to a linear map f. It turns out that the ansatz

$$\hat{f}(M \cap N) = M' \cap N \tag{6}$$

(where N is an arbitrary element of $\mathcal{M}_2$) does what is needed. In particular, if M'' is a third element of $\mathcal{M}_1$, then the isomorphisms $M \to M' \to M''$ defined by (6) obviously compose consistently.

Our remaining task is to lift the just-constructed mapping, $\hat{f} : \mathcal{P}M \to \mathcal{P}M'$, to a linear function, $f : M \to M'$. Given that for $v \in M$, the mapping $\hat{f}$ already determines the ray in M' to which v should go, the only further input needed to define $f(v)$ is its normalization. Although this seems a tiny bit of extra information, the construction via which we will obtain it is surprisingly intricate. In fact, it is not really needed for present purposes; all we really need to know is that a linear lift f exists, which could be proven more easily. If nevertheless we take the trouble to construct f explicitly, it is because doing so will introduce us to a certain type of "simple square" that will play an important role in the next section.

Fix spaces $M, M' \in \mathcal{M}_1$ as above, and let ℓ_0 be any ray in M, with $\ell'_0 = \hat{f}(\ell_0)$ being the corresponding ray in M', as given by (6). We know that f will take any point in ℓ_0 to some point in ℓ'_0. Given now some arbitrarily chosen reference vector, $v_0 \in \ell_0$, we need to decide which vector in ℓ'_0 will be $f(v_0)$, and it turns out that this decision determines f fully. Let v'_0 be the vector selected to be $f(v_0)$. The problem then is to determine

$f(v)$ when v belongs to some other ray $\ell \subseteq M$. That is, we need to figure out where $f(v)$ lies along the ray $\ell' = \hat{f}(\ell)$.

This problem admits a generic case and a couple of special cases. In the generic case, v_0, v_0' and v are all linearly independent. Consider then an arbitrary $v' \in \ell'$ and the *square*

$$\begin{pmatrix} a & b \\ c & d \end{pmatrix} = \begin{pmatrix} v_0 & v_0' \\ v & v' \end{pmatrix} \tag{7}$$

whose elements belong to the rays

$$\begin{pmatrix} \ell_0 & \ell_0' \\ \ell & \ell' \end{pmatrix}. \tag{8}$$

By construction (cf. (6)),

$$\pi_1 a = \pi_1 c, \quad \pi_1 b = \pi_1 d, \quad \pi_2 a = \pi_2 b, \quad \pi_2 c = \pi_2 d. \tag{9}$$

Consequently, the two row sums and the two column sums belong to $\mathfrak{S}$ (i.e., all four sums are simple vectors in V), but what about the overall sum, $a + b + c + d$? In the answer to this question lies the key to our construction of f. In fact (as we will prove shortly), this sum meets $\mathfrak{S}$ for precisely one point v' in the ray ℓ', and by setting $f(v) = v'$, we define f uniquely on the ray ℓ. Doing the same for the other rays in M, we will obtain a function $f : M \to M'$ which is linear, unique up to a multiplicative prefactor and whose action on rays is by definition that of $\hat{f}$.

So much for the generic case. Before turning to the special cases, observe that just from (9) alone, we can write the rays in (8) as

$$\begin{pmatrix} M \cap N & M' \cap N \\ M \cap N' & M' \cap N' \end{pmatrix}, \tag{10}$$

where $M = \pi_1 a = \pi_1 c$, $N = \pi_2 a = \pi_2 b$, $M' = \pi_1 b = \pi_1 d$, $N' = \pi_2 c = \pi_2 d$. The generic case just treated corresponded to an array (10) in which the subspaces, M, M', N, N', were all distinct, and correspondingly, the vectors, a, b, c, d, in (7) were linearly independent. The special cases we still need to treat are those in which $M = M'$ or $N = N'$.

Consider first the special case where $N = N'$ or equivalently, $\ell = \ell_0$. Here, we know the answer trivially because $v = \lambda v_0$ for some scalar λ, whence $v' = f(v) = f(\lambda v_0) = \lambda f(v_0) = \lambda v_0'$. The square in (7) thus assumes the form

$$\begin{pmatrix} a & b \\ \lambda a & \lambda b \end{pmatrix}. \tag{11}$$

The other special case is that where $M = M'$, or equivalently (since, as we know, any two elements of $\mathcal{M}_1$ are either equal or disjoint), $\ell_0' = \ell_0$. Here, f is just mapping M to itself, an obvious solution for which would be to take f to be the identity map. However, we could equally well take it to be a multiple of the identity by a scalar μ, in which case, our square would take on the appearance

$$\begin{pmatrix} a & \mu a \\ c & \mu c \end{pmatrix}, \tag{12}$$

a form that follows immediately from (11) by symmetry. For completeness, let us also record the doubly special case where $M = M'$ and $N = N'$ both hold, leading to a square of the design

$$\begin{pmatrix} a & \mu a \\ \lambda a & \lambda \mu a \end{pmatrix}, \tag{13}$$

as one sees by combining (11) with (12). All these special cases, (11)–(13), can be obtained from the generic case by forming limits. Amalgamating these special cases with the generic one, we arrive at the following definition.

Definition: A *square* (or *simple square*) is a matrix $\begin{pmatrix} a & b \\ c & d \end{pmatrix}$ of simple vectors which satisfy (9) and which in the generic case satisfy $a + b + c + d \in \mathfrak{S}$ or in the special cases take on one of the forms (11)–(13).

The reason for separating the generic from special cases in the definition is that $a + b + c + d \in \mathfrak{S}$ suffices in the generic case but not in the special cases. Of course, it holds in the latter cases too, albeit it is trivial there. It's also worth noting that given any simple square, one can multiply any row or column by a scalar without invalidating its status as a square. And for completeness, let us recall from above that the two row sums and the two column sums also belong to $\mathfrak{S}$.

This completes the description of our procedure for defining f. In order to understand why it works, let's "look behind the curtain" to see what our squares amount to when expressed in terms of vectors in $V_1 \otimes V_2$. (This should also help illuminate the rather abstract development we have been following in this section.) Recall that the four rays in (8) can also be written as the intersecting subspaces exhibited in (10). Now, by Eqs. (3) and (4), $M = V_1 \otimes \beta_0$ for some $\beta_0 \in V_2$, while $N = \alpha_0 \otimes V_2$ for some $\alpha_0 \in V_1$, and similarly, $M' = V_1 \otimes \beta$, $N' = \alpha \otimes V_2$, for some α and β. Without loss of generality, we can therefore write (7) in the form

$$\begin{pmatrix} a & b \\ c & d \end{pmatrix} = \begin{pmatrix} \alpha_0 \otimes \beta_0 & \alpha_0 \otimes \beta \\ \alpha \otimes \beta_0 & \lambda \alpha \otimes \beta \end{pmatrix}, \tag{14}$$

where λ is some unknown coefficient of proportionality. This form makes it plain that the row and column sums are indeed simple, for example, $a + b = \alpha_0 \otimes (\beta_0 + \beta)$. As for the overall sum, $a + b + c + d$, it will be the simple vector, $(\alpha_0 + \alpha) \otimes (\beta_0 + \beta)$, *provided that* $\lambda = 1$. Were $\lambda \neq 1$ on the other hand, the same sum would equal $(\alpha_0 + \alpha) \otimes (\beta_0 + \beta) + (\lambda - 1)\alpha \otimes \beta$, which according to Lemma 2, could be simple only if α were proportional to α_0 or β were proportional to β_0, meaning we'd be back in one of the special cases we disposed of earlier.

In summary, consider a square of simple vectors belonging to rays of the form exhibited in (10) with $M, M' \in \mathcal{M}_1$ and $N, N' \in \mathcal{M}_2$ and assume we are in the generic case where $M \neq M'$, $N \neq N'$. On condition that the sum of all four simple vectors is itself simple, any three of them determine the fourth uniquely. The vectors must in that case "secretly" take the form (14) with $\lambda = 1$:

$$\begin{pmatrix} \alpha_0 \otimes \beta_0 & \alpha_0 \otimes \beta \\ \alpha \otimes \beta_0 & \alpha \otimes \beta \end{pmatrix}. \tag{15}$$

Our special cases correspond to $\alpha_0 \propto \alpha$ and/or $\beta_0 \propto \beta$, and they also fit the pattern (15), which accordingly represents the universal form that a square assumes when one views it "from behind the curtain".

Returning to the task of lifting $\hat{f} : \mathcal{P}M \to \mathcal{P}M'$ to a linear isomorphism, $f : M \to M'$, we can now see that the construction of $f(v)$ following Eq. (9) does indeed do the job because it maps $M = V_1 \otimes \beta_0$ to $M' = V_1 \otimes \beta$ by carrying $\alpha \otimes \beta_0 \in M$ to $\alpha \otimes \beta \in M'$, a correspondence which is plainly linear when α varies. Of course, the fact that f is a lift of $\hat{f}$ cannot determine it uniquely because any multiple of a lift is another lift. It's thus no accident that our construction involved a free choice of reference vectors, v_0 and v_0'. A different choice, however, could only alter f by an overall factor, as follows from the general fact that any two linear isomorphisms that induce the same mapping on rays must agree up to scale.[f] For the same reason, we don't need to check our isomorphisms f for coherence. Given that they cohere on $\mathcal{P}M \to \mathcal{P}M' \to \mathcal{P}M''$, as we already know they do, they must also cohere up to scale on $M \to M' \to M''$, and that's the best we can do.

Taking as input solely the set $\mathfrak{S}$ of simple vectors in V, we have now identified with one another the members of $\mathcal{M}_1$ via isomorphisms which are unique up to scale. On one hand, we used this to derive from V-cum-$\mathfrak{S}$ a canonically given projective space that is naturally isomorphic to $\mathcal{P}V_1$ (the space of "pure states of system-1" in a quantal interpretation). On the other hand, these same identifications produce a vector space that is naturally isomorphic to V_1 itself, albeit only modulo a scaling ambiguity. The same procedure applied to $\mathcal{M}_2$ rather than $\mathcal{M}_1$ would obviously recover $\mathcal{P}V_2$ and V_2 in the same sense. Our task now is to complete the story by re-building V as the tensor product of the two vector spaces just constructed.

5. The Analysis and Synthesis of a Tensor Product

Our previous work has already led us to pay close attention to the maximal linear subspaces of $\mathfrak{S} \subseteq V = V_1 \otimes V_2$. Let us now select two such spaces, $W_1 \in \mathcal{M}_1$ and $W_2 \in \mathcal{M}_2$, and then select further a vector $w_0 \in W_1 \cap W_2$ to serve as their common "base point".[g] We have then

$$W_1 = \pi_1 w_0, \quad W_2 = \pi_2 w_0.$$

We want to demonstrate that V can be construed as the tensor product of these two spaces.

To that end, and basing ourselves on the concept of "square" introduced in Section 4, we will introduce a new bilinear product, $\overline{\otimes} : W_1 \times W_2 \to V$,

---

[f]Proof. Call the maps f and g and let x and y be any two independent vectors in their domain with $z = x + y$. By assumption, $g(x) = \lambda f(x)$ and $g(y) = \mu f(y)$, and we want to prove that $\mu = \lambda$. By rescaling either f or g if necessary, we can assume that $\lambda = 1$. But then $f(z) = f(x) + f(y)$ would lie in a different ray from $g(z) = f(x) + \mu f(y)$ unless $\mu = 1$ as well.

[g]In Sections 3 and 4, we usually used the letters M and N to denote maximal linear subspaces of $\mathfrak{S}$. The notation, W_1, W_2, here is chosen to emphasize the parallelism with V_1, V_2.

as follows. For $w_i \in W_i$ $(i = 1, 2)$, let us define $w_1 \overline{\otimes} w_2$ to be the solution of the following square:

$$
\begin{pmatrix}
w_0 & w_2 \\
w_1 & w_1 \overline{\otimes} w_2
\end{pmatrix}. \tag{16}
$$

In other words, $w = w_1 \overline{\otimes} w_2$ must satisfy the conditions

$$w \in \pi_2 w_1 \cap \pi_1 w_2$$

$$w_0 + w_1 + w_2 + w \in \mathfrak{S}.$$

As we have seen, these conditions determine $w_1 \overline{\otimes} w_2$ uniquely in the generic case where $w_1 \notin W_2$ and $w_2 \notin W_1$. In the special cases where this is not true, a scaling ambiguity remains. To supplement (16) for such cases, we can stipulate that $w_0 \overline{\otimes} w_0 = w_0$ and more generally that $w_0 \overline{\otimes} w_2 = w_2$ and $w_1 \overline{\otimes} w_0 = w_1$. These rules[h] render $w_1 \overline{\otimes} w_2$ unique. For example, if $w_1 \in W_2$, then $w_1 \in W_1 \cap W_2$, whence $w_1 = \lambda w_0$ since, as always, $\dim(W_1 \cap W_2) = 1$. Therefore, $w_1 \overline{\otimes} w_2 = (\lambda w_0) \overline{\otimes} w_2 = \lambda w_2$, exactly as in (11).

We learned in the previous section [following Eq. (15)] that $\overline{\otimes}$ would be bilinear when defined in this manner.[i] Therefore (compare definition (2) of tensor product in Section 2), it induces a linear map $\Phi : W_1 \otimes W_2 \to V$. In fact, Φ is an isomorphism. To see this, let's go back to the representation of V as $V_1 \otimes V_2$ and write $w_0 = \alpha_0 \otimes \beta_0$, $w_1 = \alpha \otimes \beta_0$, $w_2 = \alpha_0 \otimes \beta$. Then, as one sees by comparing (15) with (16), $w_1 \overline{\otimes} w_2 = \alpha \otimes \beta$. This means first of all that the simple vectors $\alpha \otimes \beta \in \mathfrak{S}$ coincide with the vectors of the form $w_1 \overline{\otimes} w_2$ for some $w_i \in W_i$ $(i = 1, 2)$. Consequently, we can build up a basis of V by choosing a basis $\{e_j \mid j = 1 \cdots \dim W_1\}$ for W_1, and a similar basis $\{f_k \mid k = 1 \cdots \dim W_2\}$ for W_2, and then taking our basis elements to be $e_{jk} = e_j \overline{\otimes} f_k$. That these e_{jk} constitute a basis for V follows from the

---

[h]In the previous section, we already introduced rules for the special cases; the rules stated here are simply their instances for the situation at hand. If we have restated them here, it is only in order to make the definition of $\overline{\otimes}$ more self-contained.

[i]One can also deduce the bilinearity of $\overline{\otimes}$ directly from the definition (16), if one proves first the following useful lemma: The set of first rows (a, b) which make a square with a fixed second row (c, d) is closed under addition and scalar multiplication and similarly for columns instead of rows. Closure under scalar multiplication we already noticed, and closure under sum can be deduced from the general square form (15). Taken together, the row and column assertions suffice to prove linearity of $\overline{\otimes}$ in both arguments.

fact that the e_j (respectively the f_k) have the form $\alpha_j \otimes \beta_0$ (resp. $\alpha_0 \otimes \beta_k$), whereby the α_j (resp. β_k) constitute a basis of V_1 (resp. V_2) if and only if the e_j (resp. f_k) constitute a basis of W_1 (resp. W_2), and furthermore $e_{jk} = e_j \overline{\otimes} f_k = \alpha_j \otimes \beta_k$.

The upshot is that a "special basis" for V (i.e., a basis of vectors $\alpha_j \otimes \beta_k$) is the same thing as a pair of bases for W_1 and W_2, modulo the familiar $GL(1)$ ambiguity that one can rescale the W_1 basis by λ if one simultaneously rescales the W_2 basis by $1/\lambda$. Recall now from Section 2 that our "second possible answer" to what constitutes a tensor product structure for V was "an equivalence class $\mathfrak{T}$ of special bases for V". We have thus demonstrated that from $\mathfrak{S} \subseteq V$, one can derive uniquely a tensor product structure in that sense. Conversely, given such a structure $\mathfrak{T}$, we immediately obtain $\mathfrak{S}$ from it as the union of all of the members of the special bases that comprise $\mathfrak{T}$. To the extent that $\mathfrak{S}$ is a simpler and more natural object than an equivalence class of special bases (and is also more intrinsic to V), we have reason to maintain that in $\mathfrak{S}$ we have an answer to the question: "What does it mean for V to be a tensor product?"

5.1. *Using "pointed vector spaces"*

The above construction began with an arbitrarily chosen "base vector" $w_0 \in V$ such that $W_1 = \pi_1 w_0$ and $W_2 = \pi_2 w_0$. The ambiguity inherent in such a choice does not impugn our demonstration of the equivalence, $\mathfrak{T} \leftrightarrow \mathfrak{S}$, but it does mean that in the procession, $(V_1, V_2) \to V\text{-cum-}\mathfrak{S} \to (W_1, W_2)$, a different choice of w_0 would produce a different pair of spaces, W_1, W_2. If desired, one could arrange for W_1 and W_2 to be unique by working with "pointed vector spaces", i.e., by equipping V_1 and V_2 with distinguished "base points", $\alpha_0 \in V_1$ and $\beta_0 \in V_2$, and then taking $v_0 = \alpha_0 \otimes \beta_0 \in V_1 \otimes V_2$ to be the base point of V. Our construction above (with w_0 taken to be v_0) would then recover the pairs (V_1, α_0) and (V_2, β_0) essentially uniquely from the triple $(V, \mathfrak{S}, v_0)$.

Remark: Interestingly, the "histories Hilbert spaces" $\mathcal{H}$ that play a role in quantum measure theory[3] automatically come with distinguished vectors $|\Omega\rangle \in \mathcal{H}$, where Ω represents the full history space (the unit of the corresponding event algebra). However, it is generally false for coupled subsystems that $\mathcal{H}$ for the composite system is the Hilbert space tensor product of the $\mathcal{H}$s for the subsystems. (Even when the vector space dimensions match, the norms in general will not.)

6. Categorical Matters (and a Shortcoming of the Functor Concept)

From a given vector space V, one can form new spaces, like the dual space V^* or the double dual V^{**}. With two vector spaces, there are other possibilities, including their direct sum, their tensor product and so forth. Although a vector space formed in one of these ways will be isomorphic to infinitely many other vector spaces, its "inner constitution" will in general be distinctive, with the result that it will to a certain extent "remember where it came from". One may say then that it *carries the structure of* a dual space, a direct sum or a tensor product, as the case may be. In each instance, one can try to identify concretely where this extra information resides, and for a vector space V that arose as a tensor product, our discussion has pointed to the set $\mathfrak{S}$ of simple vectors within V as the pertinent structure. Adopting a notation that keeps track of $\mathfrak{S}$, we may say that from an ordered pair (V_1, V_2) of vector spaces, there arises via tensor product the ordered pair $(V, \mathfrak{S})$.

A question then is to what extent the transformation $(V_1, V_2) \to (V, \mathfrak{S})$ is reversible. How perfectly does V remember where it came from, or to ask this another way, how well can we reconstruct V_1 and V_2 given V and $\mathfrak{S}$? When we dealt with pointed spaces, we discovered that (V_1, V_2) could "in essence" be recovered fully. But in the unpointed case, it appeared that although $(V, \mathfrak{S})$ is determined by (V_1, V_2), the latter could be recovered from the former only up to some sort of $GL(1)$ ambiguity. This suggests that in the pointed case, a vector space carrying the structure of a tensor product is in some sense equivalent to the factor spaces from which it arose, whereas in the unpointed case, there is only partial equivalence.

But what concept of equivalence is implicitly animating these expectations? Simple isomorphism will not do, being too narrow in one way (because by definition structures of different types like (V_1, V_2) and $(V, \mathfrak{S})$ cannot be isomorphic) and too broad in another way (because, for example, any two vector spaces of equal dimension are isomorphic). Maybe one can put the underlying thought into words by saying that "A and B are equivalent if B can be constructed from A and vice versa." I am not sure that mathematics knows any framework which really does justice to this thought but perhaps the category-theoretical concepts of functor and natural isomorphism come closest to providing one, and so it seems worth considering how they apply to the question at hand. We will take this up momentarily, but first let's see a very simple illustration of how $\mathfrak{S}$ is able to remember "where V came from".

6.1. *A small illustration: Topology remembers spin*

As a small illustration of how the set $\mathfrak{S}$ of simple tensors encodes the structure of V as a tensor product, let us return to the example of *spin-3/2 $\otimes$ spin-1* versus *spin-1/2 $\otimes$ spin-5/2*. To distinguish these two possible provenances of V, one from the other, it is enough to pay attention to the topology of $\mathfrak{S}$, for example, its dimensionality. Taking into account that an element of $\mathfrak{S}$ has by definition the form $\alpha \otimes \beta$ and that α and β are unique modulo the obvious $GL(1)$ ambiguity, we can observe that the (complex) dimensionality of $\mathfrak{S}$ is one less than the sum of the dimensionalities of the factor spaces. In our examples, this yields for $\dim(\mathfrak{S})$ the respective values, $4 + 3 - 1 = 6$ for $\mathbf{3/2} \otimes \mathbf{1}$ and $2 + 6 - 1 = 7$ for $\mathbf{1/2} \otimes \mathbf{5/2}$. In fact, it is easy to verify that this simple test works in general. If we know that V arose from the combination of two spins, then the topological dimension of $\mathfrak{S}$ determines fully what those spins were. Of course (and as we have now seen in great detail), the same information can be deduced with a bit more work from the dimensionalities of the maximal linear subspaces of $\mathfrak{S}$ or from the dimensionalities of $\mathcal{M}_1$ and $\mathcal{M}_2$ as "foliations" of $\mathfrak{S}$.

6.2. *A functorial gloss on our constructions*

Now back to categories, functors and natural isomorphisms. A *category* is basically a collection of spaces of a given type (its "objects") and of structure-preserving mappings between these spaces (its "morphisms"). For our purposes, it will be best to limit the latter to isomorphisms, i.e., to require them to be invertible. A *functor* between two categories, I and II, is a kind of black box that converts the objects and morphisms of category-I to objects and morphisms of category-II while preserving composition of morphisms. Conceptually, it is telling you that you can build spaces and mappings of type II from spaces and mappings of type I (but unfortunately, it is not telling you *how* to do so.)

The two categories of interest to us here can be denoted as VEC $\times$ VEC and TVEC, where the former is the category of pairs (V_1, V_2) and the latter[j] is the category of pairs $(V, \mathfrak{S}_V)$. A morphism in VEC $\times$ VEC will thus be a pair of linear isomorphisms, while a morphism in TVEC will be a linear isomorphism between vector spaces that preserves their respective subsets $\mathfrak{S}$. When our spaces are pointed, all these isomorphisms will of

---

[j]The "T" in TVEC is meant to suggest the word "tensor".

course also need to preserve the respective base points. Let us now describe some of our constructions in terms of functors between VEC × VEC and TVEC, concentrating for the time being exclusively on the pointed case.

The first functor of interest, which we will designate as $\otimes : \text{VEC} \times \text{VEC} \to \text{TVEC}$, is that induced by the tensor product itself. It takes a pair of vector spaces (A, B) to their tensor product, $V = A \otimes B$ equipped with its space $\mathfrak{S}_V$ of simple vectors $\alpha \otimes \beta$, and it takes a pair (f, g) of (invertible) linear functions between vector spaces to their tensor product $f \otimes g$. Conversely, given an object $(V, \mathfrak{S}_V, v_0) \in \text{TVEC}$ (where I've now indicated the base point v_0 explicitly), we saw how to locate within V the subspaces $W_1 = \pi_1(v_0)$ and $W_2 = \pi_2(v_0)$, which were certain maximal linear subsets of $\mathfrak{S}_V$. Thereby, we in effect defined a second functor, $D : \text{TVEC} \to \text{VEC} \times \text{VEC}$, that goes in the direction opposite to $\otimes$ and for which $D(V, \mathfrak{S}_V, v_0) = ((W_1, v_0), (W_2, v_0))$. Of course, one has not defined a functor fully until one tells how it acts on morphisms, but that is self-evident for D. A morphism in TVEC from $(V, \mathfrak{S}_V, v_0)$ to $(V', \mathfrak{S}'_V, v'_0)$ is nothing but an invertible linear function, $f : V \to V'$, such that $f[\mathfrak{S}_V] = \mathfrak{S}'_V$ and $f(v_0) = v'_0$. Such an f induces immediately a pair of (base point preserving) functions $f_1 : W_1 \to W'_1$ and $f_2 : W_2 \to W'_2$, and so $D(f) = (f_1, f_2)$.

Now what of the expectation that $\otimes$ and D are in essence each other's inverses? Were that literally true, we would be able to express it by writing $\otimes \circ D = 1$ and $D \circ \otimes = 1$, but unfortunately, both equations are, strictly speaking, false. Consider first the composed functor, $D \circ \otimes$. What happens when we apply it to the pair of pointed vector spaces $((A, \alpha_0), (B, \beta_0))$? Tracing through the definitions, we find $\otimes((A, \alpha_0), (B, \beta_0)) = (A \otimes B, \mathfrak{S}_{A \otimes B}, \alpha_0 \otimes \beta_0)$ and then $D(A \otimes B, \mathfrak{S}_{A \otimes B}, \alpha_0 \otimes \beta_0) = ((W_1, w_1), (W_2, w_2))$, where $W_1 = \pi_1(\alpha_0 \otimes \beta_0) = A \otimes \beta_0$, $W_2 = \pi_2(\alpha_0 \otimes \beta_0) = \alpha_0 \otimes B$, and $w_1 = w_2 = \alpha_0 \otimes \beta_0$. In other words,

$$(D \circ \otimes)((A, \alpha_0), (B, \beta_0)) = ((A \otimes \beta_0, \alpha_0 \otimes \beta_0), (\alpha_0 \otimes B, \alpha_0 \otimes \beta_0)). \quad (17)$$

While $(D \circ \otimes)((A, \alpha_0), (B, \beta_0))$ is thus not exactly identical with $((A, \alpha_0), (B, \beta_0))$, there is between them an obvious correspondence, $((A, \alpha_0), (B, \beta_0)) \longleftrightarrow (D \circ \otimes)((A, \alpha_0), (B, \beta_0))$, given by the linear isomorphisms,

$$\alpha \longleftrightarrow \alpha \otimes \beta_0 \quad \text{and} \quad \beta \longleftrightarrow \alpha_0 \otimes \beta. \quad (18)$$

The bijection (18) is an instance of what is called a *natural isomorphism* between functors, and so category theory gives us a precise way to express that D is effectively a right-inverse of $\otimes$ by saying that $\otimes \circ D$ is "naturally isomorphic" to the identity functor, a relationship which we will write as

$$\otimes \circ D \cong 1. \tag{19}$$

It is evident from its definition that the correspondence (18) establishes an isomorphism between two *objects* in VEC $\times$ VEC. If read from left to right, it is a mapping $\Psi : ((A, \alpha_0), (B, \beta_0)) \to (D \circ \otimes)((A, \alpha_0), (B, \beta_0))$, but what is it that earns Ψ the title, "natural", thereby authorizing the use of the symbol $\cong$ in (19)? It is that Ψ also induces the correct correspondence between *morphisms* by converting (f_1, f_2) into $(D \circ \otimes)(f_1, f_2)$. This is self-evident when one unpacks the definitions (cf. (20)), but even without unpacking the definitions, we could have been assured that Ψ was natural, if we had reflected that it was defined *intrinsically*, utilizing nothing more than the structures displayed in (17). Indeed, I think it would be fair to say that this possibility of being constructed from intrinsic information without the intervention of any arbitrary choices is what best expresses the intuitive meaning of "naturality".

The distinction between plain isomorphism $\simeq$ and natural isomorphism $\cong$ is perhaps most familiar in the example of dual vector spaces, where both V^{**} and V^* are isomorphic to V, but only the isomorphism between V^{**} and V is natural. Given an element $v \in V$, one can define $v^{**} \in V^{**}$ by the equation $v^{**}(f) = f(v)$, where $f \in V^*$. On the other hand, there is no way to pass deterministically from v to an element $f \in V^*$ without the aid of a basis for V, or a metric, or some such auxiliary structure.

Remark: A natural isomorphism sets up an equivalence between functors in much the same way as a similarity transformation sets up an equivalence between group representations. If R_1 and R_2 are representations of the group G related by the similarity transformation S, then $SR_1(g)S^{-1} = R_2(g)$ or equivalently, $SR_1(g) = R_2(g)S$. Now, if we replace R_1 and R_2 by functors F_1 and F_2, and the arbitrary group element g by an arbitrary morphism f, we obtain the condition for a family of invertible morphisms S to define a natural isomorphism between F_1 and F_2, namely $SF_1(f) = F_2(f)S$. Often, this last equation is represented by drawing the commutative diagram

$$
\begin{array}{ccc}
F_1 X & \overset{F_1 f}{\to} & F_1 Y \\
S_X \downarrow & & \downarrow S_Y, \\
F_2 X & \underset{F_2 f}{\to} & F_2 Y
\end{array}
$$

where X and Y are any objects in the category and $f : X \to Y$ is any morphism between them. When, as in our case, F_1 is the identity functor, the diagram for $S : 1 \to F$ simplifies to

$$
\begin{array}{ccc}
X & \xrightarrow{f} & Y \\
S\downarrow & & \downarrow S. \\
FX & \xrightarrow{Ff} & FY
\end{array}
\tag{20}
$$

One sees in this case that the functor F must be a bijection between the morphisms f and the morphisms Ff, and conversely, the fact that F is such a bijection captures to a large extent everything that the equation $F \cong 1$ means.

Having established that $D \circ \otimes \cong 1$, let us now try to demonstrate the complementary equivalence, $\otimes \circ D \cong 1$. Following the same steps as before, let us apply the functor $\otimes \circ D$ to the object $(V, \mathfrak{S}_V, v_0) \in \text{TVEC}$, obtaining first $D(V, \mathfrak{S}_V, v_0) = ((W_1, v_0), (W_2, v_0))$ and then $\otimes((W_1, v_0), (W_2, v_0)) = (W_1 \otimes W_2, \mathfrak{S}_{W_1 \otimes W_2}, v_0 \otimes v_0)$, which taken together tell us that

$$
(\otimes \circ D)(V, \mathfrak{S}_V, v_0) = (W_1 \otimes W_2, \mathfrak{S}_{W_1 \otimes W_2}, v_0 \otimes v_0).
\tag{21}
$$

Can we exhibit a natural isomorphism equating $(W_1 \otimes W_2, \mathfrak{S}_{W_1 \otimes W_2}, v_0 \otimes v_0)$ to $(V, \mathfrak{S}_V, v_0)$ and therefore $\otimes \circ D$ to the identity functor? To this question, we already have the answer in the form of the isomorphism, $\Phi : W_1 \otimes W_2 \to V$, which we constructed earlier with the aid of the intrinsically defined product $\overline{\otimes}$, and for which $\Phi(w_1 \otimes w_2) = (w_1 \overline{\otimes} w_2)$. As with Ψ before, it is straightforward to verify that Φ is natural, as indeed it had to be, given its intrinsic nature. Therefore, $\otimes \circ D \cong 1$.

We have now proven that the composition of $\otimes$ with D in either order is naturally isomorphic to the identity. Thus, category theory, by introducing the concept of natural isomorphism $\cong$ as a replacement for strict equality, has given us a way to make precise (and then to verify) the informal claims that, in the pointed case, the functor $\otimes$ is invertible and that D is its inverse.

Turn now to the unpointed case and to our expectation that it will not be possible to recover the pair (A, B) from $(A \otimes B, \mathfrak{S})$ when the spaces involved are not equipped with base points. Can we also corroborate this expectation within the categorical framework? Stated formally, the question is whether there exists a functor $D : \text{TVEC} \to \text{VEC} \times \text{VEC}$ which is a "left-inverse" to $\otimes$ in the sense that $D \circ \otimes \cong 1$. In fact, it's easy to see that no such functor can exist. Were D such a functor then, as we noted in connection with (20), the mapping, $f \mapsto (D \circ \otimes)f$, would have to be invertible for morphisms, $f : (A, B) \to (A', B')$, of the category VEC $\times$ VEC, where f is by definition a pair (g, h) of individual morphisms in VEC. This, however, is clearly impossible because the functor $\otimes$ (and therefore its composition with D if the latter existed) fails to be injective, since it maps both $f = (g, h)$ and $\tilde{f} = (\lambda g, h/\lambda)$ to the single morphism $g \otimes h = (\lambda g) \otimes (h/\lambda)$. In other words, $\otimes$ acting on morphisms is not injective but many-to-one, the "many" being parametrized by a non-zero scalar λ which embodies the same $GL(1)$ ambiguity we met with earlier. This confirms that the equation $D \circ \otimes \cong 1$ can have no solution, and *a fortiori* that the functor $\otimes$ is not invertible.

A somewhat simpler example of the same nature occurs in connection with the attempt to represent a spinor geometrically. Starting from a 2-component Weyl spinor ζ, for example, one can derive algebraically a so-called null flag F, which consists of a light-like vector together with a half-plane matched to the vector.[4] But because vectors are quadratically related to spinors, both ζ and $-\zeta$ give rise to the same flag F, whence one can recover the spinor from the flag only up to an unknown sign. (This loss of information was inevitable because spinors change sign after rotating through 2π, whereas vectors do not.) To couch these relationships in categorical language, one could introduce a category of spinor spaces and a category of spaces of null flags and a functor ϕ from the former to the latter. Like $\otimes$ above, ϕ would not be invertible because it would be $2 \to 1$ on morphisms. At best, one might be able to devise, as a kind of *right* inverse to ϕ, a "functor manqué" or "functor up to sign" going from flag spaces to spinor spaces. Its existence would proclaim that, although not fully a geometrical object, a spinor is nevertheless "geometrical up to sign".

Remark: Despite its utility, the concept of functor does not necessarily illuminate the connection between its inputs and its outputs as fully as one might have expected it to do because unlike a morphism, it is blind to the individual elements of the spaces on which it acts; by definition, it does not "look inside". Thus, if ϕ is a functor and X a space (or a mapping), and if Y is the space (or mapping) that results when ϕ acts on X, then the equation $Y = \phi(X)$ tells us *that* Y is in some sense built from X, but it tells us nothing concretely about *how* Y is built from X.[k] For example, X could be a spinor space and ϕ the above functor. Then, $Y = \phi(X)$ would be the space comprised of all the null flags derived from the spinors comprising X.

---

[k]Could Bourbaki's concept[5] of "deduction of structures" come any closer to doing this?

But if $\zeta \in X$ were an *individual* spinor in X, and if F were the individual flag derived from ζ, the rules governing functors would not allow us to write "$F = \phi(\zeta)$", even though it might seem natural to do so and even though we know perfectly well what we would mean by it!

7. Questions, Further Developments, and Connection to Quantum Field Theory

In conclusion, let me mention a few questions and possible further developments suggested by the above considerations.

The most important of the constructions introduced in Sections 4 and 5 revolve around the "foliations" $\mathcal{M}_1$ and $\mathcal{M}_2$, the corresponding mappings π_1 and π_2, the *square* concept and the product $\overline{\otimes}$ which results from these via definition (16).

An obvious question that one might ask is how these distinctive ingredients generalize to the tensor product of three or more vector spaces. One could of course just treat a threefold product like $A \otimes B \otimes C$ as an iterated pairwise product like $(A \otimes B) \otimes C$, but a more symmetric construction ought to be possible, and one might expect it to uncover some new structures that are not visible in connection with simple pairwise products like $A \otimes B$.

One might also wonder whether there was anything of interest to be learned from the study of the various symmetry types that become possible when two or more of the factor spaces are equal to each other. For example, when $V \subseteq A \otimes A$ is the subspace of symmetric tensors, two natural analogs of $\mathfrak{S}$ as used above would be the set of tensors of the form $\alpha \otimes \beta + \beta \otimes \alpha$ or even more simply, of the form $\alpha \otimes \alpha$. Or for the anti-symmetric tensor product of A with itself, the set of tensors of the form $\alpha \wedge \beta = \alpha \otimes \beta - \beta \otimes \alpha$ would be a natural analog of $\mathfrak{S}$. To what extent, and in what form, could one repeat the above discussion with one of these subsets replacing $\mathfrak{S}$? And still more generally, what might be an analog of $\mathfrak{S}$ belonging to the non-abelian symmetry types (those corresponding to more general Young tableaux) that arise as subspaces of higher products, $A \otimes A \cdots \otimes A$, and which are neither "bosonic" nor "fermionic"?

Another subcase of obvious interest is that where the vector spaces are equipped with metrics, in particular where they are Hilbert spaces. One might then expect orthonormality to play a role, but would any additional, unexpected features of interest show up?

Our discussion so far has proceeded in finite dimensions. If we want to generalize it to infinite-dimensional vector spaces, a whole raft of further questions will appear, some of which concern the definition of tensor product itself. Clearly, the subset $\mathfrak{S}$ of simple tensors $\alpha \otimes \beta$ within $V = A \otimes B$ can be defined without difficulty but will our constructions based on it also go through as before? Will they still let us recover A and B and will they still lead us as in Sections 4 and 6 (say in the pointed case) to a functor D inverse to $\otimes$? In all of this, what consequences might flow from ambiguities in the definition of $\otimes$? When A and B are Hilbert spaces, $A \otimes B$ *qua* Hilbert space is unambiguous, but when they are only Banach spaces (normed vector spaces), many different spaces $A \otimes B$ have been defined.[2] One may wonder then whether $\mathfrak{S}$ will still be able to "remember" which specific choice of $\otimes$ went into the creation of V.

Among infinite-dimensional vector spaces, the Hilbert spaces have a special significance for quantum theories. Although the ambiguity in defining $A \otimes B$ is not an issue when A and B are Hilbert spaces, it can happen in connection with quantum field theory that the notion of tensor product itself seems to be transcended. If one divides a Cauchy surface into two complementary regions, then naively one would expect the overall Hilbert space $\mathcal{H}$ of the field theory to be the tensor product of Hilbert spaces associated with the two regions, just as happens with composite systems in ordinary quantum mechanics. Unfortunately, this would conflict with the fact that the operator algebras associated with the two regions (technically with their domains of dependence in space-time) are known (for free fields) to be of "type III", this being intimately linked to the infinite entanglement entropy between the two regions. One still has operator subalgebras for the regions (so-called coupled factors), but these subalgebras cannot be interpreted as acting on the separate factors of a tensor product. One thus confronts something like a tensor product of operator algebras that does not derive from a tensor product decomposition of the underlying Hilbert space $\mathcal{H}$. (Adopting the language of "quantum systems", one might say that one is dealing with "subsystems which possess observables but lack state vectors".)[1]

In the absence of a tensor product structure for $\mathcal{H}$, the notion of simple vector is not defined and therefore neither is our subset $\mathfrak{S} \subseteq \mathcal{H}$.

---

[1]This is not quite the same as saying that a type-III factor lacks pure states. According to the definitions most commonly used in the theory of operator algebras, pure states do exist, and one could thus entertain them as generalized state vectors, since in finite dimensions, state vector = pure state. However, the pure states of a type-III factor seem to be mathematically pathological (even "ineffable"), and one should probably regard them as unphysical. See Ref. 6.

Nevertheless, one might hope that some generalization of simple vector, and some corresponding subset of $\mathcal{H}$, could serve a similar function. Simple vectors are tensors of rank 1, but the tensors of ranks, 2, 3, 4,... (or better, of finite "co-rank") also partake of the tensor product structure of V. Could it be that suitable analogs of the spaces of such tensors are able to capture the structure of coupled factors of types-III or -II and in so doing, shed light on features like the area law for entanglement entropy? Especially salient in this connection is the need for some kind of spatiotemporal cutoff to render the entropy finite.[7] Physically, such a cutoff should be frame-independent (locally Lorentz invariant), and it seems suggestive that an analog of $\mathfrak{S}$, if it could be defined, would not obviously need to refer to any arbitrarily chosen reference frame.

As a first approach to some of these questions, one could ask in finite dimensions how to relate the operator algebra framework to that in the present paper. Indeed, one might have thought to identify a tensor product structure for V, not with the family $\mathfrak{S}$ of simple vectors in V but with a pair of commuting operator subalgebras which generate the algebra $L(V)$ of all linear operators on V and which have in common only the multiples of the identity operator (like Murray–von Neumann coupled factors but without any specialization to complex numbers or self-adjointness). The advantage of such an alternative approach would be that the algebras $L(A)$ and $L(B)$ reappear bodily in $L(A \otimes B)$, whereas the spaces A and B themselves need to be excavated from $A \otimes B$ more painfully, as we have seen in great detail above. (Quantally speaking, the "observables" of a subsystem carry over to the composite system, whereas the "states" do not. But see the remark in the following.) Its disadvantage would be that an algebra of operators in V is a considerably more complicated beast than the simple subset $\mathfrak{S} \subseteq V$. Be that as it may, it's clear that the two viewpoints are related. For example, an operator acting only on A (an operator in $L(A) \otimes \mathbf{1}$) will automatically be an operator that preserves $\mathfrak{S}$, suggesting how one might derive $L(A)$ and $L(B)$ from $\mathfrak{S}$.

Remark: In the context of Quantum Measure Theory,[3] the histories Hilbert space associated with a subsystem actually does reappear as a true subspace of the histories Hilbert space of the full system, the reason being that an *event* in a subsystem is *ipso facto* an event in the full system. Moreover, this subspace carries a distinguished "base point", as remarked in Section 5. When, in addition, the overall quantum measure is the product measure (as for "non-interacting subsystems in a product state"), the full histories Hilbert space is the tensor product of the subspaces, and the afore-mentioned advantage of an approach via operator algebras disappears.

Let us return, finally, to finite dimensions and to the cone $\mathfrak{S}$ of simple vectors within V, on which most of our work has been based. We have seen how $\mathfrak{S}$ endows V with the structure of a product space, but we did not provide (or even ask for) a simple criterion that would let us recognize whether a given subset $\mathfrak{S}$ could actually play the role assigned to it. That is, we did not provide necessary and sufficient conditions for there to exist an isomorphism mapping V to a space $A \otimes B$ that would map $\mathfrak{S}$ to the set of tensors of the form $\alpha \otimes \beta$. One trivially adequate criterion is that the re-constructions undertaken in Section 5 should succeed, and in particular that the building up of the *squares* should never encounter an obstacle. But one might wish for criteria that were more self-contained and more simply stated. Given that the rays in $\mathfrak{S}$ constitute a "Segre variety", one might hope that the algebraic geometry literature would contain something of this sort.

Alternatively, rather than seeking axioms for $\mathfrak{S}$, one might instead seek axioms for the squares themselves, i.e., axioms for quadruples of vectors in V. A tensor product structure for V would then be a set of quadruples satisfying these axioms.

Acknowledgements

This research was supported in part by NSERC through grant RGPIN-418709-2012. This research was supported in part by Perimeter Institute for Theoretical Physics. Research at Perimeter Institute is supported by the Government of Canada through Industry Canada and by the Province of Ontario through the Ministry of Economic Development and Innovation.

References

1. For more detail on these definitions, see: S. Sternberg, *Lectures on Differential Geometry*, Englewood Cliffs, N.J, Prentice-Hall (1964); T. Y. Thomas, *Concepts from Tensor Analysis and Differential Geometry*, Academic Press (1961), pp. 7ff; S. M. Lane, *Homology*, Academic Press (1963), pp. 138ff; Chapter 1 of Sternberg contains examples of most of the definitions, including implicitly the definition of $V \otimes W$ as the space of bilinear mappings of $V^* \times W^*$ into $\mathbb{R}$. The time-honoured definition of a tensor in terms of transformation laws for its components is presented in Thomas.
2. A. Grothendieck, Produits tensoriels topologiques et espaces nucléaires, *Mem. Am. Math. Soc.* **16** (1955).

3. R. D. Sorkin, *Mod. Phys. Lett.* **A9** (No. 33) 3119–3127 (1994); F. Dowker and R. D. Sorkin, 'Persistence of Zero' as an intrinsic causality condition in histories-based quantum mechanics (to appear); F. Dowker, S. Johnston and S. Surya, *J. Phys.* A435053052010.

4. R. Penrose, Structure of space-time, in C. M. DeWitt and J. A. Wheeler (eds.), *Battelle Rencontres: 1967 Lectures in Mathematics and Physics.* New York: Benjamin (1968).

5. N. Bourbaki, *Theory of Sets,* Paris: Hermann (1968), Chapter IV.

6. B. Blackadar, *Operator Algebras: Theory of C*-Algebras and von Neumann Algebras* (2017). https://packpages.unr.edu/media/1224/cycr.pdf; See III.2.2.3 and the remarks in III.2.2.15.

7. R. D. Sorkin, On the entropy of the vacuum outside a horizon, in B. Bertotti, F. de Felice and A. Pascolini (eds.), *Tenth International Conference on General Relativity and Gravitation (held Padova, 4-9 July, 1983), Contributed Papers*, Vol. II, pp. 734–736, Roma, Consiglio Nazionale Delle Ricerche (1983); R. D. Sorkin, The statistical mechanics of Black Hole thermodynamics, in R. M. Wald (ed.) *Black Holes and Relativistic Stars*, University of Chicago Press (1998), pp. 177–194.

Chapter 23

Noncommutative AdS_2 II: The Correspondence Principle

Allen Stern[*,§] and Aleksandr Pinzul[†,‡,¶]

*Department of Physics, University of Alabama,
Tuscaloosa, Alabama 35487, USA
†Universidade de Brasília, Instituto de Física, Brasília,
DF 70910-900, Brasil
‡International Center of Physics, Brasília, DF C.P. 04667, Brazil,
§astern@ua.edu
¶apinzul@unb.br

Using the exact solutions to the field equation for a massive scalar field on noncommutative AdS_2, we apply the AdS/CFT correspondence principle to obtain an exact result for the associated two-point function on the conformal boundary. The answer satisfies conformal invariance and has the correct commutative limit and massless limit.

1. Introduction

The conjectured AdS/CFT correspondence principle has played a central role in theoretical physics in the past two decades. It posits a strong/weak duality between gravity in the bulk of an asymptotically anti-de Sitter (AdS) space and a conformal field theory (CFT) located at the so-called conformal boundary.[1] For obvious reasons, many practical applications of the correspondence principle utilize classical, or weak, gravity in the bulk, with the intention of exploring the strong coupling regime of the boundary theory. Since a fully consistent quantum gravity treatment of the bulk remains out of reach, it is a non-trivial task to explore other domains of the correspondence. On the other hand, the incorporation of some quantum gravity effects in the bulk is possible. This has been the motivation for our recent works.[2-5] As remarked in Ref. 6, the inclusion of some quantum gravity effects might be achieved by replacing the AdS bulk by its noncommutative analog, $ncAdS_2$. While, in general, the introduction of noncommutativity destroys the isometries of a manifold and is not unique, this is not the case for AdS_2. Therefore, $ncAdS_2$ can serve as a toy model for the introduction of quantum gravity effects in the bulk (barring the known difficulties of the AdS_2/CFT_1 correspondence that already exist in the commutative case[a]). Moreover, in Ref. 2 it was shown (i) that the star product for $ncAdS_2$, when acting on functions having a well-behaved boundary limit, reduces to the point-wise product in this limit and (ii) that noncommutative corrections to isometry generators, i.e., Killing vectors, vanish near the boundary. In other words, $ncAdS_2$ is an asymptotically AdS space and so according to the correspondence principle, a CFT should be present at its boundary.

The on-shell field theory action in the bulk, $S|_{\text{on-shell}}$, plays a central role in the explicit construction of the AdS/CFT correspondence. It generates the $n-$point connected correlation functions for operators on the boundary.[9] As the first step, it is therefore necessary to obtain the solutions to field theories in the bulk, which in our case are noncommutative. Exact solutions on $ncAdS_2$ were obtained for free massless scalar and spinor fields in Ref. 5 and massive scalar fields in Ref. 6. These exact solutions will therefore lead to exact expressions for the corresponding two-point function on the boundary. This was shown in, Ref. 5 for the case of massless fields. It is the purpose of this chapter to obtain the boundary two-point function resulting from the exact solutions to the massive field equation found in Ref. 6.

The generating function $S|_{\text{on-shell}}$ is expressed in terms of the boundary values ϕ_0 of the solutions for the bulk fields, and the prescription for computing the $n-$point connected correlation function for operators $\mathcal{O}$ located at non-coincidental points x_i on the boundary is given by

$$\langle \mathcal{O}(x_1) \cdots \mathcal{O}(x_n) \rangle = \frac{\delta^n S|_{\text{on-shell}}}{\delta\phi_0(x_1) \cdots \delta\phi_0(x_n)}\bigg|_{\phi_0=0}. \tag{1.1}$$

For the case of a 2-dimensional bulk theory, both $\mathcal{O}$ and ϕ_0 are functions of only one coordinate, the time t. Conformal invariance severely restricts

---

[a]See, for example, Refs. 7, 8.

the form of the n−point correlators. For $n = 2$, one has

$$\langle \mathcal{O}(t)\mathcal{O}(t') \rangle = C_{\Delta_+} \frac{1}{|t - t'|^{2\Delta_+}}, \tag{1.2}$$

where Δ_+ is the conformal dimension and C_{Δ_+} is a constant which can be computed using (1.1). We shall compute the two-point function that results from massive scalar fields on $ncAdS_2$ and show that it has the form given in (1.2). The result for the overall factor C_{Δ_+} differs from that of the commutative theory. By taking various limits, we can compare with the previous results.

The outline for the remainder of this chapter is as follows: We first review the calculations for C_{Δ_+} in the commutative case in Section 2 and then apply the analogous procedure to find the constant factor in the noncommutative case in Section 3. Some concluding remarks are made in Section 4.

2. Commutative Case

The analysis of the correspondence principle is more conveniently performed for the Euclidean version of AdS, or $EAdS$, due to the absence of propagating states.[b] Here, we specialize to $EAdS_2$, closely following the work in Refs. 13, 14.

The field equation for a scalar field Φ with mass m on $EAdS_2$ is

$$\Delta\Phi = m^2\Phi, \tag{2.1}$$

with Δ denoting the appropriate Laplacian. $EAdS_2$ is conveniently parametrized by Fefferman–Graham coordinates (z, t). They span the half-plane, $z > 0$, $-\infty < t < \infty$, with the conformal boundary located at $z \to 0$.[c] In terms of these coordinates, the metric tensor and Laplacian are given, respectively, by

$$ds^2 = \frac{dz^2 + dt^2}{z^2} \tag{2.2}$$

and

$$\Delta = z^2(\partial_z^2 + \partial_t^2) . \tag{2.3}$$

---

[b] For treatments of the correspondence principle relying on Lorentzian signature, see Refs. 10–12.
[c] Some care is needed to define the boundary limit. One should first assume that the boundary is located at some $z = \epsilon$ and then take the limit $\epsilon \to 0$.

The field equation (2.1) results from the standard scalar field action

$$S = \frac{1}{2} \int_{R \times R_+} dt dz \left((\partial_z \Phi)^2 + (\partial_t \Phi)^2 + \frac{m^2}{z^2} \Phi^2 \right)$$

$$= -\frac{1}{2} \int_{R \times R_+} \frac{dt dz}{z^2} \Phi (\Delta - m^2) \Phi - \frac{1}{2} \int_R dt \, \Phi \partial_z \Phi \Big|_{z \to 0}. \qquad (2.4)$$

Only the boundary term survives when evaluating the action on-shell.

General solutions to (2.1) that are well behaved away from the conformal boundary have the form

$$\Phi(z,t) = \frac{1}{2\pi} \int d\omega \, e^{i\omega t} \sqrt{z} \, a(\omega) K_\nu (|\omega|z), \quad \nu = \sqrt{m^2 + \frac{1}{4}}, \qquad (2.5)$$

where $K_\nu(x)$ denotes the modified Bessel function. To determine the behavior of the solutions near the conformal boundary, we can use[15]

$$K_\nu(x) \to \frac{1}{2} \left(2^\nu \Gamma(\nu) x^{-\nu} + 2^{-\nu} \Gamma(-\nu) x^\nu \right) \left(1 + \mathcal{O}(x^2) \right), \quad \text{as } x \to 0. \quad (2.6)$$

Then, for $0 < \nu < 1$ (corresponding to $-\frac{1}{4} < m^2 < \frac{3}{4}$), $\Phi(z,t)$ tends towards

$$\frac{1}{4\pi} \int d\omega \, e^{i\omega t} a(\omega) \left(\left(\frac{2}{|\omega|} \right)^\nu \Gamma(\nu) z^{\Delta_-} + \left(\frac{|\omega|}{2} \right)^\nu \Gamma(-\nu) z^{\Delta_+} \right), \text{ as } z = \epsilon \to 0,$$
$$(2.7)$$

where $\Delta_\pm = \frac{1}{2} \pm \nu$, with Δ_+ to be identified with the conformal dimension. So, upon keeping just the leading term,

$$\Phi(\epsilon, t) \to \phi_\epsilon(t) = \epsilon^{\Delta_-} \phi_0(t), \quad \text{as } \epsilon \to 0. \qquad (2.8)$$

Here, we shall assume Dirichlet boundary conditions. $\phi_0(t)$ is regarded as the source field on the boundary, which determines the solution for Φ in the bulk.[d] This can be made explicit by writing the coefficients $a(\omega)$ of the solution (2.5) in terms of the Fourier coefficients $\phi(\omega)$ of the boundary field $\phi_\epsilon(t) = \frac{1}{2\pi} \int d\omega e^{i\omega t} \phi(\omega)$,

$$\phi(\omega) = \sqrt{\epsilon} a(\omega) K_\nu (|\omega|\epsilon). \qquad (2.9)$$

---

[d]For other boundary conditions, see, for example Ref. 14.

The solution for $\Phi(z,t)$ in the bulk can thereby be expressed in terms of these Fourier coefficients

$$\Phi(z,t) = \frac{1}{2\pi} \int d\omega e^{i\omega t} \phi(\omega) \frac{\sqrt{z}K_\nu(|\omega|z)}{\sqrt{\epsilon}K_\nu(|\omega|\epsilon)}. \tag{2.10}$$

The evaluation of the on-shell action requires an analogous expression for $\partial_z \Phi(z,t)$. For this, we can use the identity $\frac{d}{dx}K_\nu(x) = \frac{\nu}{x}K_\nu(x) - K_{\nu+1}(x)$ to write

$$\sqrt{z}\frac{d}{dz}\left(\sqrt{z}K_\nu(|\omega|z)\right) = \left(\frac{1}{2}+\nu\right)K_\nu(|\omega|z) - |\omega|zK_{\nu+1}(|\omega|z),$$

and hence

$$\partial_z \Phi(z,t) = \frac{1}{2\pi} \int d\omega e^{i\omega t} \phi(\omega) \frac{\left(\frac{1}{2}+\nu\right)K_\nu(|\omega|z) - |\omega|zK_{\nu+1}(|\omega|z)}{\sqrt{z\epsilon}K_\nu(|\omega|\epsilon)}. \tag{2.11}$$

Substitution of (2.10) and (2.11) into the boundary term in (2.4) gives the following result for the on-shell action $S|_{\text{on-shell}}$:

$$-\frac{1}{4\pi} \int dt \int dt' \int d\omega e^{i\omega(t-t')}\phi_\epsilon(t)\phi_\epsilon(t')\frac{1}{\epsilon}\left(\frac{1}{2}+\nu-|\omega|\epsilon\frac{K_{\nu+1}(|\omega|\epsilon)}{K_\nu(|\omega|\epsilon)}\right)\Big|_{\epsilon\to 0}. \tag{2.12}$$

From (2.6), the asymptotic limit of the ratio of Bessel functions is

$$\frac{xK_{\nu+1}(x)}{K_\nu(x)} \to \frac{(2^{\nu+1}\Gamma(\nu+1)+2^{-\nu-1}\Gamma(-\nu-1)x^{2\nu+2})}{(2^\nu\Gamma(\nu)+2^{-\nu}\Gamma(-\nu)x^{2\nu})} \quad \text{as } x \to 0,$$

which to leading order is $2\nu\left(1 - \frac{\Gamma(-\nu)}{\Gamma(\nu)}\left(\frac{x}{2}\right)^{2\nu}\right)$ for $0 < \nu < 1$. Upon applying this result, along with the integral[14]

$$\frac{1}{2\pi}\int d\omega e^{-i\omega t}|\omega|^\rho = \frac{2^\rho\Gamma(\frac{\rho+1}{2})}{\sqrt{\pi}\Gamma(-\frac{\rho}{2})}\frac{1}{|t|^{\rho+1}}, \quad \rho \neq -1, -3, \ldots, \tag{2.13}$$

one gets[e]

$$S|_{\text{on-shell}} = -\frac{\nu\Gamma(-\nu)}{2^{2\nu+1}\pi\Gamma(\nu)} \int dt \int dt'\,\phi_\epsilon(t)\phi_\epsilon(t')\,\epsilon^{2\nu-1} \int d\omega\, e^{i\omega(t-t')}|\omega|^{2\nu}\bigg|_{\epsilon\to 0}$$

$$= -\frac{\nu}{\sqrt{\pi}}\frac{\Gamma(\nu+\frac{1}{2})}{\Gamma(\nu)} \int dt \int dt'\,\phi_\epsilon(t)\phi_\epsilon(t')\,\epsilon^{2\nu-1}\frac{1}{|t-t'|^{2\nu+1}}\bigg|_{\epsilon\to 0}.$$

(2.14)

In terms of the ϵ-independent source field $\phi_0(t)$ in (2.8), the result is

$$S|_{\text{on-shell}} = -\frac{\nu}{\sqrt{\pi}}\frac{\Gamma(\nu+\frac{1}{2})}{\Gamma(\nu)} \int dt \int dt'\,\frac{\phi_0(t)\phi_0(t')}{|t-t'|^{2\nu+1}}.$$

(2.15)

The two-point function for the massive scalar (1.2) is now easily recovered from (1.1), with the resulting factor C_{Δ_+} given by

$$C_{\Delta_+} = -\frac{2\nu}{\sqrt{\pi}}\frac{\Gamma(\nu+\frac{1}{2})}{\Gamma(\nu)}.$$

(2.16)

3. Noncommutative Case

We now repeat the procedure for the noncommutative case. The field equation for a massive scalar field $\hat{\Phi}$ on noncommutative $EAdS_2$ is

$$\hat{\Delta}\hat{\Phi} = m^2\hat{\Phi},$$

(3.1)

where $\hat{\Delta}$ is the noncommutative version of the Laplacian (2.3). As in Ref. 6, one can express $\hat{\Delta}$ in terms of operators $\hat{X}^a$, $a = 1,2,3$, which satisfy the $su(1,1)$ algebra

$$[\hat{X}^a, \hat{X}^b] = i\alpha\epsilon^{abc}\hat{X}_c,$$

(3.2)

and have $\hat{X}^a\hat{X}_a = -1$. The indices are raised and lowered with the metric $\text{diag}(-1,1,1)$, ϵ^{abc} is totally antisymmetric, with $\epsilon^{012} = 1$, and α is the noncommutative parameter, analogous to Planck's constant, which should

---

[e]The singular term

$$-\frac{1}{4\pi}\left(\frac{1}{2}-\nu\right)\int dt \int dt' \int d\omega\, e^{i\omega(t-t')}\frac{\phi_0(t)\phi_0(t')}{\epsilon^{2\nu}}\bigg|_{\epsilon\to 0}$$

can be ignored since the integration in ω leads to a delta function and we are interested in $t \neq t'$.

correspond to the quantum gravity scale. The Laplacian is given by

$$\frac{\alpha^2}{2}\Delta\hat{\Phi} = \hat{X}_a\hat{\Phi}\hat{X}^a + \hat{\Phi}. \tag{3.3}$$

The field equation (3.1) results from the variation of $\hat{\Phi}$ in the following action:

$$\hat{S} = -\frac{1}{2\alpha^2}\text{Tr}\left\{[\hat{X}^a, \hat{\Phi}][\hat{X}_a, \hat{\Phi}] - (\alpha m)^2\hat{\Phi}^2\right\}. \tag{3.4}$$

The trace in (3.4) can be understood as an integration over symbols of operators, and the integration domain can again be taken to be the half-plane spanned by commuting coordinates z and t. As in (2.4), the action can be split up into two terms, one of which vanishes on-shell and the other is a boundary term. Moreover, as was shown in Ref. 2, the boundary term for the noncommutative theory is *identical* to that of the commutative theory. (A conformal boundary limit can be defined for noncommutative $EAdS_2$ in terms of the representation theory for the $su(1,1)$ algebra.[2]) Thus, the on-shell action once again takes the form

$$S|_{\text{on-shell}} = -\frac{1}{2}\int_R dt\, \hat{\Phi}\partial_z\hat{\Phi}\Big|_{z\to 0}, \tag{3.5}$$

where here $\hat{\Phi}$ actually denotes the symbol of the field.

As shown in Ref. 6, (3.1) has exact solutions, which are given in terms of generalized Legendre functions. Upon restricting to fields that are well behaved in the bulk, and integrating over all frequencies ω, one has

$$\hat{\Phi} = \frac{1}{2\pi}\int d\omega\, a(\omega)e^{i\omega\hat{t}/2} P_{-\Delta_-}^{-\frac{2\kappa}{\alpha}}\left(\frac{2\hat{r}}{|\omega|\alpha}\right) e^{i\omega\hat{t}/2}, \tag{3.6}$$

where $\hat{t}$ and $\hat{r}$ are noncommutative operators, which can be used to construct $\hat{X}^a$ (see Ref. 6), and $\kappa = \sqrt{1 + \frac{\alpha^2}{4}}$. $\hat{t}$ and $\hat{r}$ satisfy the commutator $[\hat{t}, \hat{r}] = -i\alpha$.

As in the commutative case, the solution can be expressed in terms of boundary fields. The symbol of $\hat{r}^{-1}$ tends to zero in this limit, and the algebra of functions of $\hat{r}$ and $\hat{t}$ is effectively commutative near the boundary. Therefore, the symbol of the solution (3.6) has the conformal boundary limit:

$$\phi_\epsilon(t) = \frac{1}{2\pi}\int d\omega\, e^{i\omega t} a(\omega)\, P_{-\Delta_-}^{-\frac{2\kappa}{\alpha}}\left(\frac{2}{|\omega|\alpha\epsilon}\right), \qquad \text{as } \epsilon \to 0. \tag{3.7}$$

From Ref. 6, this tends to $\epsilon^{\Delta_-}\phi_0(t)$ as $\epsilon \to 0$ for $\nu > 0$, just as in the commutative case (2.8). We shall again assume Dirichlet boundary conditions, with $\phi_0(t)$ regarded as the source field. Re-expressing the solution (3.6) in terms of the Fourier transform $\phi(\omega)$ of the boundary field $\phi_\epsilon(t)$,

$$\phi(\omega) = a(\omega) \, P_{-\Delta_-}^{-\frac{2\kappa}{\alpha}}\left(\frac{2}{|\omega|\alpha\epsilon}\right), \tag{3.8}$$

we get

$$\hat{\Phi} = \frac{1}{2\pi} \int d\omega \, \phi(\omega) e^{i\omega\hat{t}/2} \frac{P_{-\Delta_-}^{-\frac{2\kappa}{\alpha}}\left(\frac{2\hat{r}}{|\omega|\alpha}\right)}{P_{-\Delta_-}^{-\frac{2\kappa}{\alpha}}\left(\frac{2}{|\omega|\alpha\epsilon}\right)} e^{i|\omega|\hat{t}/2}. \tag{3.9}$$

Once again we need to compute the derivative of the solution with respect to the radial coordinate. This, in general, will introduce operator ordering ambiguities. However, such ambiguities are not a concern for the computation of the on-shell action, since we only need the result in the conformal boundary limit where the coordinates effectively commute. So, let us choose the radial derivative to be $\frac{i}{\alpha}[\hat{t}, \hat{\Phi}]$. Then, the symbol of $\frac{i}{\alpha}[\hat{t}, \hat{\Phi}]$ in the boundary limit is $-z^2 \partial_z \hat{\Phi}|_{z\to 0}$. After applying the identity[16]

$$\frac{dP_\rho^\mu(x)}{dx} = \frac{1}{1-x^2}\left((\rho+1)xP_\rho^\mu(x) - (-\mu+\rho+1)P_{\rho+1}^\mu(x)\right), \tag{3.10}$$

we get the following result for $\partial_z \hat{\Phi}|_{z=\epsilon}$ as $\epsilon \to 0$:

$$-\frac{1}{4\pi}\int d\omega e^{i\omega t}\phi(\omega)\alpha|\omega|\left((\Delta_- -1)\left(\frac{2}{|\omega|\alpha\epsilon}\right) + \left(\Delta_+ + \frac{2\kappa}{\alpha}\right)\frac{P_{\Delta_+}^{-\frac{2\kappa}{\alpha}}\left(\frac{2}{|\omega|\alpha\epsilon}\right)}{P_{-\Delta_-}^{-\frac{2\kappa}{\alpha}}\left(\frac{2}{|\omega|\alpha\epsilon}\right)}\right).$$

The substitution of this result in the on-shell action (3.5) yields

$$S|_{\text{on-shell}} = \frac{1}{8\pi}\int dt \int dt' \int d\omega e^{i\omega(t-t')}\phi_\epsilon(t)\phi_\epsilon(t')$$

$$\times \alpha|\omega|\left((\Delta_- -1)\left(\frac{2}{|\omega|\alpha\epsilon}\right) + \left(\Delta_+ + \frac{2\kappa}{\alpha}\right)\frac{P_{\Delta_+}^{-\frac{2}{\alpha}\kappa}\left(\frac{2}{|\omega|\alpha\epsilon}\right)}{P_{-\Delta_-}^{-\frac{2}{\alpha}\kappa}\left(\frac{2}{|\omega|\alpha\epsilon}\right)}\right)\Bigg|_{\epsilon\to 0}. \tag{3.11}$$

As with the commutative answer, the term that yields a delta function after integration in ω can be dropped. To evaluate the remaining term, we need the asymptotic behavior of $yP_{\nu+\frac{1}{2}}^\mu(1/y)\,/\,P_{\nu-\frac{1}{2}}^\mu(1/y)$ as $y \to 0$, which is

$$\frac{2\nu}{\nu-\mu+\frac{1}{2}}\left(1 - \frac{\Gamma(-\nu)\Gamma(\nu-\mu+\frac{1}{2})}{2^{2\nu}\Gamma(\nu)\Gamma(-\nu-\mu+\frac{1}{2})}y^{2\nu}\right), \quad \text{for } 0 < \nu < 1.$$

Applying this to (3.11) gives

$$S|_{\text{on-shell}} = \frac{1}{4\pi} \int dt \int dt' \int d\omega e^{i\omega(t-t')} \phi_\epsilon(t) \phi_\epsilon(t') \frac{1}{\epsilon}$$

$$\times \frac{2\nu(\Delta_+ + \frac{2\kappa}{\alpha})}{\nu + \frac{2\kappa}{\alpha} + \frac{1}{2}} \left(1 - \frac{\Gamma(-\nu)\Gamma(\nu + \frac{2\kappa}{\alpha} + \frac{1}{2})}{2^{2\nu}\Gamma(\nu)\Gamma(-\nu + \frac{2\kappa}{\alpha} + \frac{1}{2})} \left(\frac{|\omega|\alpha\epsilon}{2} \right)^{2\nu} \right) \Bigg|_{\epsilon \to 0}.$$

$$(3.12)$$

The integral over ω can be performed, again using (2.13), with the result for the on-shell action being

$$-\frac{\nu}{\sqrt{\pi}} \frac{\Gamma(\nu + \frac{1}{2})\Gamma(\nu + \frac{2\kappa}{\alpha} + \frac{1}{2})}{\Gamma(\nu)\Gamma(-\nu + \frac{2\kappa}{\alpha} + \frac{1}{2})} \int dt \int dt' \phi_\epsilon(t) \phi_\epsilon(t') \frac{1}{\epsilon} \left(\frac{\alpha\epsilon}{2} \right)^{2\nu} \frac{1}{|t - t'|^{2\nu+1}} \Bigg|_{\epsilon \to 0}.$$

In terms of the ϵ-independent source fields ϕ_0, this becomes

$$S|_{\text{on-shell}} = -\left(\frac{\alpha}{2} \right)^{2\nu} \frac{\nu}{\sqrt{\pi}} \frac{\Gamma(\nu + \frac{1}{2})\Gamma(\nu + \frac{2\kappa}{\alpha} + \frac{1}{2})}{\Gamma(\nu)\Gamma(-\nu + \frac{2\kappa}{\alpha} + \frac{1}{2})} \int dt \int dt' \frac{\phi_0(t)\phi_0(t')}{|t - t'|^{2\nu+1}}.$$

The conformal answer for the two-point function (1.2) is then once again recovered after using (1.1). Now, the overall factor is

$$C_{\Delta_+}^{\text{nc}} = -\left(\frac{\alpha}{2} \right)^{2\nu} \frac{2\nu}{\sqrt{\pi}} \frac{\Gamma(\nu + \frac{1}{2})\Gamma(\nu + \frac{2\kappa}{\alpha} + \frac{1}{2})}{\Gamma(\nu)\Gamma(-\nu + \frac{2\kappa}{\alpha} + \frac{1}{2})}. \qquad (3.13)$$

The ratio of the noncommutative result to the commutative result (2.16) is given by the simple expression

$$\mathcal{R}^{(2)}(\alpha, \nu) = \left(\frac{\alpha}{2} \right)^{2\nu} \frac{\Gamma(\nu + \frac{2\kappa}{\alpha} + \frac{1}{2})}{\Gamma(-\nu + \frac{2\kappa}{\alpha} + \frac{1}{2})}. \qquad (3.14)$$

One can check various limits. The ratio smoothly goes to one in the commutative limit $\alpha \to 0$. It reduces to κ in the massless case, $\nu = \frac{1}{2}$, which agrees with the result found in Ref. 5.[f]

---

[f] If one does a leading order expansion in α^2, one gets

$$\mathcal{R}^{(2)}(\alpha, \nu) = 1 + \frac{\nu}{12} \left(\frac{13}{4} - \nu^2 \right) \alpha^2 + \mathcal{O}(\alpha^4).$$

This disagrees with the result found in Ref. 3, which is due to the fact that a different regularization scheme was used in that approach. The answers agree on the other hand, if one re-evaluates integrals in Ref. 3 with the regularization scheme used here. Both regularizations give the same result in the massless case. For a discussion of the different regularizations, see the appendix in Ref. 13.

We also note that the result found here agrees with that found in Ref. 17 for the massless case, but it differs when $\nu \neq \frac{1}{2}$.

4. Concluding Remarks

It had previously been conjectured that the boundary two-point function associated with field theory on a noncommutative AdS_2 bulk satisfies the constraints of conformal invariance, i.e., it has the form (1.2). This was previously shown to be true up to leading order in the noncommutativity parameter α. In this chapter, we proved that the conjecture is true to all orders in α, at least for the scalar field. A similar proof should be possible for spinors. The result means that the noncommutativity of the bulk only affects the overall normalization of the boundary two-point function.

It is a non-trivial step to see whether or not these results can be extended to the case of $n(> 2)$−point correlators on the boundary. This will require analyzing a fully interacting field theory on the noncommutative bulk. If conformal invariance is satisfied for the $n(> 2)$−point boundary correlators, then, once again, only the overall factor is effected by the noncommutative bulk. We would then have an expression for the ratio $\mathcal{R}^{(n)}(\alpha, \nu)$ of the overall factor for the noncommutative correlator with that of the commutative correlator. If the ratio is such that $\sqrt[n]{\mathcal{R}^{(n)}(\alpha, \nu)} = \sqrt{\mathcal{R}^{(2)}(\alpha, \nu)}$, it would suggest that the only effect that bulk noncommutativity has on the boundary is to renormalize the conformal operators $\mathcal{O}$. If instead this turns out not to be the case, we would conclude that the couplings between boundary operators pick up quantum gravity corrections as well. Only a preliminary investigation in this direction has been undertaken so far. A cubic interaction term was introduced to the massive scalar field on noncommutative $EAdS_2$ in Ref. 3. The analysis there was quite non-trivial because only a perturbative solution (in α) was available, and one needs to perturb in the other parameter, the coupling constant λ, as well. Although there were strong indications that the leading order correction to 3−point function satisfied the constraints of conformal invariance, we were not able to write down an explicit expression for the leading order correction, in either of the parameters. The fact that we now have an exact answer for the free theory could be quite beneficial for obtaining the 3−point function, since then we would only have to perform an expansion in one parameter, i.e., λ. The only obstacle to proceeding with the analysis is the construction of a meaningful Green's function on the noncommutative bulk. We hope to report on this construction in a future work.

References

1. J. M. Maldacena, The large N limit of superconformal field theories and supergravity, *Adv. Theor. Math. Phys.* **2** (1998) 231–252. doi: 10.1023/A:1026654312961.

2. A. Pinzul and A. Stern, Non-commutative AdS_2/CFT_1 duality: The case of massless scalar fields, *Phys. Rev. D* **96**(6) (2017) 066019. doi: 10.1103/PhysRevD.96.066019.

3. F. R. de Almeida, A. Pinzul and A. Stern, Noncommutative AdS_2/CFT_1 duality: The case of massive and interacting scalar fields, *Phys. Rev. D* **100**(8) (2019) 086005. doi: 10.1103/PhysRevD.100.086005.

4. F. Lizzi, A. Pinzul, A. Stern and C. Xu, Asymptotic commutativity of quantized spaces: The case of $\mathbb{CP}^{p,q}$, *Phys. Rev. D* **102**(6) (2020) 065012. doi: 10.1103/PhysRevD.102.065012.

5. A. Pinzul and A. Stern, Exact solutions for scalars and spinors on quantized Euclidean AdS2 space and the correspondence principle, *Phys. Rev. D* **104**(12) (2021) 126034. doi: 10.1103/PhysRevD.104.126034.

6. A. Pinzul and A. Stern, Noncommutative AdS_2 I: Exact solutions, *see the contribution to this volume* (2023).

7. A. Strominger, AdS(2) quantum gravity and string theory, *JHEP* **01** (1999) 007. doi: 10.1088/1126-6708/1999/01/007.

8. J. Maldacena and D. Stanford, Remarks on the Sachdev-Ye-Kitaev model, *Phys. Rev. D* **94**(10) (2016) 106002. doi: 10.1103/PhysRevD.94.106002.

9. E. Witten, Anti-de Sitter space and holography, *Adv. Theor. Math. Phys.* **2** (1998) 253–291. doi: 10.4310/ATMP.1998.v2.n2.a2.

10. V. Balasubramanian, P. Kraus and A. E. Lawrence, Bulk versus boundary dynamics in anti-de Sitter space-time, *Phys. Rev. D* **59** (1999) 046003. doi: 10.1103/PhysRevD.59.046003.

11. V. Balasubramanian, P. Kraus, A. E. Lawrence and S. P. Trivedi, Holographic probes of anti-de Sitter space-times, *Phys. Rev. D* **59** (1999) 104021. doi: 10.1103/PhysRevD.59.104021.

12. D. Marolf, States and boundary terms: Subtleties of Lorentzian AdS/CFT, *JHEP* **05** (2005) 042. doi: 10.1088/1126-6708/2005/05/042.

13. D. Z. Freedman, S. D. Mathur, A. Matusis and L. Rastelli, Correlation functions in the CFT(d)/AdS(d+1) correspondence, *Nucl. Phys. B* **546** (1999) 96–118. doi: 10.1016/S0550-3213(99)00053-X.

14. P. Minces and V. O. Rivelles, Scalar field theory in the AdS/CFT correspondence revisited, *Nucl. Phys. B* **572** (2000) 651–669. doi: 10.1016/S0550-3213(99)00833-0.

15. H. Bateman and A. Erdélyi, *Higher Transcendental Functions*, vol. 2,*California Institute of Technology. Bateman Manuscript project*, McGraw-Hill, New York, NY (1955).

16. I. S. Gradshteyn and I. M. Ryzhik, *Tables of Integrals, Series, and Prodcts*, 7th edn. Elsevier/Academic Press, Amsterdam (2007).

17. B. Ydri, The AdS_θ^2/CFT_1 correspondence and noncommutative geometry I: A QM/NCG correspondence, *IJMPA* **37**(13) (2022) 2250077.

https://doi.org/10.1142/9789811270437_0024

Chapter 24

Quantum Probability as the Sum of Holonomies of the Canonical One-Form

C. G. Trahern

cgtrahern@gmail.com

We reconsider the representation of the quantum transition amplitude as a sum over continuous (in time) histories. Continuous histories may be decomposed into three sets: those histories corresponding to paths in configuration space which are globally (or piece-wise) differentiable with respect to time and the remaining paths which are not differentiable. Under reasonable assumptions, we show that the contribution from the first two sets of paths can be expressed in terms of unitary, anti-Zeno propagators. We can then show that both the transition amplitude and probability are expressible as the sum of holonomies of the canonical one-form on the set of differentiable paths in phase space connecting the initial and final states.

1. Introduction

Transition probability in quantum mechanics is defined for closed systems via the Born rule in terms of the norm squared of the amplitude

$$\langle q_f|U(t)|q_i\rangle, \tag{1.1}$$

where $|q_i\rangle$ and $|q_f\rangle$ are states in the system Hilbert space representing points q_i, q_f in configuration space and $U(t) = exp(-iHt)$, where H is the Hamiltonian for the system. As is well known, Eq. (1.1) may be recast as a sum over histories in Hilbert space (and in turn, after evaluation, in terms of paths in the classical configuration space) by insertion of multiple resolutions of the identity at intervening times defined by partitioning the time interval $(0, T)$ in intervals δt. Each term of this sum is a so-called history operator[2] defined in terms of products of Heisenberg picture projection operators. The limiting form of this sum as δt tends to zero and the number of insertions of the identity tends to ∞ correspond to a set of paths of which the majority are not differentiable with respect to time. However, there *are* subsets which contain differentiable paths. Consequently, the amplitude corresponding to a sum of paths can be defined in terms of contributions from these three sets: those paths which are globally at least once differentiable (or piece-wise differentiable) with respect to time, and the remainder which are not.

The amplitude's contributions from the differentiable paths each correspond to evaluation of unitary Anti-Zeno (AZ) propagators[1] for each path or in the case of piece-wise differentiable path, to a product of AZ propagators for each differentiable segment of the path. The contribution from a non-differentiable or discontinuous 'path' corresponds to a product of amplitudes which in the limit that $\delta t \to 0$ are distributions, $\delta(q_{j+1} - q_j)$. When the initial and final states are not equal, these jump discontinuities evaluate to zero. Thus, as the non-differentiable paths' contributions to the amplitude all contain products of subfactors which are discontinuous in the limit, the contribution from the entire set to the overall amplitude is zero on a term-by-term basis.

Since both the globally and piece-wise differentiable sets can be expressed in terms of unitary AZ propagators in the limit, we will argue that the evaluation of these AZ propagators is a pure phase, $exp(iS(q))$, where $S(q)$ is the classical action functional, and q is a path connecting the initial and final points (q_i, q_f) in the classical configuration space for the globally differentiable set, and for the subpaths connecting each segment (q_{j+1}, q_j) for the piece-wise differentiable set. One can lift the path q from configuration space to phase space and write $S()$ in term of the canonical one form $\theta = p \wedge dq - H \wedge dt$, $S = \int_\gamma \theta$, where γ is a path in phase space connecting (q_i, q_f). (For the piece-wise differentiable set of paths, the total action evaluates to a sum of actions for each differentiable segment.)

Taking the norm square of Eq. (1.1), the probability is then expressed as a sum of terms, each of which is the action evaluated on one of the differentiable (or piece-wise differentiable) loops starting from the initial point, passing through the final point and returning to the initial point. Each term will be referred to as a holonomy of the canonical one-form.

We elaborate on these points in the rest of this chapter. In Section 2, we review the history formulation of quantum mechanics.[3–5] In Section 3, we discuss the anti-Zeno propagator as presented by Balachandran and Roy.[1] In Section 4, we discuss the separation of paths into the differentiable and non-differentiable sets in more detail. In Section 5, we present the main result of this chapter. Finally, we state some conclusions and comments in Section 6, touching upon what changes occur if the configuration space has non-trivial topology or homotopy.

2. History Formalism of Quantum Mechanics

The history formulation of quantum mechanics has developed in several stages. Initially,[3-5] it was developed to provide criteria, via the decoherence conditions, to define probabilities corresponding to statements about the results of measurement in terms of a classical physical description through sequences of events described by products of projection operators at a set of discrete times. More recently, the history formulation has been used in quantum information theory (QI) where statements about histories of systems also include the possibility of interference between histories, and the decoherence conditions fail. The history formulation has thus expanded to include most of the intellectual terrain of quantum and classical physical descriptions. We adopt the expanded view of the history formulation here.

Castellani has a good exposition of the formulation of history operators in Ref. [2]. A history operator is written in terms of a product sequence of Heisenberg picture projection operators, $P(t_i) = U(t_i)^\dagger P_i U(t_i)$, with the (t_i) a discrete set of times at which the events or *ideal* measurements occur, and $U(t_i) = exp(-iHt_i)$. The transition amplitude in Eq. (1.1) can be expressed in terms of sums of such histories (where interference of histories clearly may occur) by insertions of a discrete number of the resolution of identity[2] at the discrete set of times (t_i).

Normally, a history as defined in terms of discrete sequences of events conforms to actual experimental practice since continuous measurements are admittedly not performable in the laboratory. However, continuous histories, i.e., the limit of continuous sequences of events, should also be discussed as these correspond to our actual experience of the world. We do not experience the world in a stroboscopic way, with gaps (say, of the 'awareness of nothing') separating perceptions of events, but as a continuous stream of perceptions of the external world whether that stream of events arises from some neurally processed coarse-graining of the set of discrete events or not.

Consequently, as an aside, we believe the interpretation of projection operators as representing *ideal* measurements should be reconsidered.

We suggest that projection operators would better be interpreted as reflecting the knowledge a system has of itself. Then, the knowledge that we, as observers of other systems, may have of another system can arise only from interaction and subsequent correlation with those systems, and our corresponding knowledge of those systems only reflects the knowledge we have of our own correlated state via the projection operators that describe 'us' as quantum systems. The necessity of interaction and correlation between observers and systems cannot be ignored by interpreting projection operators as *ideal measurements* in one way so we can ignore the interaction with observers but then limiting histories to only discrete sets of events because that is more realistic.

If projection operators are understood as the knowledge a system may have of itself, then continuous sequences of projections can be used to describe our experience. We necessarily interact with a discrete subset of the event stream of the observed system, and this corresponds to experimental practice. But both system and observer may experience themselves and the world in a continuous manner as can be described by continuous sequences of projections onto their respective Hilbert spaces. Using Hilbert space to describe experience may seem somewhat outlandish, but it is not an unreasonable idealization to do so.

If one only inserts the resolution of identity in Eq. (1.1) at a discrete set of times, then the amplitude is a sum of histories defined over sequences of events at those times. Taking the limit where resolutions of the identity are inserted at all times results in the amplitude being expressed as a sum over continuous (in time) histories. The histories defined over continuous sequences of projections or events are the ones of interest in this paper. When the path in Hilbert space defined by the sequence of projections is differentiable (more precisely, weakly analytic[1]) with respect to time, those histories can be expressed in terms of anti-Zeno propagators.

3. Anti-Zeno Propagation

The AZ propagator was perhaps named 'in opposition' to the original Zeno propagation[6] because unlike Zeno evolution where the initial state does not change or remains in a subspace of Hilbert space very close to the initial state under continuous projections, in AZ propagation, the initial state is transformed by continuously different projections, tracing out the aforementioned path in Hilbert space to a different final state. With two rather non-restrictive assumptions, this continuous sequence of projections or continuous history is unitary as was first observed by von Neumann in Ref. [7]. This result may appear paradoxical as 'projective measurement' is usually associated with the 'collapse of the wave function', i.e., with non-unitary evolution.

In the previous section, we stated that continuous sequences of projections can be used to describe our experience of the world. AZ propagators show under what conditions such continuous sequences of events correspond to unitary evolution of a system experiencing those events.

Two assumptions are made in deriving this form of unitary evolution. First, the initial (and in the case here, also the final) state satisfies the conditions $P|q_i\rangle = |q_i\rangle$, where P is the initial projection in the history, and $\langle q_f| P_f = \langle q_f|$ where $P_f \neq P$ is the final projection in the history. Second, the continuous path of projections, $P(t) = V(t)^\dagger P\ V(t)$, where $V(t)$ is unitary with $V(0) = 1$, $V(T)^\dagger P\ V(T) = P_f$ is at least once differentiable (or weakly analytic) in t. The first and second assumptions mean that when the path in Hilbert space is differentiable, then the AZ propagation is unitary.[1]

Under these assumptions, the amplitude

$$\langle q_f|K(T)|q_i\rangle \tag{3.1}$$

where $K(t)$ is the anti-Zeno propagator[1]

$$K(t) = exp(-iHt)A(T,0), \ \ A(T,0) = \lim_{n\to\infty} \prod_{i=1}^{n} U(t_i)^\dagger P(t_i)\ U(t_i) \tag{3.2}$$

for a history defined by $P(t)$, $t_i = T(i-1)/(n-1)$, is one of the terms in the sum over histories representation of the amplitude Eq. (1.1) for one of the globally differentiable paths. The set of globally differentiable paths all correspond to evolution by unitary anti-Zeno propagators for each path in the set $\{P(t)\}$.

Being unitary operators, AZ propagators have pure phases as eigenvalues. When the matrix element of an AZ propagator is evaluated along a specific path, this phase should be, and will be assumed to be, $exp(iS(q))$, where $S()$ is the classical action functional and q is the path in configuration space corresponding to the continuous sequence of projected states, $P(t)$.

Since this path $q(t)$ is differentiable, we can lift it to phase space and write the amplitude as an integral of the canonical one-form $\theta = p \wedge dq - H \wedge dt$, where $p = \partial L/\partial \dot{q}$, and L is the classical Lagrangian on the configuration space and $\dot{q}$ is the velocity. We write the action $S(q)$ in terms of its phase space equivalent form $S(q,p) = \int_\gamma \theta$, where γ is a path in phase space with q_i, q_f as fixed endpoints, and $p(t)$ determined from $q(t)$.

4. Contributions from the Sets of Paths

We have distinguished three sets of paths arising from taking the continuous limit of history operators when the number, N, of consecutive insertions of 'events' in a history $N \to \infty$: the set of globally differentiable paths, the set of piece-wise differentiable paths and the set of non-differentiable paths, the latter necessarily containing many jump discontinuities in the limit.

4.1. *The set of globally differentiable paths*

The cardinality of the set of globally differentiable paths is notably *stable* with respect to the limit of infinite insertions of the identity. The number of globally differentiable paths connecting initial and final states for the interval $(0, T)$ once determined does not change. If another resolution of identity is inserted, one event from that insertion will occur on the path and the rest of the events create new paths which contribute either to the piece-wise or non-differentiable sets. The contribution of the set of globally differentiable paths to the amplitude is in this sense fixed and is not perturbed by further insertions of the identity. Of course, the set of globally differentiable paths does not truly 'exist' until the limit is taken, but its stability property, as defined here, can be inferred *a priori*. Thus, for each globally differentiable path, a single AZ propagator can be assigned to its history operator.

4.2. *The set of piece-wise differentiable paths*

The cardinality of the set of piece-wise differentiable paths is, on the other hand, unstable in the limit. With each insertion of the identity, new paths are generated corresponding to the many projection operators added. These will either contribute to new piece-wise differentiable segments or add to the non-differentiable paths' set.

The history operator for each piece-wise differentiable path can be written as a product of AZ propagators, one for each segment. The paths corresponding to these segments are unspecified (i.e., not necessarily straight lines) other than by being differentiable. There is also an implicit dependence on the inflection points in each such path where the path fails to be globally differentiable. (As another side note, the transition amplitude cannot depend on these inflection points, so it must be the case that the total contribution to the amplitude from the sum of terms from the piece-wise differentiable paths is zero.)

4.3. The set of non-differentiable paths

Finally, the set of non-differentiable paths is composed of all the paths not contained in the other two sets. Each term in the non-differentiable path set is characterized by the fact that it contains at least one factor such as

$$\langle q_{j+1}|U(\delta t)|q_j\rangle. \tag{4.1}$$

As $\delta t \to 0$, this factor becomes a distribution $\delta(q_{j+1} - q_j)$. Since $q_{j+1} \neq q_j$, each such factor is zero. So, in fact, all the non-differentiable terms evaluate to 0 before the sum is taken. This result may appear surprising given that the cardinality of this set dwarfs that of the other two sets combined. However, this just reflects the absence of contributions from jump discontinuities to the amplitude.

Summarizing, the quantum amplitude has contributions, in principle, only from the set of globally differentiable and piece-wise differentiable paths.

5. Quantum Amplitudes/Probability as Sums of Holonomies

Having argued that terms associated with the non-differentiable paths do not contribute to the amplitude, we ignore them and re-write Eq. (1.1) as a sum of histories as follows:

$$exp(iS(\gamma_{cl}))\left(1 + \sum_{\gamma_i} exp(iS(\gamma_i \cup \gamma_{cl}^{-1}))\right) = exp(iS(\gamma_{cl}))\left(1 + \sum_{\gamma_l} Hol(\gamma_l)\right), \tag{5.1}$$

$$Hol(\gamma_l) = exp(iS(\gamma_l)), \tag{5.2}$$

where we have separated the contribution from the classical trajectory, γ_{cl}, as an overall factor, and γ_l is then defined as the loop composed of the path γ_i and the inverse of the classical solution path, γ_{cl}^{-1}. $Hol()$ is the holonomy defined here as the classical action evaluated on the loop γ_l.

Even though we pulled out the term corresponding to the classical trajectory above, we could have chosen any of the globally differentiable paths to define the overall factor. However, as the classical trajectory has special status, we will keep it in focus.

Taking the norm square of Eq. (1.1), the probability becomes (up to overall normalization)

$$| \langle q_f|U(t)|q_i \rangle |^2 = \sum_{\gamma,\gamma'} exp(iS^*(\gamma'))exp(iS(\gamma)) = \sum_{\gamma,\gamma'} exp(iS(\gamma \cup \gamma'^{-1})).$$

$$(5.3)$$

The latter term is the sum over all loops starting at q_i, passing through q_f and returning to q_i. Thus, each term is given in terms of the action defined on a loop in configuration space. Once this path is lifted to phase space as discussed earlier, the term becomes a holonomy of the canonical one-form θ. The probability is thus given in terms of the sum of holonomies of the canonical one-form.

Both amplitude and probability can be seen to be given in terms of a sum of holonomies of the canonical one-form for all closed loops connecting initial and final points.

Remembering that $exp(iS(\gamma))$ transforms under canonical transformations by $S \to S+\Delta$, where Δ depends only on the difference of values of the canonical transformation at the end points of the path, we see that the probability, defined on a set of loops starting and returning to q_i, is invariant. This can also be seen by examining the change of S by canonical deformations leaving the loop fixed. We can write S over a loop γ_l in terms of a bounding disk, D, as $S(\gamma_l) = \int_D d\theta$, where $\gamma_l = \partial D$, $d\theta = \omega - dH \wedge dt$, and ω is the symplectic form. The difference in action where the disc is deformed keeping the loop γ_l fixed is $\Delta S = \int_{D'} d\theta - \int_D d\theta = \int_{\partial V} d\theta = \int_V d^2\theta = 0$, where $\partial V = D' - D$ is a closed volume in phase space bounded by the two disks. As $\Delta S = 0$, S is invariant under deformations, i.e., canonical transformations, of D that leave the boundary $\gamma_l = \partial D = \partial D'$ fixed.

6. Conclusions

We have shown (with many assumptions along the way) how the quantum transition amplitude and probability can be expressed in terms of a sum of holonomies of the canonical one-form. It appears from this analysis that the vast majority of contributions to the path representation of the amplitude cancel one against another, with the term corresponding to the classical trajectory being the only survivor — a result that is already well known for quadratic Lagrangians.[8] Even the contributions from all differentiable paths other than the classical solution must cancel one against the other given that we know the path integral representation of the amplitude by other means. It is worth emphasizing that the two sets of paths (the globally and piece-wise differentiable sets) independently and internally cancel one against the other leaving only the one term $exp(iS(q_{cl}))$ defined by the classical solution q_{cl} connecting q_i and q_f.

If the configuration space contains topological "obstructions", i.e., where parts of configuration space are inaccessible to the system, e.g., in the cases of single slit diffraction, multi-slit interference, magnetic field confined to an impenetrable solenoidal region due to the Aharonov–Bohm effect or other kinds of topological issues in configuration space, there will likely be multiple solutions to the classical equations of motion due to the multiple connectedness of the configuration space.[9–11] In these cases, the amplitude Eq. (1.1) will be represented by a sum of terms, one for each classical solution, and each term will be multiplied by the factor containing the sum of holonomies with respect to that classical solution.

$$\sum_{\gamma_{cl_k}} exp(iS(\gamma_{cl_k})) \left(1 + \sum_{\gamma_i} exp(iS(\gamma_i \cup \gamma_{cl_k}^{-1}))\right) = \sum_{\gamma_{cl_k}} exp(iS(\gamma_{cl_k}))F(Hol_k),$$

$$(6.1)$$

where

$$F(Hol_k) = \left(1 + \sum_{\gamma_i} exp(iS(\gamma_i \cup \gamma_{cl_k}^{-1}))\right). \qquad (6.2)$$

In principle, the sum of holonomy factors, $F(Hol_k)$, may differ from unity, but experience indicates that most, if not all, of the physics is described just by the interference of the leading overall factor defined in terms of the classical solutions. Thus, $F(Hol_k)$ most likely always equal 1, i.e., the sum of holonomies of paths defined with respect to any classical solution always cancel one against each other in the overall sum. Consequently, when calculating the probability, it will only be the sum of the holonomies of the various classical trajectories, e.g.,

$$Hol_{jk} = exp(iS(\gamma_{cl_j} \cup \gamma_{cl_k}^{-1})), \qquad (6.3)$$

that determines the shape of the interference pattern in the probability distribution.

Finally, we have used the word 'holonomy' for the expression $exp(iS(\gamma_l))$, where S is the classical action and γ_l is a differentiable loop in configuration space. Holonomy is usually, perhaps strictly, used to refer to the exponential of the closed loop integral of a connection one-form in gauge theory. Even though the canonical one-form θ is not a connection one-form in that sense, its behavior under canonical transformations is similar, i.e., $\theta \to \theta + df$, where f is the function used to generate the canonical transformation. The author hopes the reader will excuse the abuse of technical language to refer to the exponential of integrals of θ as holonomies.

Acknowledgements

The author would like to thank Bal for the enjoyable and collegial relationship over the years and for his support during the years in Syracuse working toward the completion of the degree. I am also very grateful for all the time we spent together doing and thinking about physics, politics, surviving winters in Syracuse and enjoying life outside of physics. The author would also like to thank V.P. Nair for his kind help with the "LaTeXnical" aspects of this chapter and for his long time friendship.

References

1. A. P. Balachandran and S. M. Roy, *Int. J. Mod. Phys.* **A17** (2002) 4007–4024.
2. L. Castellani, History operators in quantum mechanics, arXiv:1810.03624 [quant-ph].
3. R. B. Griffiths, Consistent histories and the interpretation of quantum mechanics, *J. Stat. Phys.* **36** (1984) 219.
4. M. Gell-Mann and J. B. Hartle, Classical equations for quantum systems, *Phys. Rev. D* **47** (1993) 3345.
5. M. Gell-Mann and J. B. Hartle, Quantum mechanics in the light of quantum cosmology, arXiv:1803.04605 [gr-qc].
6. B. Misra and E. C. G. Sudarshan, *J. Math. Phys.* **18** (1977) 756; C. B. Chiu, B. Misra and E. C. G. Sudarshan, *Phys. Rev.* **D16** (1977) 520; *Phys. Lett.* **B117** (1982) 34; K. Kraus, *Found. Phys.* **11** (1981) 547.
7. J. Von Neumann, *Mathematical Foundations of Quantum Mechanics*, Princeton University Press (1955), p. 364; E. P. Wigner, *Foundations of Quantum Mechanics*, B. d'Espagnat (ed.), Academic, NY (1971), formulae 14 and 14(a), p. 16.
8. J. S. Schulman, *Techniques and Applications of Path Integration*, John Wiley & Sons (1981), pp. 31–41.

9. L. S. Schulman, A path integral for spin, *Phys. Rev.* **176**(5) 1558–1569.

10. J. S. Dowker, Quantum theory on multiply connected spaces, *J. Phys. A* **5** 936–943.

11. A. Mouchet, Path integrals in a multiply-connected configuration space (50 years after), arXiv:2010.01504 [quant-ph].

Chapter 25

Noncommutative Geometry and Super-Chandrasekhar White Dwarf

T. R. Govindarajan[*,§], Surajit Kalita[†,¶]
and Banibrata Mukhopadhyay[‡,‖]

[*]*The Institute of Mathematical Sciences,
Taramani, Chennai 600113, India*
[†]*High Energy Physics, Cosmology & Astrophysics Theory Group,
Department of Mathematics & Applied Mathematics,
University of Cape Town, Cape Town 7701, South Africa*
[‡]*Department of Physics, Indian Institute of Science,
Bangalore 560012, India*
[§]*trg@imsc.res.in*
[¶]*surajit.kalita@uct.ac.za*
[‖]*bm@iisc.ac.in*

The geometry at the infinitesimal level cannot be described by the conventional axioms of geometry was the famous quote of Riemann himself. He went ahead to suggest that the physics of measuring process of length will affect it. Classical general relativity relies heavily on Riemannian geometry. We argue that the noncommutative geometry naturally fits this scheme at extremely dense astrophysical systems. In the last two decades, several over-luminous type Ia supernovae have been observed, which predict possible white dwarfs violating Chandrasekhar limit significantly. We provide a model using squashed fuzzy sphere to describe such systems by allowing extra mass to be accumulated.

1. Introduction

In 1930s, Bronstein,[1] explained the conflict between quantum mechanics and general relativity and argued that gravitational dynamics prohibits measurement of arbitrarily small space-time distances. He also concluded that classical Riemannian geometry should be modified to account for this. Bernhard Riemann[2] in his famous habilitation lecture quoted

> *"..it seems that empirical notions on which the metrical determinations of space are founded, the notion of a solid body and a ray of light cease to be valid for the infinitely small. We are therefore quite at liberty to suppose that the metric relations of space in the infinitely small do not conform to hypotheses of geometry; and we ought in fact to suppose it, if we can thereby obtain a simpler explanation of phenomena.....*
> *...................."*
>   *Bernhard Riemann*

Noncommutative (NC) geometry[3] can fit in this framework. But the question that raises is what is the scale beyond which such effects will be seen. The general assumption is that the Planck length would be the most appropriate. We will argue that in extreme density astrophysical situations, there could be models which bring out these features. This is because the geometry is determined by the matter present, whereas geometry determines the dynamics of matter. NC geometry is generally specified by some length parameter. The parameters that are fundamentally available are the Planck length as well as the Compton wavelength of the particle(s) constituting the matter. In extreme dense cases when interparticle distances are close to their Compton wavelength along with gravitational forces, we can anticipate interplay between them.

A white dwarf (WD) is the end state of a star with mass $(10 \pm 2)\,M_\odot$. Chandrasekhar showed that the maximum mass of a carbon–oxygen non-magnetized and non-rotating WD is about $1.4\,M_\odot$. Above this mass limit, they burst out to produce type Ia supernovae (SNeIa) with nearly similar luminosities. But we now have several over-luminous SNeIa, such as SN2003fg, SN2006gz, SN2007if, SN2009dc and many more. This suggests that they originated from super-Chandrasekhar WDs with mass $2.1 - 2.8\,M_\odot$.[4] Several models, using rotation, magnetic fields, modified gravity, generalized Heisenberg uncertainty principle, to name a few, have been proposed to explain these masses. Unfortunately, no super-Chandrasekhar WDs have been observed directly so far, and it is impossible to conclude their origin.

The motivation for our proposal comes from the fact that white dwarf gets stabilized after exhausting nuclear fusion process in a star due to two forces: 1. gravitational pull of the mass of the star and 2. Fermi–Dirac statistical pressure due to the electron gas. When electrons get extremely close to each other such that interelectron separation is less than the Compton wavelength of the electrons, the NC geometry becomes effective. One of the important outcomes of the NC geometry, realized through Drinfeld twist, is that fermions behave less of fermions.[5] This reduces the outward pressure and allows more matter to accumulate at the core and makes WD more massive. We realize this scenario through the squashed fuzzy sphere model.

We show that the squashed fuzzy sphere algebra provides the energy dispersion relation for single particles, which alters the equation of state (EoS) of the degenerate electrons present in a WD. This results in the increasing mass of the WD beyond the Chandrasekhar mass limit. We also explain that the length scale below which noncommutativity becomes important depends on the interelectron separation. Hence, the inference of super-Chandrasekhar WDs may indirectly predict the existence of noncommutativity in high-density regimes.

2. Length Scale for NC Geometry in White Dwarf

For a typical WD, the radius is ~ 6000 km and mass $\mathcal{O}(1\ M_\odot)$ possessing an average density 10^6 g/cm^3. The mean inter-electron distance therein can be easily estimated to be of the $\mathcal{O}(10^{-10}$ cm$)$. This mean distance is of the order of the Compton wavelength of electrons which is 2.4×10^{-10} cm. As the star gets compressed more due to gravitational pull, the density increases and interelectron distance decreases. Hence, the modifications in the statistical pressure take place due to noncommutativity. The next scale is nuclear radius or Compton wavelength of neutrons which is $\mathcal{O}(10^{-13}$ cm$)$. The stability of the very massive stars can be achieved at scales smaller than Compton wavelength of the electrons. The reason is the meaning of single particle will be lost due to pair creations when the mean distance gets smaller than Compton wavelength. Exchange statistics through noncommutativity should be modified with this input. Uncertainties in measurement of distance between particles in quantum theory was undertaken by Wigner himself in the 1960s, and Salekar and Wigner[6] argued that the uncertainty in length scale for a system has to be $\delta \approx (L L_P^2)^{1/3}$, where L is the interelectron distance and L_P is the Planck length, and one needs to consider δ as the quantum measurement of length. Hence, for a WD, one can expect to observe a reasonable noncommutativity effect depending on the interelectron separation, even though it is far from the Planck scale. The mean distance between electrons becomes much less than the Compton wavelength at extremely high densities. Certain features of noncommutativity like the notion of statistics, which needs localization of particles, get distorted and can affect physical systems, for example, stable mass limit of the WD. Similar distortions take place even in quantum Hall effect which we will mention in discussions.

3. Fuzzy Sphere and Squashed Fuzzy Sphere

The fuzzy sphere[7] is similar to the angular momentum algebra where the position coordinates take the role of the angular momentum variables. In an N-dimensional irreducible representation of SU(2) group, the commutation relation for the coordinates of fuzzy sphere, X_i $(i = 1, 2, 3)$, is given by

$$[X_j, X_l] = \frac{ik\hbar}{R}\epsilon_{jlm}X_m, \tag{3.1}$$

where $k = \frac{2R^2}{\hbar(N^2-1)}$ and R is the radius of the sphere. The quantum mechanics has interesting connections to the quantum Hall effect which has been exploited by several authors. See, for example, Nair's work.[8]

A squashed fuzzy sphere[9] is obtained from the fuzzy sphere by projecting all the points to an equatorial plane (see Fig. 1). The equatorial plane can be in any direction, though we have taken $X_1 - X_2$ plane.

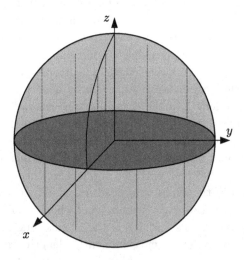

Fig. 1. Squashed fuzzy sphere.

The commutation relation for the squashed fuzzy sphere is easy to obtain, given by

$$[X_1, X_2] = \frac{ik\hbar}{R}\sqrt{R^2 - X_1^2 - X_2^2}. \tag{3.2}$$

As expected, noncommutativity vanishes on the surface of the sphere.

The Dirac operator[9] for the squashed fuzzy sphere is defined as

$$\rlap{/}{D} = \frac{c}{k}\left(\sigma_1 \otimes [X_1, \cdot] + \sigma_2 \otimes [X_2, \cdot]\right), \tag{3.3}$$

where c is the speed of light and $\sigma_{1,2}$ are Pauli matrices. Squaring the Dirac operator, we can easily get the energy eigenvalues:

$$E_{l,m}^2 = \frac{2\hbar\, c}{k\sqrt{N^2 - 1}}\{l(l+1) - m(m \pm 1)\}, \tag{3.4}$$

where l, m are quantum numbers. Here, $l = 0, \ldots, N - 1$ and $m = -l, \ldots, +l$.

We can convert the commutation relations to that between polar and azimuthal angles. Since there is no particular direction we need to assign for equatorial plane, squashed fuzzy sphere has rotational symmetry about the chosen plane. However, there is no noncommutativity in the radial direction. Let p_r be the radial momentum, and the energy dispersion relation in the squashed fuzzy sphere can be obtained as[10]

$$E^2 = p_r^2 c^2 + m_e^2\, c^4 \left[1 + \{l(l+1) - m(m \pm 1)\}\frac{2\hbar}{m_e^2 c^2\, k\, \sqrt{N^2 - 1}}\right]. \tag{3.5}$$

In the large angular momentum limit (that is continuum limit), we get[10]

$$E^2 = p_r^2 c^2 + m_e^2 c^4\left(1 + \frac{4\nu\hbar}{m_e^2 c^2\, k}\right), \quad \nu = \mathcal{Z}^{0+}. \tag{3.6}$$

Role of NC parameter is played by $1/k$. We can model the WD by a sequence of concentric squashed fuzzy spheres such that at the surface of each spheres, the above relation is valid. Nevertheless, $k \propto R^2$, and hence $1/k$ or the strength of noncommutativity decreases from center to the surface.

4. Equation of State for Degenerate Electron Gas

We can rewrite the energy dispersion by comparing with magnetic Landau level problem as[11]

$$E^2 = p_r^2 c^2 + m_e^2 c^4 \left(1 + 2\nu \frac{B}{B_c} \right), \qquad (4.1)$$

where $B_c = \frac{m^2 c^3}{e\hbar}$ and $B = \frac{2c}{ek}$. B_c is known as critical (Schwinger) magnetic field. The grand canonical partition function for this system is given by

$$\mathcal{Z} = \frac{4V}{\beta \, k \, h^2} \int_0^{p_{r,max}} dp_r \sum g_\nu \log(1 + z e^{-\beta E}), \qquad (4.2)$$

where $z = e^{\beta\mu}$ is the fugacity, $\beta = \frac{1}{k_B T}$ and μ is the chemical potential. The pressure $\mathcal{P}$ can be worked out by standard procedure (for details, see Ref. 10). It is given by

$$\mathcal{P} = \frac{2\rho^{2/3}}{\zeta \, h \, \mu_e^{2/3} m_p^{2/3}} \sum_0^{\nu_{max}} g_\nu \mathcal{F}(\nu), \qquad (4.3)$$

where

$$\mathcal{F} = p_F E_F - \left(m_e^2 c^3 + \frac{4\nu\hbar c}{k} \right) \log \left(\frac{E_F + p_F c}{\sqrt{m^2 c^4 + \frac{2\nu\hbar c^2}{k\pi}}} \right), \qquad (4.4)$$

where E_F stands for the Fermi energy.

We also need the Fermi momentum p_F in terms of volume (or density). It is

$$p_F = \frac{\zeta \, h}{4\mu_e^{1/3} m_p^{1/3} (2\nu_{max} + 1)} \rho^{1/3}. \qquad (4.5)$$

These two equations determine the EoS for degenerate electron gas in squashed fuzzy sphere.[11]

5. Modified Chandrasekhar Limit

We now have to balance the pressure due to degenerate electron gas providing inputs for extreme density through noncommutativity (squashed fuzzy sphere) with gravitational force using TOV (Tolman–Oppenheimer–Volkoff) equations:[12, 13]

$$\frac{d\mathcal{M}}{dr} = 4\pi \, r^2 \, \rho, \tag{5.1}$$

$$\frac{d\mathcal{P}}{dr} = -\frac{G}{r^2} \left(\rho + \frac{\mathcal{P}}{c^2} \right) \left(\mathcal{M} + \frac{4\pi r^3 \mathcal{P}}{c^2} \right) \left(1 - \frac{2G\mathcal{M}}{r \, c^3} \right)^{-1}, \tag{5.2}$$

where $\mathcal{M}$ is the mass of the star inside a sphere of radius r and G is Newton's gravitational constant. Figure 2 indicates how the mass changes as a function of radius. It also exhibits the variation of the mass as we change the central density ρ_c of the WDs. This is obtained by solving the TOV equations and the EoS. For comparison, we also plot the same curves for Chandrasekhar EoS. We expect that the effect of noncommutativity should not be significant at lower densities and the usual statistical mechanics, governed by usual Fermi statistics, should be sufficient. Therein the occupied energy levels increases, the effect of noncommutativity decreases and all the curves merge with the Chandrasekhar curve. To compare EoS at low density, we find $\zeta = 1.51$ for $\nu_{max} = 0$, while $\zeta = 156.56$ for $\nu_{max} = 50$. We get the noncommutativity is significant if $\mathcal{L} \leq 1.11 \times 10^{-10}$ cm for $\nu_{max} = 0$, whereas $\mathcal{L} \leq 1.09 \times 10^{-11}$ cm for $\nu_{max} = 50$. These correspond

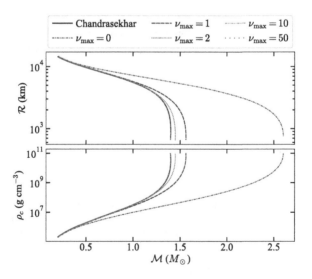

Fig. 2.   Mass–radius relations of WDs for squashed fuzzy sphere.

to the central density $\rho_c \approx 10^7 \, \mathrm{g/cm^3}$ and $\rho_c \approx 1.1 \times 10^{11} \, \mathrm{g/cm^3}$, respectively. Hence, we do not see any effect of noncommutativity in the WDs with $\rho_c \leq 2 \times 10^{10} \, \mathrm{g/cm^3}$ for $\nu_{max} = 50$, and we have Chandrasekhar's original mass–radius relation.

6. Neutron Drip

As density increases and relativistic electrons are very close to the nuclei, neutronization process starts where inverse beta decay results in proton converting to neutron. As neutron density increases, energy levels of neutron near rest energy of the neutron. When the system consists of ions rather than free protons and neutrons, at high density, electron will become neutron combining with proton and will 'drip' out. This is known as neutron drip when we get neutron fluid core in the star.[14]

The modified dispersion relation for a squashed fuzzy sphere is different compared to that of general relativity. Hence, one can expect that the property of neutron drip alters in the presence of noncommutativity. At the neutron drip density, protons and electrons combine to form fluid neutrons. Above this density, the electron degenerate pressure is no longer sufficient.

The neutron drip density ρ_D is given by

$$\rho_D = \frac{n_e \, M(A, Z)/Z + \epsilon_e - n_e m_e c^2}{c^2}, \tag{6.1}$$

where $M(A, Z)$ is the energy of ion with atomic number Z and mass number A. ϵ_e is the electron energy density at zero temperature, which is given by

$$\epsilon_e = m_e \, c^2 \, \frac{4\pi\theta_D}{\lambda_e^3} \sum_0^{\nu_{max}} g_\nu (1 + \nu\theta_D) \Psi \left(\frac{x_F(\nu)}{\sqrt{1 + 2\nu\theta_D}} \right), \tag{6.2}$$

where

$$x_F = \sqrt{(2230.31 - 2\nu\theta_D)}, \quad \Psi(z) = \frac{1}{2} z \sqrt{1 + z^2} + \frac{1}{2} \log \left(z + \sqrt{1 + z^2} \right) \tag{6.3}$$

and $\theta_D = 2\hbar/m_e^2 c^2 k$. Figure 3 shows the variation of ρ_D with respect to θ_D.

We note that before the linear regime, the drip density oscillates about the density, which is approximately $3 \times 10^{11} \, \mathrm{g/cm^3}$ drip density in the absence of noncommutativity. It is also evident from the figure that the onset drip density increases linearly after a certain θ_D. This is because, beyond certain θ_D, the quantized electrons reside only in the ground state. To summarize, ρ_D changes with the increase in noncommutativity, and

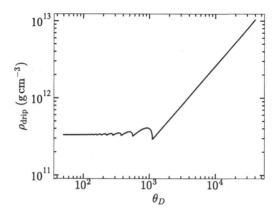

Fig. 3. Variation of neutron drip density as a function of θ_D.

hence above this density, we can no longer use the EoS of degenerate electrons.

7. Reminiscences and Conclusions

Prof. A. P. Balachandran was instrumental in introducing one of the authors (TRG) to NC Geometry and Quantum Space-time. About four decades back, APB (as he is affectionately called) came to Madras University where TRG was a graduate student. We worked together along with another graduate student named Vijayalakshmi on using monopole algebra in a novel way.[15] This was used to describe particles with spin. Interesting thing was that it had a NC algebraic structure in-built into it. Two decades later, we worked on Dirac operators on fuzzy spheres and realized that it had the potential to evade fermion doubling obstruction,[16] normally present in lattice type discretization of space-time. Later, we continued our research on several fundamental questions. It is interesting that TRG could return to the fuzzy sphere geometry to understand the super-Chandrasekhar white dwarfs much later. Fuzzy sphere algebra has been used extensively to understand the quantum mechanics of Hall effect. The length scale of magnetic length provides scale for NC geometry.[8] Another author (BM) also had a chance to meet APB in several occasions, though briefly, and to get illuminated by his general discussion on noncommutativity, CPT theories, etc.

Noncommutativity is a fundamental property of matter and geometry together. In the presence of noncommutativity, the statistical behavior of the constituent particles changes. This is because of the reason that in noncommutativity, the fermions behave less of a conventional fermion. It is evident that apart from the ground level, each energy levels comprise of two up-spin and two down-spin electrons. In the conventional picture of quantum mechanics, we know that two electrons with the same spin cannot occupy a common energy state due to Pauli's exclusion principle. However, since more than one electron with the same spin occupy the common energy state in fuzzy noncommutativity, the effective repulsive pressure between the fermions gets reduced. On the other hand, the Chandrasekhar mass limit arises due to the balance of effective fermion pressure and gravitational attraction. If the fermions' repulsive force reduces, it can collapse more and extra mass can be accumulated to form the super-Chandrasekhar WDs.

Our work is primarily based on semiclassical gravity, not quantum gravity, and hence the scale of noncommutativity is expected to be different from the Planck scale. In the context of specific dynamics of quantum mechanics, gravity and statistical mechanics, the NC scale gets complicated. It is evident that if $L < L_{\text{eff}}$, the noncommutativity is important. Indeed, Salecker and Wigner[6] already pointed it out as a limitation of quantum measurement of lengths, and the new uncertainty in length scale appears to be $\delta \approx (LL_P^2)^{1/3}$. Hence, the length scale of uncertainty in a WD, which is a low-energy system, depends both on the Planck length and Compton wavelength of the electron.

The inferences of super-Chandrasekhar WDs are indirect, as none of them have been detected from direct observations. Similarly, there is no observational evidence for noncommutativity either. Gravitational waves can be one of the prominent tools to detect such massive WDs directly. We have shown that WDs governed by NC geometry can emit gravitational radiation for a long duration even if they possess a minimal surface magnetic field.[17] In this way, one can estimate both the masses and radii of the WDs; thereby comparing with the theoretical mass–radius curves, we might have indirect observational evidence of noncommutativity.

References

1. M. Bronstein, General relativity and gravitation, **44** (2012) 267–283.
2. B. Riemann, Translated by W.K.Clifford, *Nature* **8** (1873) 183.
3. A. Connes, *Noncommutative Geometry*, Academic Press (1994).
4. D. A. Howell, M. Sullivan, P. E. Nugent, R. S. Ellis, A. J. Conley, D. LeBorgne, R. G. Carlberg, J. Guy, D. Balam, S. Basa, D. Fouchez, I. M. Hook, E. Y. Hsiao, J. D. Neill, R. Pain, K. M. Perrett and C. J. Pritchet,

The type Ia supernova SNLS-03D3bb from a super-Chandrasekhar-mass white dwarf star, *Nature* **443** (2006) 308.

5. A. P. Balachandran, T. R. Govindarajan, G. Mangano, A. Pinzul, B. A. Qureshi and S. Vaidya, *Phys. Rev. D* **75** (2007) 045009.

6. H. Salecker and E. P. Wigner, *Phys. Rev.* **109** (1958) 571.

7. J. Madore, Classical and quantum gravity, **9** (1992) 69.

8. V. P. Nair, *Noncommutative Mechanics, Landau Levels, Twistors, and Yang-Mills Amplitudes*, Vol 1., Berlin, Heidelberg, Springer Berlin Heidelberg 97 (2006).

9. S. Andronache and H. C. Steinacker, *J. Phys. A* **48** (2015) 295401.

10. S. Kalita, T. R. Govindarajan and B. Mukhopadhyay, *Int. J. Mod. Phys. D* **30** (2021) 2150101.

11. S. Kalita, T. R. Govindarajan and B. Mukhopadhyay, Proceedings of 16th Marcel Grossmann Meet WSPC (2021) gr-qc/:2111.05878.

12. R. C. Tolman, *Phys. Rev.* **55** (1939) 364.

13. J. R. Oppenheimer and G. M. Volkof, *Phys. Rev.* **55** (1939) 374.

14. M. V. Vishal and B. Mukhopadhyay, *Phys. Rev. C* **89** (2014) 065804.

15. A. P. Balachandran, T. R. Govindarajan and B. Vijayalakshmi, *Phys. Rev. D* **18** (1978) 1950.

16. A. P. Balachandran, T. R. Govindarajan and B. Ydri, *Mod. Phys. Lett. A* **15** (2000) 1279.

17. S. Kalita, B. Mukhopadhyay, T. Mondal and T. Bulik, *ApJ* **896** (2020) 69.

Chapter 26

Chiral Anomaly in $SU(N)$ Gauge Matrix Models and Light Hadron Masses

S. Vaidya[*,§], N. Acharyya[†,¶] and M. Pandey[‡,‖]

[*] Centre for High Energy Physics, Indian Institute of Science,
Bengaluru 560012, India
[†] School of Basic Sciences, Indian Institute of Technology Bhubaneswar,
Jatni, Khurda, Odisha 752050, India
[‡] School of Theoretical Physics, Dublin Institute for Advanced Studies,
Dublin 4, D04 C932, Ireland
[§] vaidya@cts.iisc.ac.in
[¶] nirmalendu@iitbbs.ac.in
[‖] mpandey@stp.dias.ie

The $SU(N)$ Yang–Mills matrix model admits self-dual and anti-self-dual instantons. The Euclidean Dirac equation in an instanton background has n_+ positive and n_- negative chirality zero modes. We show that the index $(n_+ - n_-)$ is equal to a suitably defined instanton charge. Further, we show that the path integral measure is not invariant under a chiral rotation and relate the non-invariance of the measure to the index of the Dirac operator. Axial symmetry is broken anomalously to a finite group. For N_f fundamental fermions, this residual symmetry is $\mathbb{Z}_{2N_f}$, whereas for adjoint quarks, it is $\mathbb{Z}_{4N_f}$.

1. Introduction

In a quantum theory, the anomalies account for the failure to preserve certain symmetries of its classical counterpart. The most well-known example of such an anomaly occurs in gauge theories with massless Dirac fermions Ψ. The classical action is invariant under the vector rotation $\Psi \to e^{i\alpha}\Psi$ as well as the axial rotation $\Psi \to e^{i\alpha\gamma^5}\Psi$, but only the former is preserved in the quantum theory while the later is anomalously broken.[1,2] This is the *axial anomaly*, which manifests in the non-conservation of the axial current $j_A^\mu = \bar\Psi\gamma^\mu\gamma^5\Psi$ in the quantum theory. Rather than $\partial_\mu j_A^\mu = 0$, we have $\partial_\mu j_A^\mu \sim \epsilon^{\mu\nu\rho\sigma}F_{\mu\nu}F_{\rho\sigma}$.

Fujikawa[3, 4] showed that the axial anomaly may be understood by examining the Euclidean path integral $\int \mathcal{D}\bar\Psi\mathcal{D}\Psi e^{-S_E^F}$. Expanding Ψ and $\bar\Psi$ in the basis of eigenfunctions of $\slashed{D}$ shows that although S_E is invariant under axial rotations, the measure $\mathcal{D}\bar\Psi\mathcal{D}\Psi$ is not. The change in the measure yields $\partial_\mu j_A^\mu \sim \epsilon^{\mu\nu\rho\sigma}F_{\mu\nu}F_{\rho\sigma}$.

We demonstrate here that the quantum $SU(N)$ gauge matrix model coupled to N_f massless quarks also exhibits an anomalous breaking of axial symmetry. Our demonstration of the axial anomaly closely follows the Euclidean approach: there exist zero modes of the Euclidean Dirac operator in the background of (anti-)self-dual instanton gauge configurations, and the index of $\slashed{D}$ is non-zero.

The surprise here is that the quantum $SU(N)$ gauge matrix model, at least at first sight, is vastly different from $SU(N)$ gauge field theory. Even the structure of instanton configurations is very different. In the gauge field theory, distinct self-dual instantons are labeled by the instanton charge $\pi_3(SU(N)) = \mathbb{Z}$, whereas in the corresponding matrix model, they are labeled by a finite set of integers. But even though the number of distinct self-dual instantons is finite, it is enough to disturb the balance between left- and right-handed fermion zero modes, leading to the anomalous breaking of axial symmetry.

The matrix model discussed here was first presented in Refs. 5–7 and has been shown to be an excellent candidate for an effective low-energy approximation of $SU(N)$ Yang–Mills theory on $S^3 \times \mathbb{R}$. In particular, a numerical investigation of its spectrum gave remarkably accurate predictions for the light glueball and hadron masses.[8, 9] To serve as a correct low-energy approximation of Yang–Mills theory, however, this quantum mechanical model must also exhibit the axial anomaly, and this work provides this important conceptual support.

A (slightly) longer version of this work has appeared in Ref. 10.

2. Brief Review of the Matrix Model

In the matrix model of $SU(N)$ gauge theory, the dynamical degrees of freedom are elements of $\mathcal{M}_N$, the set of all $3 \times (N^2 - 1)$-dimensional real matrices M_{ia}, and gauge transformations act as $M \to MS(h)^T$, $h \in SU(N)$. The configuration space of the model is the base space of principle bundle $\text{Ad}\,SU(N) \to \mathcal{M}_N \to \mathcal{M}_N/\text{Ad}\,SU(N)$. This fiber bundle being twisted[11,12] lies at the heart of the Gribov problem in Yang–Mills theory.

With $A_i = M_{ia}T_a$ (T_as are the generators of $SU(N)$ in the fundamental representation) and introducing $A_0 \equiv M_{0a}T_a$ to define a gauge-covariant time derivative $D_t A_i = \dot{A}_i - i[A_0, A_i]$, the Yang–Mills Lagrangian is given by

$$L_{YM} = g^{-2}\text{Tr}[(D_t A_i)^2 - V(A)], \quad V(A) = B_i B_i, \qquad (2.1)$$

where $B_i \equiv B_{ia}T_a = -A_i - \frac{i}{2}\epsilon_{ijk}[A_j, A_k]$ is the chromomagnetic field.

Under a gauge transformation $h(t) \in SU(N)$, $A_i \to hA_i h^{-1}$ and $A_0 \to hA_0 h^{-1} - \dot{h}h^{-1}$. It is easy to check that L_{YM} is indeed gauge invariant.

The quarks Ψ are Grassmann-valued matrices depending only on time and transform in the fundamental representation of color $SU(N)$ and in the spin-$\frac{1}{2}$ representation of spatial rotations. The Lagrangian with minimally coupled massless quarks is given by $L = L_{YM} + L_F$, where $L_F = [\bar{\Psi}\left(i\gamma^0 D_t + \gamma^i A_i - \frac{3}{2}\gamma^5\gamma^0\right)\Psi]$ is the gauge-covariant Dirac Lagrangian on S^3.[7]

Performing a Wick rotation $t \to -i\tau$ and $A_0 \to iA_0$ in the Lagrangian L gives us the Euclidean action $S_E = S_E^{YM} + S_E^F$:

$$S_E^{YM} = \frac{1}{g^2}\int d\tau\,\text{Tr}\,[\mathbb{E}_i \mathbb{E}_i + V(A)], \quad S_E^F = \int d\tau\,\Psi^\dagger i\slashed{D}\Psi, \qquad (2.2)$$

where $\mathbb{E}_i(\equiv \mathbb{E}_i^a T_a) = \frac{\partial A_i}{\partial \tau} - [A_0, A_i]$ is the chromoelectric field, the potential $V(A)$ is as in (2.1) and $\slashed{D}$ is the Dirac operator: $i\slashed{D} = \frac{\partial}{\partial \tau} - A_0 - \gamma^0\gamma^i A_i - \frac{3}{2}\gamma^5$.

The classical action S_E is, indeed, invariant under the axial rotation $\Psi \to e^{i\alpha\gamma^5}\Psi$ and hence, has $U(1)_A$ symmetry. However, as we demonstrate in the following, the quantum effects do not preserve this symmetry.

3. Instantons

Classical vacuum configurations of (2.1) are given by those A_i which satisfy $V(A) = 0$, i.e., $[A_i, A_j] = i\epsilon_{ijk}A_k$. This has solutions $A_i = L_i$, where L_i are generators of the Lie algebra of $SU(2)$. Thus, classically, there are multiple degenerate vacua, which correspond to the matrices A_i forming a general N-dimensional representation of L_i.

The tunneling between degenerate classical minima is captured by instantons, the finite action solutions of equations of motion of the action S_E^{YM} in (2.2). This action is extremized when $\mathbb{E}_i = \pm B_i$, whose solutions with plus (minus) sign give the self-dual (anti-self-dual) instantons.

Using the gauge freedom, the self-dual equation can be transformed to the temporal gauge $A_0 = 0$ to read

$$\frac{dA_i}{d\tau} = \pm \left(-A_i - \frac{i}{2}\epsilon_{ijk}[A_j, A_k] \right). \tag{3.1}$$

Substituting the ansatz

$$A_i = \phi(\tau)L_i^{(1)} + \left(1 - \phi(\tau)\right)L_i^{(2)}, \quad [L_i^{(1)}, L_j^{(2)}] = 0 \tag{3.2}$$

into (3.1), where L_is are generators of $SU(2)$, we obtain

$$\frac{\partial \phi_s}{\partial \tau} = -\phi_s(1 - \phi_s), \quad \frac{\partial \phi_a}{\partial \tau} = \phi_a(1 - \phi_a), \tag{3.3}$$

which have solutions

$$\phi_s(\tau) = \frac{1}{1 + e^{(\tau-\tau_0)}}, \quad \phi_a(\tau) = \frac{1}{1 + e^{-(\tau-\tau_0)}}, \tag{3.4}$$

where s denotes the self-dual instanton and a denotes the anti-self-dual instanton. In general, $L_i^{(\alpha)}$ is a direct sum of $r_0^{(\alpha)}$ irreps with each irrep $L_i^{(\alpha),r}$ with $j_r^{(\alpha)} = \frac{N_r^{(\alpha)}-1}{2}$.

If we transform the solutions by the curve $h(\tau)$ to obtain A_μ^h, we obtain instanton solutions that go between $L_i^{(1,2)}$ and $h_b L_i^{(2,1)} h_b^{-1}$, where $h_b = h(-\infty)$ for the self-dual and $h_b = h(\infty)$ for the anti-self-dual solution. This does not affect any of the subsequent arguments, since the instanton charge is gauge-invariant.

The instanton number, defined as $\mathcal{T} = c \int_{-\infty}^{\infty} d\tau \text{Tr}\, \mathbb{E}_i B_i$, is the integral of a total derivative. The normalization factor c is fixed to 4 so that $\mathcal{T}$ is an integer. For the(anti-)self-dual instantons,

$$\mathcal{T}_s = -\mathcal{T}_a = \frac{2}{3} \sum_{\alpha=1}^{2} \epsilon_{\alpha\beta} \sum_{r=1}^{r_0^{(\alpha)}} j_r^{(\alpha)}(j_r^{(\alpha)} + 1)(2j_r^{(\alpha)} + 1). \tag{3.5}$$

The form (3.5) is not very enlightening as there is no simple relation between the $\{j_r^{(\alpha)}\}$ and N. Fortunately, for the (anti-)self-dual instanton, we can construct another charge that is expressible only in terms of $r_0^{(\alpha)}$, the number of irreps at $\tau = \pm\infty$. Defining $\mathbb{E}'_i \equiv \frac{d\phi_{s/a}}{d\tau}(e_i^{(2)} - e_i^{(1)})$ with $e_i^{(\alpha)} = \bigoplus_{r=1}^{r_0^{(\alpha)}} \frac{3}{(j_r^{(\alpha)}+1)(2j_r^{(\alpha)}+1)} L_i^{(\alpha),r}$, the new charge is given by

$$\mathcal{T}_{\text{new}} = 4 \int d\tau \ \text{Tr} \ \mathbb{E}'_i B_i = \pm(r_0^{(2)} - r_0^{(1)}). \tag{3.6}$$

The charge $\mathcal{T}_{\text{new}}$ is still an integral over a total τ-derivative and hence a topological invariant. It depends only on $(r_0^{(2)} - r_0^{(1)})$, rather than the Casimirs or other labels of each individual irrep. As we will demonstrate in the following section, $\mathcal{T}_{\text{new}}$ for a (anti-)self-dual instanton is *equal* to the index of the Dirac operator in that instanton background.

4. Index of $\not{D}$ in Instanton Background

The Hermitian Dirac operator obeys $\{\not{D}, \gamma^5\} = 0$ and consequently, for every eigenfunction ψ_n of $\not{D}$ with a non-zero eigenvalue λ_n, there is an eigenfunction $\gamma^5\psi_n$ with eigenvalue $-\lambda_n$. The eigenfunctions with zero eigenvalue (the zero modes) can also be arranged to be eigenfunctions of γ^5:

$$\not{D}\chi_k^{\pm} = 0, \quad \gamma^5\chi_k^{\pm} = \pm\chi_k^{\pm}, \quad k = 1, 2, \ldots n_{\pm}, \tag{4.1}$$

but n_+ and n_- need not be equal. The index of the Euclidean Dirac operator $\not{D}$ is defined as the difference ind $\not{D} = n_+ - n_-$.

In the Weyl basis, the Euclidean Dirac operator is

$$\not{D} = \begin{pmatrix} 0 & \mathcal{L} \\ \mathcal{L}^\dagger & 0 \end{pmatrix}, \quad \mathcal{L} \equiv -i\left(\frac{d}{d\tau} - A_0 + \sigma_i A_i + \frac{3}{2}\right). \tag{4.2}$$

We solve for the zero modes $\Psi_{\alpha A}$ of $\not{D}$ in the temporal gauge. For a generic background gauge field $A_0^h = -\frac{dh}{d\tau}h^{-1}$, the corresponding zero mode is obtained via the gauge transformation $\Psi^h = u(h)\Psi$. In any case, $n_{\pm}$ and hence the index do not change under the gauge transformation and we can make our entire argument in temporal gauge.

For the (anti-)self-dual instanton, $\sigma_i A_i(\tau) + 3/2$ is a $2N$-dimensional Hermitian matrix whose eigenvalues $\xi_i(\tau)$ are time-dependent but the eigenvectors are not. So, $\mathcal{L}$ can be diagonalized and $\not{D}$ brought to the form

$$\not{D} = \begin{pmatrix} 0 & \mathcal{L}_d \\ \mathcal{L}_d^\dagger & 0 \end{pmatrix}, \quad \mathcal{L}_d = -i\left(\frac{d}{d\tau} + \Sigma\right), \tag{4.3}$$

where $\Sigma = \text{diag}(\xi_1(\tau), \xi_2(\tau), \ldots \xi_{2N}(\tau))$. In the new basis, the positive and negative chirality zero modes are of the form $\chi^+ = (0, \psi^+)^T$ and $\chi^- =$

$(\psi^-, 0)^T$ respectively, where the $\psi^\pm$ satisfy $\mathcal{L}_d\psi^+ = 0$ and $\mathcal{L}_d^\dagger\psi^- = 0$. The index is given by ind $\slashed{D} = \dim \text{Ker}(\mathcal{L}_d) - \dim \text{Ker}(\mathcal{L}_d^\dagger)$.

For a Dirac operator of the form (4.3), there is a neat way to determine the index from $\xi_i(\tau \to \pm\infty)$ using Callias' index theorem:[13]

$$\text{ind } \slashed{D} = \frac{1}{2} \sum_{i=1}^{2N} \left[\text{sgn}\, \xi_i(\tau \to -\infty) - \text{sgn}\, \xi_i(\tau \to \infty) \right]. \qquad (4.4)$$

Let us apply (4.4) to the simple case when $A_i = \phi_{s/a}(\tau)L_i$, i.e., when $L_i^{(2)} = 0$ and $L_i^{(1)} = L_i$. Say L_i consists of r_0 irreducible blocks with each block of dimension $N_r \equiv (2j_r + 1)$. In each block, $\sigma_i A_i + 3/2$ is a $2N_r$-dimensional matrix and has $(N_r - 1)$ degenerate eigenvalues $\lambda_{1,s/a} \equiv \frac{1}{2}[3 - \phi_{s/a}(\tau)(N_r + 1)]$ and $(N_r + 1)$ degenerate eigenvalues $\lambda_{2,s/a} \equiv \frac{1}{2}[3 + \phi_{s/a}(\tau)(N_r - 1)]$. As $\lambda_{2,s/a} > 0$ has the same sign at both $\tau = -\infty$ and $\tau = \infty$, their contribution in (4.4) vanishes and consequently the index in (anti-)self-dual background is

$$\text{ind } \slashed{D}_s = \frac{1}{2} \sum_{r=1}^{r_0} (N_r - 1) \left[1 - \frac{2 - N_r}{|2 - N_r|} \right] = -\text{ind } \slashed{D}_a. \qquad (4.5)$$

For each block $N_r > 2$, the term in the square brackets above evaluates to 2. There exist $N_r - 1$ normalizable zero modes of $\slashed{D}$ with one chirality (+1 for the self-dual instanton background and −1 for anti-self-dual one) and none with the opposite chirality. For blocks with $N_r = 1$, the index evaluates to 0 and consequently there are no normalizable zero modes.

For the blocks with $N_r = 2$, it seems at first sight that the index (4.5) is undefined. For this case, one eigenvalue of Σ vanishes at either $\tau = -\infty$ or $+\infty$, depending on whether the background is self-dual or anti-self-dual. Thus, for large $|\tau|$, the Dirac operator resembles that of a free particle. The corresponding zero mode is not strictly normalizable but is delta function normalizable: it is a zero-energy resonance or a threshold state.

In fact, the explicit solutions of $\slashed{D}\psi_{s/a,n}^\pm = 0$ ($n = 1, 2, \ldots, N_r$) yield $(N_r - 1)$ normalizable zero modes with positive (negative) chirality for $N_r \geq 3$ and one zero energy resonance for $N_r = 2$. On the other hand, there are $(N_r + 1)$ solutions with positive (negative) chirality and $2N_r$ solutions with negative (positive) chirality which are non-normalizable.

As we shall see in the following, both zero modes and zero-energy resonance states have non-zero contribution to the axial anomaly. Therefore, to correctly take into account all contributions, we must extend the definition of ind $\slashed{D}$ to include zero modes as well as zero-energy resonances with positive and negative chiralities, as in Ref. 14. More generally, for the ansatz (3.2), we find that ind $\not{D}_s = -$ind $\not{D}_a = r_0^{(2)} - r_0^{(1)}$, where $r_0^{(a)}$ is the number of irreps in $L_i^{(a)}$. Thus, the index is *equal* to $\mathcal{T}_{\text{new}}$ defined in (3.6). This yields the matrix model version of the Atiyah–Singer index theorem

$$\text{ind } \not{D} = \mathcal{T}_{\text{new}}. \tag{4.6}$$

It is important to recognize that $\mathcal{T}_{\text{new}}$ is a quantity computed from the pure gauge sector, while the index of the Dirac operator counts the difference between the number of zero modes of opposite chiralities. It is remarkable that there exists such a simple relation between the two: *a priori*, there is no reason to expect this equality. Moreover, this relation between the charge and the index can be suitably adapted for adjoint fermions as well, hinting towards its universal nature in the context of the matrix model.

5. Non-invariance of Fermion Measure

We adapt Fujikawa's method to demonstrate the axial anomaly and its relation to ind $\not{D}$. Specifically, we show that under a $U(1)_A$ transformation, the measure of the fermionic path integral is not invariant. The Jacobian of the transformation gives the integrated anomaly.

In the Euclidean fermionic path integral $\int \mathcal{D}\bar{\Psi}\mathcal{D}\Psi e^{-S_E^F}$, we expand the fermionic field Ψ as $\Psi = \sum a_n \Phi_n$, and $\bar{\Psi} = \sum_n b_n \Phi_n$, where Φ_ns are eigenfunctions of $\not{D}$ with non-zero eigenvalues, zero modes as well as zero-energy resonances. In this basis, the fermionic path integral measure is given by $d\mu \equiv \mathcal{D}\bar{\Psi}\mathcal{D}\Psi = \prod_{n,m} da_n db_m$. Under a $U(1)_A$ rotation with an infinitesimal α, the transformed Ψ' can be expanded as $\Psi' = \sum a_n' \Phi_n$ $\bar{\Psi}' = \sum_n b_n' \Phi_n$, where the coefficients transform linearly as $a_n' = C_{mn} a_m$ and $b_n' = C_{mn} b_m$.

As a consequence, the path integral measure transforms as $d\mu \to [\det C]^{-2} d\mu$. The Jacobian $[\det C]^{-1} = e^{-i \int d\tau \, \alpha \mathcal{A}(\tau)}$, where the anomaly function $\mathcal{A}(\tau)$ is defined as $\mathcal{A}(\tau) = \sum_n \Phi_n^\dagger \gamma^5 \Phi_n$. In $\mathcal{A}(\tau)$, the summation is over an infinite number of modes and hence divergent. We introduce a gauge-invariant regulator $e^{-\beta \not{D}^2}$ and formally take the limit $\beta \to 0$ at the end: $\mathcal{A}(\tau) = \lim_{\beta \to 0} \sum_n \Phi_n^\dagger \gamma^5 e^{-\beta \not{D}^2} \Phi_n$.

We already saw that for non-zero eigenvalues, the eigenfunctions Φ_n and $\gamma^5 \Phi_n$ of $\not{D}$ are orthogonal and so do not contribute to the summation. However, the zero modes and the zero-energy scattering states can have non-zero contribution to $\mathcal{A}(\tau)$.

Setting α = constant, the integrated anomaly is

$$\int d\tau \, \mathcal{A}(\tau) = \int d\tau \left[\sum_{k=1}^{n_+} \chi_k^{+\dagger} \gamma^5 \chi_k^+ + \sum_{k=1}^{n_-} \chi_k^{-\dagger} \gamma^5 \chi_k^- \right]. \qquad (5.1)$$

Using (4.1), we get $\int d\tau \, \mathcal{A}(\tau) = (n_+ - n_-) = \text{ind } \slashed{D}$.

Thus, in a background gauge configuration where the Dirac operator has a non-zero index, the fermion measure is not invariant under axial transformations, and axial symmetry is anomalously broken.

Because ind $\slashed{D}$ is always an integer, $e^{\mp 2i\alpha \int d\tau \, \mathcal{A}(\tau)} = 1$, when $\alpha = n\pi$ for any N (here, $n \in \mathbb{Z}$). This means that under axial rotations with $\alpha = n\pi$, the anomaly vanishes. Thus, the $U(1)_A$ is anomalously broken to $\mathbb{Z}_2$, the residual axial symmetry.

The Dirac operator is diagonal in flavor, and we simply obtain N_f copies of the spectrum of $\slashed{D}$ for a single flavor. The axial symmetry is now broken to $\mathbb{Z}_{2N_f}$.

6. Adjoint Weyl Fermion

The adjoint Weyl fermions are relevant for supersymmetric gauge matrix models.[15] We show that axial symmetry is anomalously broken in this case too.

The Euclidean fermionic action with an adjoint Weyl fermion λ can be written as

$$S_E^F = \int dt \, \lambda^\dagger \left(\partial_\tau - A_0 + \sigma^i \mathcal{F}_i + \frac{3}{2} \right) \lambda, \qquad (6.1)$$

where $\mathcal{F}_i = M_{ia} G_a$ and $G_a = -i f_{abc}$ are the $SU(N)$ generators in the adjoint representation. As before, we choose the temporal gauge.

With $A_i = \phi_{s/a} L_i^{(1)} + (1 - \phi_{s/a}) L_i^{(2)}$ takes the form $\mathcal{F}_i = \phi_{s/a} \mathcal{J}_i^{(1)} + (1 - \phi_{s/a}) \mathcal{J}_i^{(2)}$, where $\mathcal{J}_i^{(\alpha)}$ are the representations of $SU(2)$ obtained by embedding $L_i^{(\alpha)}$ in $(N^2 - 1)$ dimensions. It is straightforward to show that $\mathcal{J}_i$ is given by one singlet representation removed from the following direct sum:

$$\bigoplus_{r,q} [(N_r + N_q - 1) \oplus (N_r + N_q - 3) \oplus \cdots \oplus (|N_r - N_q| + 1)]. \qquad (6.2)$$

Thus, $\mathcal{J}_i^{(\alpha)}$ is always a direct sum of $q_0^{(\alpha)}$ irreps $\mathcal{J}_i^{(\alpha),q}$ of dimensions $\mathcal{N}_q^{(\alpha)}$, with $\sum_{q=1}^{q_0^{(\alpha)}} \mathcal{N}_q^{(\alpha)} = N^2 - 1$. We can again define the spin in each block as

$j_q^{(\alpha)} = \frac{\mathcal{N}_q^{(\alpha)}-1}{2}$. Just as before, we can now define $\mathcal{E}'_i = \frac{d\phi_{s/a}}{d\tau}(e_i^{(2)} - e_i^{(1)})$ and $\mathcal{B}_i = -\mathcal{F}_i - \frac{i}{2}\epsilon_{ijk}[\mathcal{F}_j, \mathcal{F}_k]$, where $e_i^{(\alpha)} = \bigoplus_{q=1}^{q_0^{(\alpha)}} \frac{3}{(j_q^{(\alpha)}+1)(2j_q^{(\alpha)}+1)} \mathcal{J}_i^{(\alpha),q}$. The instanton charge for the embedding is given by

$$\mathcal{T}_{\text{new}}^{\text{adj}} = \int d\tau \, \text{Tr} \, \mathcal{E}'_i \mathcal{B}_i = q_0^{(2)} - q_0^{(1)}. \tag{6.3}$$

The index calculation is exactly as in (4.4) and (4.5), by replacing $N_r^{(\alpha)}$ by $\mathcal{N}_q^{(\alpha)}$ with $\sum_q \mathcal{N}_q^{(\alpha)} = N^2 - 1$. By correctly taking into account the zero modes as well as the zero-energy resonances, the index is now given by ind $\slashed{D} = q_0^{(2)} - q_0^{(1)} = \mathcal{T}_{\text{new}}^{\text{adj}}$.

We observe that all the even-dimensional blocks occur an even number of times. This is understandable because such representations can come only from the cross terms in (6.2) (for which $r \neq q$), which occur twice in the direct sum. So, all distinct eigenvalues occur an even number of times, i.e., are at least doubly degenerate. This degeneracy can also be seen in the spectrum of the Dirac operator: for every eigenvector φ_n of $\slashed{D}$ with eigenvalue λ_n, there exist another eigenvector $(\sigma_2 \otimes \mathbf{1})\varphi^*$ with the same eigenvalue. Consequently, the index is always an even integer, and so the anomaly $e^{-2i\alpha \, \text{ind}\slashed{D}}$ takes value 1 when $\alpha = \frac{n\pi}{2}$, $n \in \mathbb{Z}$. Therefore, a single adjoint Weyl fermion breaks the $U(1)_A$ axial symmetry to a residual $\mathbb{Z}_4$ subgroup. For N_f flavors, the residual symmetry is $\mathbb{Z}_{4N_f}$.

7. Discussion

Though the gauge matrix model is very different from the corresponding gauge field theory, it nevertheless retains important non-perturbative features of the field theory like the axial anomaly. In the usual discussion of the axial anomaly in non-abelian gauge field theories, only the irreducible connections are considered, and it is the instanton number of such connections that is related to the fermion zero modes. *A priori*, there is no reason that the residual axial symmetry should match with the corresponding result in the field theory, and it is surprising that it matches for the case of fundamental fermions.

Our result on the anomaly provides a strong conceptual support to the numerical investigations of the matrix model. Figure 1 displays the results of predictions.[9]

In addition to reproducing the masses of light hadrons with surprising accuracy, the numerics also show that the pseudo-scalar mesons are much

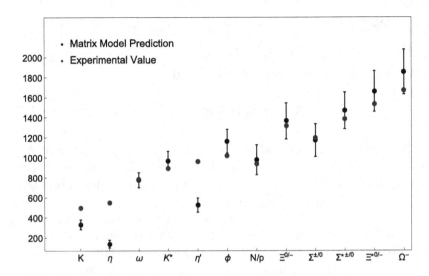

Fig. 1. Matrix model predictions of light hadron masses.

lighter than their scalar counterparts. Furthermore, it also finds the η'-meson to be considerably heavier than the η-meson. The result on the axial anomaly presented in this chapter serves to strengthen the position of the $SU(3)$ gauge matrix model as an effective low-energy approximation of QCD.

Finally, we remark that the axial anomaly is present for any $SU(N)$ gauge group, and there is no reason to expect that it is washed out in the large N limit.

Acknowledgments

It is with great pleasure that we dedicate this chapter to A. P. Balachandran on the occasion of his 85th birthday. Bal has been an important influence on us in shaping our understanding (and taste!) of physics. We take this opportunity to wish him many more years of productive activity.

References

1. S. L. Adler, *Phys. Rev.* **177** (1969) 2426–2438.
2. J. S. Bell and R. Jackiw, *Nuovo Cim. A* **60** (1969) 47–61.
3. K. Fujikawa, *Phys. Rev. Lett.* **42** (1979) 1195–1198.
4. K. Fujikawa, *Phys. Rev. D* **21** (1980) 2848 [erratum: *Phys. Rev. D* **22** (1980) 1499].

5. A. P. Balachandran, S. Vaidya and A. R. de Queiroz, *Mod. Phys. Lett. A* **30**(16) (2015) 1550080.

6. A. P. Balachandran, A. de Queiroz and S. Vaidya, *Int. J. Mod. Phys. A* **30**(09) (2015) 1550064.

7. M. Pandey and S. Vaidya, *J. Math. Phys.* **58**(2) (2017) 022103.

8. N. Acharyya, A. P. Balachandran, M. Pandey, S. Sanyal and S. Vaidya, *Int. J. Mod. Phys. A* **33**(13) (2018) 1850073.

9. M. Pandey and S. Vaidya, *Phys. Rev. D* **101**(11) (2020) 114020.

10. N. Acharyya, M. Pandey and S. Vaidya, *Phys. Rev. Lett.* **127**(9) (2021) 092002.

11. I. M. Singer, *Commun. Math. Phys.* **60** (1978) 7–12.

12. M. S. Narasimhan and T. R. Ramadas, *Commun. Math. Phys.* **67** (1979) 121–136.

13. C. Callias, *Commun. Math. Phys.* **62** (1978) 213–234.

14. D. Bolle, F. Gesztesy, H. Grosse, W. Schweiger and B. Simon, *J. Math. Phys.* **28** (1987) 1512–1525.

15. V. Errasti Díez, M. Pandey and S. Vaidya, *Phys. Rev. D* **102**(7) (2020) 074024.

Chapter 27

The QM/NCG Correspondence

Badis Ydri

Department of Physics, Badji-Mokhtar Annaba University,
Annaba, Algeria
ydribadis@gmail.com
In Honor of Bal on the Occasion of His 85th Birthday

The correspondence between quantum mechanics and noncommutative geometry is illustrated in the context of the noncommutative AdS_θ^2/CFT_1 duality where CFT_1 is identified as conformal quantum mechanics. This model is conjectured here to describe the gauge/gravity correspondence in one dimension.

1. A QM/NCG Correspondence

In this chapter, we will focus on the case of two dimensions with Euclidean signature where the positive curvature space is given by a sphere $\mathbb{S}^2$ with isometry group $SO(3)$ and the negative curvature space is given by a pseudo-sphere $\mathbb{H}^2$ with isometry group $SO(1,2)$ (which is Euclidean AdS^2). The space $\mathbb{S}^2 \times AdS^2$ is in fact the near-horizon geometry of extremal black holes in general relativity and string theory. The information loss problem in four dimensions on $\mathbb{S}^2 \times AdS^2$ is then reduced to the information loss problem in two dimensions on AdS^2.

The quantization of these two spaces yields the fuzzy sphere $\mathbb{S}_N^2$[5] and the noncommutative pseudo-sphere AdS_θ^2[6–8] respectively which enjoy the same isometry groups $SO(3)$ and $SO(1,2)$ as their commutative counterparts. The fuzzy sphere is unstable and suffers collapse in a phase transition to Yang–Mills matrix models (topology change or geometric transition), whereas the noncommutative pseudo-sphere can sustain black hole configurations (by including a dilaton field) and also suffers collapse in the form of information loss process (quantum gravity transition).

Explicitly, the noncommutative AdS_θ^2 is given by the following embedding and commutation relations:

$$-\hat{X}_1^2 + \hat{X}_2^2 + \hat{X}_3^2 = -R^2, \quad [\hat{X}^a, \hat{X}^b] = i\kappa\epsilon^{ab}{}_c\hat{X}^c, \tag{1.1}$$

$$\hat{X}^a = \kappa K^a. \tag{1.2}$$

The K^a are the generators of the Lie group $SO(1,2) = SU(1,1)/\mathbb{Z}_2$ in the irreducible representations of the Lie algebra $[K^a, K^b] = i\epsilon^{ab}{}_c K^c$ given by the discrete series $D_k^\pm$ with $k = \{1/2, 1, 2/3, 2, 3/2, \ldots\}$.[12]

By substituting the solution (1.2) in the constraint equation in (1.1), the value of the deformation parameter κ is found to be quantized in terms of the $su(1,1)$ pseudo-spin quantum number $j \equiv k - 1$ as follows:

$$\frac{R^2}{\kappa^2} = k(k-1). \tag{1.3}$$

The commutative limit is then defined by $\kappa \longrightarrow 0$ or $k \longrightarrow \infty$.

The fact that we have for AdS^2 two disconnected one-dimensional boundaries makes this case very different from higher-dimensional anti-de Sitter space-times and is probably what makes the AdS^2/CFT_1 correspondence the most mysterious case among all examples of the AdS/CFT correspondence. For example, see Ref. 11.

In, Ref. 1 the difficulty of the AdS^2/CFT_1 correspondence is traced instead to the fact that the conformal quantum mechanics residing at the boundary is only quasi-conformal, and as a consequence, the theory in the bulk is only required to be quasi-AdS. In other words, we will take the opposite view and start or assume that the CFT_1 theory on the boundary is really given by the dAFF conformal quantum mechanics.[9, 10]

In the dAFF conformal quantum mechanics, the $so(2,1) = su(1,1)$ Lie algebra generators P (translations), D (scale transformations) and K (special conformal transformations) are realized in terms of a single degree of freedom $q(t)$ with scaling dimension Δ. The corresponding conjugate momentum is given by $p(t) = \dot{q}(t)$ and we start from the standard Heisenberg algebra

$$[q, p] = i. \tag{1.4}$$

The $so(2,1)$ Lie algebra is given explicitly by

$$[D, P] = -iP, \quad [D, K] = iK, \quad [K, P] = -2iD. \tag{1.5}$$

A straightforward calculation gives then

$$P = \frac{p^2}{2} + V(q), \quad V(q) = \frac{g}{q^2}, \tag{1.6}$$

$$D = tP + \frac{\Delta}{2}(qp + pq), \quad \Delta = -\frac{1}{2}, \tag{1.7}$$

$$K = -t^2 P + 2tD + \frac{1}{2}q^2. \tag{1.8}$$

The coupling constant g is completely fixed by conformal invariance. Indeed, we compute the Casimir operator

$$-C = \frac{1}{2}(PK + KP) - D^2 = \frac{g}{2} - \frac{3}{16} \equiv \mp k(k - 1). \tag{1.9}$$

The minus sign corresponds to Lorentzian AdS^2 (relevant to the dAFF quantum mechanics), whereas the plus sign corresponds to Euclidean AdS^2 (relevant to the noncommutative AdS_θ^2).

It is clearly observed that the Lorentz group $SO(1,2)$ is the fundamental unifying symmetry structure underlying the following commutative, non-commutative and quantum spaces:

(i) the AdS^2 space-time (classical geometry and gravity),
(ii) the noncommutative AdS_θ^2 space (or first quantization of the geometry),
(iii) the quantum theory on the boundary (conformal quantum mechanics),
(iv) the IKKT-type Yang–Mills matrix models defining quantum gravity (or second quantization of the geometry).

In particular, the algebra of quasi-primary operators on the boundary is seen to be a subalgebra of the operator algebra of noncommutative AdS_θ^2. This leads us to the conclusion/conjecture that the theory in the bulk must be given by noncommutative geometry and not by classical gravity, i.e., it is given by AdS_θ^2 and not by AdS^2. We end up therefore with an AdS_θ^2/CFT_1 correspondence where the CFT_1 is given by the dAFF conformal quantum mechanics. See Ref. 1 for more details.

This is then a correspondence or duality between quantum mechanics (QM) on the boundary and noncommutative geometry (NCG) in the bulk which provides a concrete model for the AdS^{d+1}/CFT_d correspondence[4] in one dimension (see Figure 1).

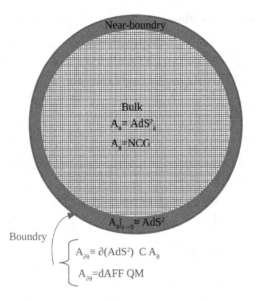

Fig. 1. The QM/NCG correspondence: The bulk is given by the algebra associated with noncommutative AdS^2_θ while the boundary is given by a subalgebra thereof. The behavior of near-boundary is precisely that of commutative AdS^2.

In summary, the main argument underlying this QM/NCG duality consists in the following main observations:

(i) Both noncommutative AdS^2_θ and the dAFF conformal quantum mechanics enjoy the same symmetry structure given by the group $SO(1,2)$. However, dAFF conformal quantum mechanics is only quasi-conformal in the sense that there is neither an invariant vacuum state nor strictly speaking primary operators.[9,10] Analogously, noncommutative AdS^2_θ is only quasi-AdS as it approaches AdS^2 only at large distances (commutative limit).

(ii) Asymptotically, AdS^2_θ is an AdS^2 space-time, i.e., it has the same boundary.[8] And furthermore, the algebra of quasi-primary operators on the boundary (which defines in the same time the geometry of the boundary and the dAFF quantum mechanics) is in some sense a subalgebra of the operator algebra of noncommutative AdS^2_θ.[1]

(iii) Metrically, the Laplacian operator on the noncommutative AdS^2_θ shares the same spectrum as the Laplacian operator on the commutative AdS^2 space-time.[7] The boundary correlation functions computed using the quasi-primary operators reproduce the bulk AdS^2 correlation functions.[10]

2. Noncommutative Black Holes and Matrix Models

The matrix model describing quantum gravitational fluctuations about the noncommutative AdS^2_θ black hole is given by a 4−dimensional Yang–Mills matrix model with mass deformations which are invariant only under the subgroup of $SO(1,2)$ pseudo-rotations around $D_4 \equiv \Phi$ (dilaton field). This model is given explicitly by the action[2]

$$
S[D] = N\mathrm{Tr}\bigg(-\frac{1}{4}[D_\mu, D_\nu][D^\mu, D^\nu] + 2i\kappa D^2[D^1, D^3] + \beta D_a D^a
$$

$$
+ 2i\kappa_1 \hat{\Phi}[D^1, D^3] - \frac{1}{2}\kappa_1^2 D_1 D^1 - \frac{1}{2}\kappa_1^2 D_3 D^3 \bigg). \tag{2.1}
$$

We have then two parameters κ and $\kappa_1 = 2\pi T\kappa$ or equivalently κ and T. We will fix the value of the parameter κ_1 so that an increase in the actual temperature T of the black hole corresponds to a decrease in the noncommutativity parameter (or inverse gauge coupling constant) κ. We are interested in the values $\kappa \neq 0$ and $\beta = 0$.

The pure noncommutative AdS^2_θ geometry appears as a classical (global) minimum of the matrix model (2.1) with $\kappa \neq 0$ and $\kappa_1 = 0$, i.e., $T = 0$.

Yang–Mills matrix models are generally characterized by two phases: (1) a Yang–Mills phase and (2) a geometric phase due to the Myers terms, i.e., the cubic terms in (2.1).

The first cubic term in (2.1) is responsible for the condensation of the geometry (AdS^2_θ space), whereas the second cubic term in (2.1) is responsible for the stability of the noncommutative black hole solution in the AdS^2_θ space. There should then be two critical points $\kappa_{1\mathrm{cr}}$ and $\kappa_{2\mathrm{cr}}$ corresponding to the vanishing of the two cubic expectation values $\langle i\mathrm{Tr}D^2[D^1, D^3]\rangle$ and $\langle i\mathrm{Tr}\Phi[D^1, D^3]\rangle$, respectively. By assuming now that $\kappa_1 > \kappa$, we have $\kappa_{2\mathrm{cr}} > \kappa_{1\mathrm{cr}}$ and we have the following three phases (a gravitational phase, a geometric phase and a Yang–Mills phase):

- For $\kappa > \kappa_{2\mathrm{cr}}$, we have a black hole phase where $\hat{\Phi} \propto \hat{X}_2$. This is the gravitational phase in which we have a stable noncommutative AdS^2_θ black hole.
- For $\kappa_1 < \kappa < \kappa_2$, the second cubic term effectively vanishes and condensation is now only driven by the first cubic term. In other words, although we have in this phase $\hat{\Phi} \sim 0$, the three matrices D_a still fluctuate about the noncommutative AdS^2_θ background given by the matrices $\hat{X}_a$. This is the geometric phase.

- As we keep decreasing the temperature, i.e., increasing κ below κ_1, the AdS^2_θ background evaporates in a transition to a pure Yang–Mills matrix model. In this Yang–Mills phase, all four matrices become centered around zero.

This is the Hawking process as it presents itself in the Yang–Mills matrix model, i.e., as an exotic line of discontinuous transitions with a jump in the entropy, characteristic of a first-order transition, yet with divergent critical fluctuations and a divergent specific heat characteristic of a second-order transition.

An alternative way of recasting Hawking process as a phase transition starts from dilaton gravity in two dimensions which can be rewritten as a nonlinear sigma model.[17] For constant dilaton configurations, i.e., for $\Phi = \text{constant}$, this action reduces to the potential term[2]

$$V = m^2 \int d^2x \sqrt{\det g}\left(M - N(\Phi)\right)^2, \quad N(\Phi) = \int^\Phi d\Phi' V(\Phi'). \quad (2.2)$$

The limit $m^2 \longrightarrow \infty$ is understood. The mass functional M is constant on the classical orbit given by the ADM mass of the AdS^2 black hole, viz. $M = 2\pi^2 T^2/\Lambda$.[18] For the case of the Jackiw–Teitelboim action, we have $V = 2\Lambda^2\Phi$ and $N = \Lambda^2\Phi^2$. The noncommutative analog of this Jackiw–Teitelboim potential term is a real quartic matrix model[19]

$$V = N_H \text{Tr}(\mu\hat{\Phi}^2 + g\hat{\Phi}^4), \quad N_H \equiv 2k - 1. \quad (2.3)$$

The regularization issues are discussed in the following section. The parameters μ and g are given explicitly by

$$\mu = -2\pi M \cdot \left(\frac{m^2}{k^2}\right), \quad g = \pi\Lambda^2 \cdot \left(\frac{m^2}{k^2}\right). \quad (2.4)$$

The one-cut solution (disordered phase) in which the dilaton field $\hat{\Phi}$ fluctuates around zero occurs for all values $M \leq M_*$, where

$$M_* = \frac{\Lambda}{\sqrt{\pi}}\frac{k}{m}. \quad (2.5)$$

The large mass limit $m^2 \longrightarrow \infty$ is then required to be correlated with the commutative limit $k \longrightarrow \infty$. This critical value M_* is consistent with the black hole mass $M = 2\pi^2 T^2/\Lambda$. In other words, the dilaton field is zero when the mass parameter M of the matrix model is lower than the black hole mass, i.e., the black hole has already evaporated and we have only pure AdS_θ^2 space in this phase.

Above the line $M = M_*$, we have a two-cut solution (non-uniform ordered phase) where the dilaton field $\hat{\Phi}$ is proportional to the idempotent matrix γ (which satisfies $\gamma^2 = 1$). In this two-cut phase, we have therefore a non-trivial dilaton field corresponding to a stable AdS_θ^2 black hole. The transition at $M = M_*$ between the the one-cut (AdS_θ^2 space) and two-cut phases (AdS_θ^2 black hole) is third order.

3. Phase Structure of the Noncommutative $\text{AdS}_\theta^2 \times \mathbb{S}_N^2$

The quantum gravitational fluctuations about the noncommutative geometry of the fuzzy sphere $\mathbb{S}^2$, the noncommutative pseudo-sphere $\mathbb{H}^2$ and the noncommutative near-horizon geometry $\text{AdS}_\theta^2 \times \mathbb{S}_N^2$ are given by IKKT-type Yang–Mills matrix models with additional cubic Myers terms which are given explicitly by the actions

$$S_\text{S}[C] = N_\text{S} \text{Tr}_\text{S} L_\text{S}[C], \quad \mathbb{S}^2, \tag{3.1}$$

$$S_\text{H}[D] = N_\text{H} \text{Tr}_\text{H} L_\text{H}[D], \quad \mathbb{H}^2, \tag{3.2}$$

$$S_\text{HS}[D, C] = N_\text{H} N_\text{S} \text{Tr}_\text{H} \text{Tr}_\text{S} \left(L_\text{S}[C] + L_\text{H}[D] - \frac{1}{4}[D_a, C_b][D^a, C_b] \right), \quad \mathbb{H}^2 \times \mathbb{S}^2. \tag{3.3}$$

The Lagrangian terms are given in terms of three $(2l+1) \times (2l+1)$ Hermitian matrices C_a (where $N_\text{S} = 2l + 1 \equiv N$) and three Hermitian operators D_a by the equations

$$L_\text{S}[C] = -\frac{1}{4}[C_a, C_b]^2 + \frac{2i}{3} \alpha \epsilon_{abc} C_a C_b C_c + \beta_S C_a^2, \tag{3.4}$$

$$L_\text{H}[D] = -\frac{1}{4}[D_a, D_b][D^a, D^b] + \frac{2i}{3} \kappa f_{abc} D^a D^b D^c + \beta_H D_a D^a. \tag{3.5}$$

The coefficients ϵ_{abc} and f_{abc} provide the structure constants of the rotation group $SO(3) = SU(2)/\mathbb{Z}_2$ and the pseudo-rotation group $SO(1,2) = SU(1,1)/\mathbb{Z}_2$, respectively.

The solutions of the equations of motion which follow from the actions (3.1), (3.2) and (3.3) are precisely given by the fuzzy sphere $\mathbb{S}_N^2$, the noncommutative pseudo-sphere $\mathbb{H}_\theta^2$ and the noncommutative near-horizon geometry $\text{AdS}_\theta^2 \times \mathbb{S}_N^2$ configurations. These noncommutative configurations are given explicitly by

$$C_a = \phi_S L_a, \quad \phi_S = \alpha \varphi_S, \tag{3.6}$$

$$D_a = \phi_H K_a, \quad \phi_H = \kappa \varphi_H, \tag{3.7}$$

$$C_a = \phi_S L_a \otimes 1_H, \quad D_a = 1_S \times \phi_H K_a. \tag{3.8}$$

These matrix models (3.1), (3.2) and (3.3) exhibit emergent geometry transitions from a geometric phase (the sphere, the pseudo-sphere and the near-horizon geometry of a Reissner–Nordstrom black hole) to a pure Yang–Mills matrix phase with no background geometrical structure.

This fundamental result is confirmed in the case of the sphere by Monte Carlo simulations of the Euclidean Yang–Mills matrix model (3.1). But in the other cases, we have only at our disposal the effective potential. Indeed, the Monte Carlo method cannot be applied to the matrix models (3.2) and (3.3) in their current form for two main reasons: first, the Lorentzian signature of the embedding space-time (these noncommutative spaces are in fact branes residing in a larger space-time) and second, the Hilbert spaces $\mathcal{H}_k$ corresponding to the noncommutative operator algebras are actually infinite-dimensional (which is required by the underlying $SO(1,2)$ symmetry).

However, the matrix models (3.2) and (3.3) can be regularized by means of a simple cutoff which consists in keeping only the $2k - 1$ states in the Hilbert spaces turning therefore the infinite-dimensional operator algebras into finite-dimensional matrix algebras. By assuming also that the AdS space-time has the same radius as the sphere, we have the natural identification

$$N_{\mathrm{H}} = 2k - 1 \equiv N. \tag{3.9}$$

The one-loop effective potential around the pseudo-sphere background (3.7) is computed from the action (3.2) using the background field method. We obtain the result (with $\tilde{\kappa}^4 = 4\kappa^4 k(k-1)$ and $\tau_H = -\beta_H/\kappa^2$)

$$\frac{2V_H}{N^2} = \tilde{\kappa}^4 \left[\frac{1}{4}\varphi_H^4 - \frac{1}{3}\varphi_H^3 + \frac{1}{2}\tau_H\varphi_H^2 \right] + \log\varphi_H^2, \quad T_H = \frac{1}{\tilde{\kappa}^4} = g_H^2. \tag{3.10}$$

The one-loop effective potential around the sphere background (3.6) is obtained from this result with the substitutions $k \longrightarrow -l$, $\varphi_H \longrightarrow \varphi_S$, $\kappa \longrightarrow \alpha$, $\beta_H \longrightarrow -\beta_S$, $\tau_H \longrightarrow \tau_S$, $g_H \longrightarrow g_S$ and $T_H \longrightarrow T_S$.

The description of the phase structure of the noncommutative pseudo-sphere $\mathbb{H}_\theta^2$ using the effective potential (3.10) is therefore formally identical to the fuzzy sphere case treated in detail in Refs. 13–16.

The sphere/pseudo-sphere solution $\varphi_{S,H} = \varphi_-(\tau_{S,H}) \neq 0$ is found to be stable only in the regime $0 < \tau_{S,H} < 2/9$. This is the fuzzy or non-commutative geometry phase. The Yang–Mills solution $\varphi_{S,H} = \varphi_0 = 0$ (matrix phase) is found to be the ground state or global minimum of the system in the regime $2/9 < \tau_{S,H} < 1/4$. In fact, the coexistence curve between the geometric phase (low-temperature T_H) and the matrix phase (high-temperature T_H) asymptotes to the line $\tau_{S,H} = 2/9$.

We turn now to the case of the noncommutative near-horizon geometry $\mathrm{AdS}_\theta^2 \times \mathbb{S}_N^2$. We can immediately state the possible phases of the noncommutative near-horizon geometry $\mathrm{AdS}_\theta^2 \times \mathbb{S}_N^2$ in the space $(\tau_S, \tau_H, \bar{\alpha}, \bar{\kappa})$ as follows:[3]

- a Yang–Mills matrix phase with no background geometrical structure which is expected at high temperature (both $\bar{\alpha}$ and $\bar{\kappa}$ approach zero) or $2/9 < \tau_{S,H} < 1/4$,
- a 2–dimensional geometric fuzzy sphere phase ($\bar{\alpha}$ approaches infinity, $\bar{\kappa}$ approaches zero and $0 < \tau_{S,H} < 2/9$),
- a 2–dimensional geometric noncommutative pseudo-sphere phase ($\bar{\alpha}$ approaches zero, $\bar{\kappa}$ approaches infinity and $0 < \tau_{S,H} < 2/9$),
- a 4–dimensional geometric noncommutative near-horizon geometry $\mathrm{AdS}_\theta^2 \times \mathbb{S}_N^2$ phase at low temperature (both $\bar{\alpha}$ and $\bar{\kappa}$ approach infinity and $0 < \tau_{S,H} < 2/9$).

In order to simplify the calculation of the one-loop effective potential around the near-horizon geometry background (3.8), we will consider the following case:

- First, in order to have a single unified temperature T_{HS}, we must have a single unified gauge coupling constant $g_{\mathrm{HS}}^2 = 1/T_{\mathrm{HS}}$, viz.

$$\alpha = \kappa. \tag{3.11}$$

- The geometric quantum entanglement between the sphere and the pseudo-sphere sectors (through the third term in the action (3.3)) is essential for the emergence of a four-dimensional space. However, there exists a special case where this geometric quantum entanglement can be partially removed while keeping the background emergent geometry four-dimensional. This is given by the case

$$\tau_S = \tau_H \iff \beta_S = -\beta_H. \tag{3.12}$$

- With this choice (3.12), one can check that the order parameters φ_S and φ_H must solve identical equations of motion. A simplified model is then obtained by setting

$$\varphi_S = \varphi_H \equiv \varphi. \tag{3.13}$$

The one-loop effective potential around the near-horizon geometry background (3.8) computed from the action (3.3) using the background field method is then given by

$$\frac{V_{HS}}{2N_S^2 N_H^2} = 2\bar{\kappa}^4 \left[\frac{1}{4}\varphi^4 - \frac{1}{3}\varphi^3 + \frac{1}{2}\tau_H \varphi^2 \right] + \log \varphi^2, \quad \bar{\kappa}^4 = \frac{\tilde{\kappa}^4}{4N}. \tag{3.14}$$

This effective potential is of the same form as the pseudo-sphere effective potential (3.10) with the substitution $\tilde{\kappa}^4 \longrightarrow 2\bar{\kappa}^4$.

In the special case given by (3.11), (3.12) and (3.13), the $2-$dimensional geometric phases disappear and we end up with a single phase transition from a noncommutative near-horizon geometry $\mathrm{AdS}_\theta^2 \times \mathbb{S}_N^2$ phase to a Yang–Mills matrix phase as the temperature $T_{\mathrm{HS}} = 1/\bar{\kappa}^4 = g_{\mathrm{HS}}^2$ is increased.

References

1. B. Ydri, [arXiv:2108.13982 [hep-th]].
2. B. Ydri, [arXiv:2109.00380 [hep-th]].
3. B. Ydri and L. Bouraiou, [arXiv:2109.01010 [hep-th]].
4. J. M. Maldacena, *Adv. Theor. Math. Phys.* **2** (1998) 231–252.
5. J. Hoppe, Ph.D. Thesis, (1982). J. Madore, *Class. Quant. Grav.* **9** (1992) 69.
6. P. M. Ho and M. Li, *Nucl. Phys. B* **590** (2000) 198.
7. D. Jurman and H. Steinacker, *JHEP* **1401** (2014) 100.
8. A. Pinzul and A. Stern, *Phys. Rev. D* **96**(6) (2017), 066019.
9. V. de Alfaro, S. Fubini and G. Furlan, *Nuovo Cim. A* **34** (1976) 569.
10. C. Chamon, R. Jackiw, S. Y. Pi and L. Santos, *Phys. Lett. B* **701** (2011) 503–507.
11. A. Strominger, *JHEP* **01** (1999) 007.
12. V. Bargmann, *Ann. Math.* **48** (1947) 568.
13. P. Castro-Villarreal, R. Delgadillo-Blando and B. Ydri, *Nucl. Phys. B* **704** (2005) 111.
14. T. Azuma, S. Bal, K. Nagao and J. Nishimura, *JHEP* **05** (2004) 005.

15. R. Delgadillo-Blando, D. O'Connor and B. Ydri, *Phys. Rev. Lett.* **100** (2008) 201601.
16. R. Delgadillo-Blando and D. O'Connor, *JHEP* **11** (2012) 057.
17. M. Cavaglia, *Phys. Rev. D* **59** (1999) 084011 [hep-th/9811059].
18. M. Cadoni and P. Carta, *Mod. Phys. Lett. A* **16** (2001) 171–178.
19. Y. Shimamune, *Phys. Lett. B* **108** (1982) 407.